International Conference on Engineering, Technologies, and Applied Sciences (ICETsAS 2019)

IOP Conference Series: Materials Science and Engineering
Volume 874

Bengkulu, Indonesia
17 - 18 October 2019

ISBN: 978-1-7138-3175-4
ISSN: 1757-8981

Printed from e-media with permission by:

Curran Associates, Inc.
57 Morehouse Lane
Red Hook, NY 12571

Some format issues inherent in the e-media version may also appear in this print version.

Institute of Physics
Dirac House, Temple Back
Bristol BS1 6BE UK

Phone: 44 1 17 929 7481
Fax: 44 1 17 920 0979

techtracking@iop.org

Additional copies of this publication are available from:

Curran Associates, Inc.
57 Morehouse Lane
Red Hook, NY 12571 USA
Phone: 845-758-0400
Fax: 845-758-2633
Email: curran@proceedings.com
Web: www.proceedings.com

TABLE OF CONTENTS

COMPUTER SCIENCE

IOP Conf. Series: Materials Science and Engineering **874** (2020) 011001 doi:10.1088/1757-899X/874/1/011001

Preface

University of Bengkulu, based on a mandate given by BKS PTN bagian barat (Coordination body of Indonesian state university in West Region), organized International Conference on Engineering Technology and Applied Science (ICETSaS) 2019 on 17-18 October 2019 in Bengkulu, Indonesia. The theme of ICETSAS 2019 is **"Coordinating Science, Technology and Culture for Sustainable and Resilient Future".** The conference has been held successfully thanks to support from Indonesian Ministry of Research, Technology, and Higher Education, Bank Bengkulu, PT Telkom, PT Sentra Ardi Purna and many others. Selected papers from the conference is published in this book.

The conference was honored by the presence of two keynote speakers i.e. **Prof. Haslan Abu Hassan, Ph.D.** from Universiti Sains Malaysia and **Prof. Dr. Ing. Yulfian Aminanda, C-Eng.** from Universiti Teknologi Brunei. Other speakers were also presenting many topics related to Geographic Information System, Cloud Computing, Web/Mobile Apps, Artificial Intelligent, Image Processing, Internet of Things, Control and Robotic System, Electronic and Instrumentation, Computer and Information System, Power Energy and Telecommunication.

The review process indicates that many submitted papers are weak in term of language even though the content is interesting. Therefore, the organizing committee allow the papers to be presented first. After that, there is some assistance in improving the language of the paper to increase the quality of the paper. Papers which have been improved is then published in this proceeding. Relawan Jurnal Indonesia has been supporting the organizing committee in the publication process.

We would like to thank the members of advisory boards (Prof. Drs. Suharno, M.Sc., Ph.D, university of Lampung and Dr. Ridwan Nurazi, M.Sc., University of Bengkulu), member of steering committee (Drs. Boko Susilo, M.Kom., University of Bengkulu ; Irza Sukmana, S.T, M.T, Ph.D., University of Lampung), and many others which cannot be mentioned here.

Organizing Committee

Helmizar, S.T, M.T, Ph.D.

IOP Conf. Series: Materials Science and Engineering **874** (2020) 011002 doi:10.1088/1757-899X/874/1/011002

Peer review statement

All papers published in this volume of *IOP Conference Series: Materials Science and Engineering* have been peer reviewed through processes administered by the proceedings Editors. Reviews were conducted by expert referees to the professional and scientific standards expected of a proceedings journal published by IOP Publishing.

IOP Conf. Series: Materials Science and Engineering **874** (2020) 012001 doi:10.1088/1757-899X/874/1/012001

Degradation, swelling profile, and gel fraction of synthetic coral scaffold incorporated PRP or PRF

Erlina Sih Mahanani[1] Maulida Nurlaeli[2] Woro Winanti[2] Monica Hafzi Azizi[2] Hammia Zanzabiela[2]

[1]Department of Dental Biomedic, School of Dentistry, Faculty of Medicine and Helath Science, Universitas Muhammadiyah Yogyakarta, Indonesi, Coresponding email: erlina.sih@umy.ac.id
[2]Dental Hospital, Faculty of Medicine and Heath Science, Universitas Muhammadiyah Yogyakarta, Indonesia

Abstract. Platelet rich plasma (PRP) and platelet rich fibrin (PRF) are sourced from human blood and have role in bone recovery. Both of them could be incorporated into scaffold for bone regeneration. Synthetic coral scaffold is a natural mimicking of sea coral designed for bone regeneration. It should have bio degradability and bio absorbability that influence new bone formation process. The purpose of this study is to investigate the degradation, swelling profile and gel fraction of synthetic coral scaffold incorporated with PRP or PRF. Synthetic coral scaffold consists of gelatin, calcium carbonate, and sodium nitrate as dispersant agent. It is divided into 3 groups, incorporated PRP, PRF, and no-incorporation. Scaffold was soaked in phosphate buffer saline (PBS) and incubated in 37° C for 24 hours. Scaffold weight was measured in every 30 minutes to observe the swelling profile and gel fraction. Degradation profile was observed after 1, 3, 6, 24, 48, 72, and 96 hours of soaking. Acceleration of degradation was measured after soaking with HC1 1N for 1, 3, 6, 24, 48, 72, 96 hours until the scaffold ran out. The result shows significant differences among PRP, PRF, and control group. PRP incorporation has slow degradation, good swelling and higher gel fraction that presents the strengthening of scaffold structure.

1. Introduction

Bone as supporting tissue forms structures and protects vital organs of the body. Bone can be damaged due to trauma or malignancy. Bone damage must be treated immediately to prevent local or systemic damage [1]. Various alternative treatments can be done to repair bone damage, such as bone graft to accelerate bone damage recovery. Bone graft is a bone tissue substitution derived from autologus, artificial, synthetic, or natural substitute material. Bone grafting may be done because the bone has ability to fully regenerate in adequate space. As bones that grow naturally, bone tissue generally replaces all bone grafts to produce an area containing new bone [2].

Tissue engineering aims to regenerate tissue and create biological substitutes for deformed or damaged tissue and organs that require cells. Tissue engineering technology is currently developed rapidly to facilitate tissue regeneration as needed [3]. Tissue engineering technology involves three factors that greatly influence the success of tissue reconstruction. The three main factors are signals, cells, and scaffolding [4].

The development of tissue engineering has been very advanced especially in the field of dentistry. The principle application of tissue engineering has expanded in various branches of dentistry, such as periodontics, maxillofacial surgery, and dental implants. Signal is an important component that can be obtained from growth factor. Growth factor can be obtained from PRP (Platelet Rich Plasma) and PRF [3]. PRP is the result of separating whole blood which is centrifuged to a certain level and in the form of a plasma layer consisting of platelet rich and poor plasma, and red blood cells. PRP contains many growth factors and other components that are very helpful for healing and regeneration process [2]. The difficulties encountered in making PRP cause evolution and simplification that result in PRF (Platelet Rich Fibrin), which has all the useful components for recovery and healing [5]. PRF contains more white blood cells that are important for wound healing process [6], and, it is the result of PRP development, namely the second generation of PRP. The process of PRF manufacture is easier and simpler than PRP because no biochemical ingredients are added, such as thrombin and anticoagulants [7]. PRF is also an autologous agent similar to PRP because it derives from the patient's own body [8]. Platelet rich fibrin are able to support bone regeneration, because it contains a lot of platelets, fibrin matrix and growth factor [9].

Besides acting as signals, PRP and PRF also have property as scaffold themselves because PRP and PRF contain fibrin. Fibrin can be used as a scaffold material because fibrin is an autologous scaffold that does not cause toxic or excessive immune reactions. It derives from the individual itself. The mechanical properties of fibrin are limited, and thus the utilization is limited. However, fibrin can be combined with other scaffold materials so that it can strengthen the structure of the scaffold itself and the time of growth factor release in accordance with the scaffold degradation [10]. PRF is a fibrin matrix containing cytokines, namely, platelets that contain growth factors and can function as resorbing membranes [9].

Scaffold has an important role in the success of tissue engineering. Various biological properties must be present in scaffolding, such as biocompatibility, biodegradation, high strength and high porosity [11]. Biocompatible means the ability of scaffold of being received by the body and degradable when the formation of new cells has been achieved. The non-degradable scaffold inside the tissue can initiate some infections in the body. The ideal velocity of scaffold degradation is equal to the velocity of tissue formation [12]. The mechanical properties of the scaffold are influenced by several factors such as original rigidity of polymer chains, types of molecular crosslinking, crosslinking and swelling solidities as the result of hydrophilic or hydrophobic balance. One of the design parameters in tissue engineering that determines its success is the mechanical properties of the scaffold. It is intended that the scaffold must be able to provide and maintain space for developing cells. In addition, the scaffold mechanical properties also influence genes adhesion and expression from the cells [10]. The biocompatible properties are influenced by the hydrophilic characteristic from constituent material of the scaffold. The good biocompatible properties of a scaffold can be shown by the ability to absorb liquid medium by the scaffold membrane without dissolving it in the liquid medium or referred to as swelling ability. This relates to the swelling profile of a scaffold which has an important role in supporting cell nutrition distribution during tissue regeneration process [13]. The swelling profile on scaffold is one of its characteristics. The swelling profile represents the process that occurs in scaffold when swelling after the fluid enters the scaffold and after the maximum swelling is obtained, it will burst [14]. Scaffold swelling ability will work optimally over large areas [15].

Synthetic coral scaffold is designed to resemble natural sea coral, has porosity, and can provide space for cells to attach and grow [16]. It also has a degradation profile that meets the criteria as a scaffold for bone tissue regeneration. This can also be incorporated with PRP for the operation of tissue engineering system [17]. The additional platelet concentrated into the scaffold material in bone regeneration serves to stimulate the proliferation and differentiation of osteoblast cells and osteoprogenitor around the bone damage area. PRP and PRF have poor mechanical properties, low stability and very fast degrees of

IOP Conf. Series: Materials Science and Engineering **874** (2020) 012001 doi:10.1088/1757-899X/874/1/012001

degradation. Mastication also disrupts bone formation process. The addition of scaffold material is expected to provide mechanical strength and maintain the stability of bone graft material during bone recovery process. The time of scaffold degradation becomes quite long so that the recovery phase occurs slowly and can run more optimally [18].

2. Material and method

2.1. Scaffold preparation

Synthetic coral scaffold was developed from gelatin and calcium carbonate with a ratio of gelatin: calcium carbonate 5:5 and 10% gelatin as a control. Sodium citrate was used as a dispersant in a small composition. The solution was molded into a thick film-like form, and freeze dried afterward. After the manufacturing process, physical crosslinking was carried out to strengthen the structure [16].

2.2. Preparation of PRP and PRF

PRP and PRF were obtained from 5 human whole blood donors who had agreed to ethical clearances. Blood was collected in vacutainer with EDTA and then centrifuged for 2 times. First with a velocity of 1200 rpm for 10 minutes, buffy coat was taken, transferred to a new vacutainer, and then centrifuged at 3500 rpm for 10 minutes to obtain PRP.

As for the PRF isolation, whole blood was put into plain vacuntainer (without anticoagulants) and was immediately centrifuged to avoid blood clots. It was centrifuged for 10 minutes at a velocity of 1600 rpm until it formed 3 layers, namely, PPP, fibrin clot (PRF) and red blood cells. The fibrin clot or PRF layer was placed in the center of the tube. Fibrin clot was transferred to a new vacutainer and was put in the refrigerator for 24 hours.

2.3. Incorporation into synthetic coral scaffold

The incorporation of PRP and PRF into artificial synthetic coral was done by immersing the scaffold in each microtube which has been filled with PRP and PRF for 10 minutes until it was absorbed into the scaffold. Scaffold was removed from the microtube, and the remaining droplets of PRP and PRF were drained. The amount of PRP and PRF used for incorporation was 100 ul for swelling profile observation and 40 ul for profile degradation.

2.4. Swelling observation

After the incorporation of PRP or PRF, firstly, scaffold with incorporated PRP/PRF was weighed as W_0 wet. Each scaffold was put in a pot and filled with a 3 ml PBS solution until the scaffold is completely immersed. It was then incubated at 37 ° C with a period of 30 minutes, 1, 2, 4, 6 and 24 hours [19]. Each period of time, each sample was lifted and drained with a filter and then wet scaffold was weighed (W_w). After weighing the wet scaffold, it was put back into the pot. This step was repeated until the last time period (24 hours).

The percentage of swelling in observing the swelling profile was obtained by formula:

- Swelling (%)= $\frac{W_w - W_0}{W_0} \times 100$

  W_0 = the initial weight of the scaffold before immersion.
  W_w = the wet weight of the scaffold after immersion.

2.5. Degradation observation

Each scaffold used for degradation observations before incorporation of PRP / PRF weighed 10 mg. Three scaffold groups was immersed into each microtube which was filled with 1.3 ml PBS and then incubated at 37 ° C for 1, 3, 6, 24, 48, 72 and 96 hours. At each time period, it was centrifuged at 3500 rpm for 2 minutes so that the scaffold was at the bottom of microtube. The supernatant was taken and measured its optical density with the UVmini-1240 SHIMADZU spectrophotometers (wavelength 280 nm). Microtube containing scaffold deposits were filled with PBS and incubated again. This was done repeatedly until the last measurement period.

The degradation acceleration was observed by replacing the immersing solution with 1N HCl and by measuring as before until the scaffold was perfectly degraded.

After getting the absorbance value for each time interval, scaffold degradation percentage then was calculated by the following formula: [20].

- $\% \text{ scaffold degradation X period} = \dfrac{\text{X period absorbance value}}{\text{100\% absorbance value}} \times 100\%$

3. Result

3.1. Swelling Profile

The time needed for incorporation was 10 minutes. The 0-minute period was the time for scaffold to dry before the incorporation of PRP or PRF and its immersion in PBS. The 10th minute period was the time for scaffold to immerse in PBS after incorporation for 10 minutes. This caused scaffold incorporated by PRP and PRF to gain significant weight compared to the control scaffold. The treatment of PRP or PRF incorporation shows the swelling process in the scaffold. The 30th minute time period, as shown in Figure 1, shows that the ratio of swelling profiles of PRP incorporated scaffold was higher compared to the swelling profile ratio of other scaffolding groups. After 30 minutes, the swelling profile in all groups decreased and in 24 hours it appeared to be a rather sharp decrease in swelling profile except the PRP incorporation group.

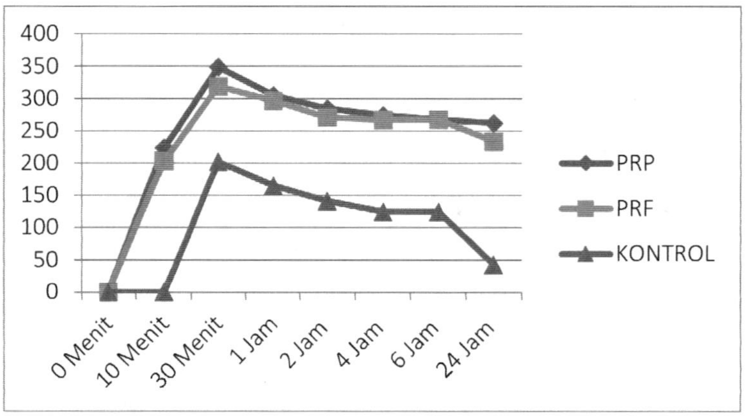

Figure 1. Swelling profile

Table 1. Gel fraction statistical analysis

Test of Homogeneity of Variety		Anova			Post Hoc	
Levene Statistic	Sig.	Mean Square		Sig.	Sample	Sig.
		Between Group	Within Group		PRP - PRF	.068
1.531	.294	128.251	13.071	.013	PRP - control	.036
					PRF - control	.858

Table 1 shows that the percentage of gel fraction in the PRP incorporation group with control group has a significant difference, whereas the difference is not significant in PRF group with PRP and PRF with the control group ($p < 0.05$).

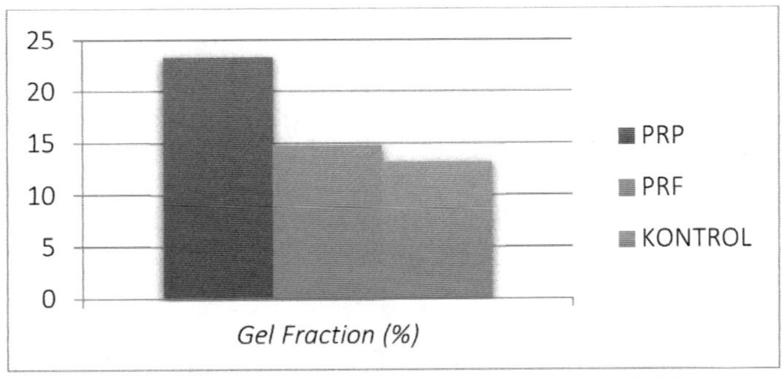

Figure 2. Gel fraction percentage of scaffold

The highest percentage of gel fraction is the PRP incorporation group shown in Figure 2. This shows that this group has a structure and mechanical properties that are stronger than other of other scaffolding groups.

3.2. Degradation Profile

Based on Figure 3 scaffold with incorporation of PRP and control (without incorporation) is degraded longer than scaffold with incorporation of PRF. Scaffold with incorporation of PRF on acceleration degradation using 1N HCl solution is degraded 100% for the first time at the 216th hour compared to the other 2 groups.

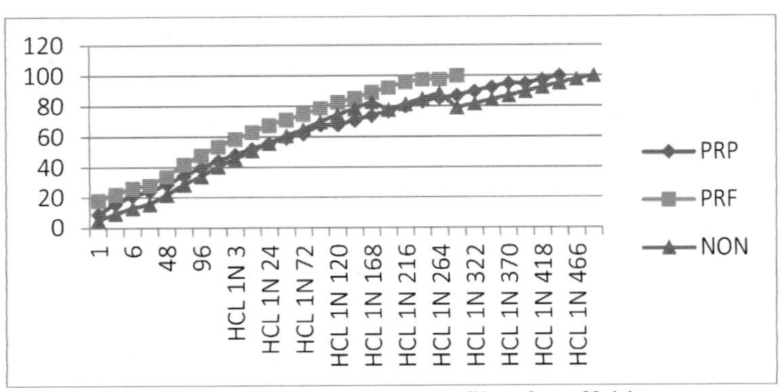

Figure 3. Degradtion Profile of scaffold

4. Discussion

The PRP scaffold group and the PRF scaffold group had undergone a swelling process previously during the incorporation process in the first 10 minutes. It was 223.95% in the PRP scaffolding group and 203.54% in the PRF scaffold group. Meanwhile, the control scaffold group had not undergone a swelling process because of no addition of PRP or PRF. The first 30 minutes period, the three sample groups continued to swell so that the percentage of swelling still continued to increase. PRP scaffold increased as much as 38.41% and an increase of 37.71% for the PRF scaffold of the initial swelling profile ratio means that scaffold swelling capabilities were still increasing. This research is in accordance with the research conducted by Wahyudi and Nurwadji, which states that swelling ability or swelling profile ratio will increase dramatically at the beginning of immersion time [21]. The increasing of the swelling profile ratio at the beginning plays a role in facilitating the process of cell migration and the formation of 3D structures. The next period, the scaffold swelling profile begins to decrease slowly with the PRP scaffold still at the highest position, but the value does not differ much from the PRF scaffold, while the control scaffold has the lowest swelling profile ratio. The next period, on the 4th and 6th hours of the scaffold swelling profile ratio tends to maintain its position until the 24th hour of immersion. According to Park et al., the swelling profile ratio will begin to decrease along with the increasing of immersion time, and after that the scaffold will experience an equilibrium phase [22]. Equilibrium phase is a phase in which the scaffold tends to maintain its weight. Based on research conducted by Zhang et al. [23], the swelling profile ratio of PVA/HA composite hydrogel is divided into 4 stages, namely increasing rapidly, decreasing, decreasing slowly and then balanced or entering the equilibrium phase, decreasing swelling ratio followed by increase in HA content in scaffold.

The significant difference between PRP and PRF is the presence of thrombin and calcium chloride which are anticogulant agents. This agent is important in helping in the fibrin polymerization process because there is no addition of thrombin and calcium chloride in PRF. The formation of fibrin matrix runs slowly, in contrast to PRP which is the addition of thrombin and calcium chloride the fibrin matrix formation process can run quickly. The fibrin matrix formation phase plays an important role in the formation of 3-dimensional structures of fibrin network [24]. Calcium ion concentration has a role in the change of fibrinogen to fibrin, could stabilize the structure of fibrinogen and accelerates fibrin formation by acting as a thrombin cofactor. The influence of the presence of thrombin in physical characters fibrin networks such as fiber diameter, mass/length, density, porosity and permeability can cause changes in cell adhesion and migration [25]. The presence of PRP activation by anticoagulants can help releasing the growth factors and cytokines from platelets and leukocytes which help stimulate the process of fibrinogen decomposition into fibrin fibers to form a network structure that can support the process of cell proliferation and differentiation. In addition, with the double centrifuge in the process of making PRP besides enriching platelet content, it is also beneficial to maintain the integrity of PRP so that it becomes more stable [26]. Research conducted by Rowe et al. mentions that differences in thrombin concentration added to fibrin hydrogel can affect the morphological structure of fibrin fibers in the 3D structure of it [27]. Whereas, low thrombin concentrations will increase bundle size fibrin so that the pores of fibrin hydrogel are larger and manage to increase cell proliferation and differentiation. It also can affect the structure of mechanical strength.

After 24 hours of immersion, the scaffold will undergo a process of breaking the structure or the so-called gel fraction process. Gel fraction was measured by drying the scaffold which is not soluble in PBS at a temperature of 500 in the oven and weighed every 24 hours until the weight is stable. The gel fraction process begins with a decrease in the mechanical properties of the scaffold which is characterized by a decrease in the weight of the scaffold. There is a difference in the gel fraction value of the scaffold group incorporated with PRP, a scaffold incorporated with PRF and control scaffold. According to a research by Park et al. [22], it is stated that the percentage of gel fraction would increase along the increasing time period and decrease in the higher PEG concentration in the PVA based Hydrogel scaffold. In addition, the gel fraction value is proportional to the tensile / mechanical strength of the scaffold. The formula used to calculate the percentage of gel fraction shows that the percentage of gel fraction is proportional to the final dry weight of the scaffold. It means the higher the percentage of gel fraction, the less weight the scaffold dissolves during the immersion process which shows the mechanical strength of the scaffold structure [28]. The results shows that among the percentage of gel fraction between 3 groups, PRP has the highest percentage of gel fraction which is indicated by the percentage of scaffold incorporated with PRP that is higher than the scaffold incorporated with PRF and control scaffolding. This shows that the scaffold incorporated with PRP has stronger mechanical strength compared to the scaffold that is incorporated with the PRF and control scaffolding.

The velocity of scaffold degradation must be in accordance with the velocity of tissue formation [29]. The degradation process is part of the release of molecules signal. The slower the process of scaffold degradation, the slower the molecules signal release to the location of bone reconstruction [30]. In line with the research that has been done, it is assumed that PRP releases the molecule signal more slowly than the PRF so that it has a slower degradation process. PRP is an autologous plasma that has a platelet concentration up to 1,000,000 / µl. The normal number of platelets in the blood is 150,000 / µl to 350,000 / µl. PRP has many growth factors which one of its functions is to stimulate collagen synthesis and to accelerate homeostasis responses in injuries so as to stimulate bone regeneration and wound healing processes [31]. PRP is a platelet-rich plasma derived from autologous or derived from the same body that is placed in the plasma and contains many growth factors. Growth factors found in PRP are PDGFαα, PDGFββ, PDGFαβ, TGF-β, TGF-β2, VEGF, and EGF [32]. Matsui and Tabata, states that platelet activation occurs when in contact with gelatin which can release various growth factors in PRP [3]. PRP contains a lot of platelets. PRP activation by thrombin or CaCl2 platelets releases α-granules and hydrogel which contain many growth factors including TGF-β [34]. TGF-β contained in PRP can increase collagen production and matrix minerals [31].

The combination of scaffold and PRP has several advantages. First, the porous structure of scaffold can stabilize the formation of fibrin network before PRP is activated by forming a 3D environment for growing and differentiating cells. Second, it controls and extends in releasing several growth factors, including TGF-β [33]. Based on the discussion, it can be assumed that one of the factors which slows down the release of growth factors in PRP is due to the porous structure of the scaffold. Anticoagulants are substances that can prevent blood clots by binding the calcium or by inhibiting thrombin formation to convert fibrinogen to fibrin in the clotting process [34]. PRP formation requires anticoagulants so that in line with the discussion. It is assumed that more calcium is found in the incorporation of PRP than in the incorporation of PRF. PRP requires fibrinogen, thrombin and calcium to form the fibrin network so that the growth factors derived from platelets can be released. The second step in PRP preparation results in activating platelets and fibrin network formation. This occurrence is similar to the natural clotting process where fibrinogen is broken down by thrombin and it is responsible for the hemostasis process. The breakdown of fibrinogen will produce a three-dimensional fibrin network. Calcium can stabilize fibrinogen structures, accelerate fibrin formation by acting as a thrombin co-factor, which is able to protect fibrinogen degradation. It can also inhibit and prolong the polymerization process of fibrin fibers. Fibrin networks are influenced by the fibrin polymerization reaction [25]. In this study, researcher used the artificial coral scaffold which contains calcium and anticoagulants. They can bind calcium, so, it can be assumed that the amount of calcium derived from artificial coral scaffolds and those bound by anticoagulants can increase fibrin network formation, stabilize fibrinogen structure and slow down the fibrinogen degradation. Overall, it can slow down the process of scaffold degradation.

Fibrin fibers produced from PRP are thinner and denser[35]. It is in line with the research of Perez et al., which states that thinner fibrin fibers can slow down clotting time [25]. It can be estimated that the increase in mechanical strength of the scaffold incorporated with PRP occurs because calcium contained in PRP and bound by anticoagulants and calcium contained in artificial coral scaffold can strengthen the structure by forming fibrin network and by slowing down the degradation process of fibrinogen. It causes PRP incorporated scaffolds becomes slower in degradation.

5. Conclusion

There was significant difference among PRP, PRF, and control group. PRP incorporation has good swelling profile, slow degradation profile, and higher gel fraction that presents strengthening of scaffold structure.

6. References

[1] Munar, M. D. L. (2006)., Study of the Physical and Biological Properties of Three Dimensional Hydroxyapatite and α-Tricalcium Phosphate Foam with Interconnected Pores, Tesis, Section of Biomaterial Engineering Division of Oral Rehabilitation Graduate School of Dental Science Kyushu University, Fukuoka.

[2] P. Kumar, B. Vinitha, and G. Fathima, Bone grafts in dentistry, J Pharm Bioallied Sci. 5(suppl 1) (2013) S125–S127.

[3] M. Matsui, and Y. Tabata, "Enhanced angiogenesis by multiple release of platelet-rich plasma contents and basic fibroblast growth factor from gelatin hydrogels," Acta Biomaterialia, 8(5), 2012, 1792–1801.

[4] Baeyens, W., Glineur, R. & Avrard, L. (2010). The use of platelet concentrates: platelet-rich plasma (PRP) and platelet-rich fibrin (PRF) in bone reconstruction prior to dental implant surgery. pp. 521-527.

[5] Camelia, M. N. dan Masulili, S. L. C., 2011. Platelet-Rich Plasma sebagai Pendekatan Perawatan Periodontal Regeneratif. MKGI, pp. 119-126.

[6] Miron RJ, Zucchelli G3, Pikos MA4, Salama M5,6,7, Lee S8, Guillemette V9, Fujioka-Kobayashi M, 2017, Use of platelet-rich fibrin in regenerative dentistry: a systematic review, Clin Oral Investig. Jul;21(6):1913-1927.

[7] Dohan, D. M., Choukroun, J., Diss, A., Dohan, S. L., Dohan, A. J. J., Mouhyi, J., and Gogly, B. (2006). Platelet-Rich Fibrin (PRF): A Second-Generation Platelet Concentrate. Part I : Technological Concepts and Evolution. Mosby, Inc, 101, pp. E37-44. Doi: 10.1016/j.tripleo.2005.07.008.

[8] Tengkawan, M., Oktawati, S., and Djais, A. I., Burhanuddin, D. P. (2013). Penanganan Resesi Gingiva Miller Klas I-II Menggunakan Platelet-Rich Fibrin dan Subepithelial Connective Tissue Graft Treating Gingivalrecessionmiller ' S Class I – Iiusing Platelet-Rich Fibrin and Subepithelial Connective Tissue Graft. Dentofacial, 12(3), pp. 169-174.

[9] Damayanti, M. M., dan Yuniarti. (2016). Review Jurnal : Pengaruh Penyembuhan Luka Pascaekstraksi Gigi. Prosiding SnaPP2016 Kesehatan, 2(1)(Pissn 2477-2364 | Eissn 2477-2356), pp. 34-39.

[10] Lee, K. Y., and Mooney, D. J. (2001). Hydrogels for Tissue Engineering. American Chemical Society, 101(7).

[11] Kitamura, C., Nishihara, T., Terashita, M., Tabata, Y., Jimi, E., Washio, A., 2011. Regeneration Approaches for Dental Pulp and Periapical Tissue with Growth Factor, Biomaterials and Laser Irradiation. Polymers,3, 1776-1793.Doi : 10.1021/cr000108x.

[12] V.V. Gaikwad, A.B. Patil, M. V. Gaikwad, Scaffold for Drug Delivery in Tissue Engineering, International journal of Pharmaceutical Science and Nanotchnology. 1 (2008) 113-122.

[13] Dhirisma, F., dan Sari, D. P. (2014). Formulation of a Membrane-Based Porous Hidrogel Combination Of HPMC (Hydroxy Propyl Metyl Cellulose) and Gelatin By Using Foaming Gas Method and Determination of Physical- Mechanical Characteristics. Karya Tulis Ilmiah strata satu, Universitas Muhammadiyah Yogyakarta.

[14] Y. Pan, S. Dong, Y. Hao, T. Chu, C. Li, and Z. Zhang, "Demineralized Bone matrix gelatin as scaffold for tissue engineering," Journal of Microbiology, 4(9), 2010, 865–70.

[15] W. Wattanutchariya, and W. Changkowchai, Characterization of Porous Scaffold from Chitosan-Gelatin/Hydroxyapatite for Bone Grafting, Proceedings of IMECS. (2014) 2.

[16] E.S. Mahanani, I. Bachtiar, and I.D. Ana, Human Mesenchymal Stem Cells Behavior on Synthetic Coral Scaffold, Key Eng Mater. 696 (2015) 205-211.

[17] E.S. Mahanani, N Farda, I. Tejaning Asih, N Khairunissa, The Effects of Platelet Rich Plasma Incorporation Towards Swelling Profile and Gel Fraction of Synthetic Coral Scaffold, 2018 1st International Conference on Bioinformatics, Biotechnology, and Biomedical Engineering (BioMIC), (2018) 978-1-5386-7599-1/18, IEEE.

[18] Rustam, A., Tatengkeng, F., Fahruddin, A. M., dan Djais, A. I. (2017). Kombinasi Perancah Silk-Fibroin Dari Kepompong Ulat Sutera (Bombyx Mori) dan Konsentrat Platelet Sebagai Inovasi Terapi Regenerasi Tulang Alveolar Silk- Fibroin Scaffold from Silkworm Cocoon (Bombyx Mori) and Platelet Concentrate- Based Mineralized Plasmatic Matrix as Innovation in Alveolar Bone Regeneration Therapy. Makassar Dent J., 6(3), pp. 107-115.
[19] Y. Li, Y. Liu, S. Li, S., G. Liang, C. Jiang, and Q. Hu, "Novel Control of Gel Fraction and Enhancement of Bonding Strength for Constructing 3D Architecture of Tissue Engineering Scaffold with Alginate Tubular Fiber," Journal of Bioscience and Bioengineering Vol. 121, No. 1, 2016, 111-116.
[20] Saito T and Tabata Y, 2012, Preparation of gelatin hydrogels incorporating low-molecular-weight heparin for anti-fibrotic therapy, Acta Biomater. Feb;8(2):646-52.
[21] Wahyudi, Ivan. A., and Nurwadji, Lea. M. (2014). The Effect of Non Freeze-Dried Hydrogel-CHA on Fibroblast Proliferation. Journal of dentistry Indonesia, 22(3), pp. 89-93. Doi : 10.14693/jdi.v2li3.226.
[22] Park, J. –S., Kim, H, -A., Choi, J. – B., Gwon, H. –J., Shin, Y. –M., Lim, Y. –M., Khil, M. S., And Nho, Y. –C. (2012). Effect of Annealing and The Addition of PEG in The PVA Based Hydrogel by Gamma Ray. Radiation Phisics And Chemistry, 81, pp. 857-860. Doi : 10.1016/j.radphyschem.2012.02.005.
[23] Zhang, D. –K., Wang. D. –G., Duan, J. –J., And Ge, S. –R. (2009). Research on the Long Time Swelling Properties of Poly (Vinyl Alcohol)/ Hydroxylapatite Composite Hydrogel. Journal Of Bionic Engineering, 6, pp. 22-28. Doi : 10.1016/SI672-6529(08)60093-1.
[24] Giannini, S., Cielo, A., Bonanome, L., Rastelli, C., Derla, C., Corpaci, F., and Falisi, G. (2015). Comparison Between PRP, PRGF, and PRF : Light and Shadows in Three Similar but Different Protocols. European Rewiew For Medical And Pharmacological Sciences, 19, pp. 927-930.
[25] Perez, Amanda, G, M., Rodrigues, Ana, A., Luzo, Angela, C, M., Lana, Jose, F, S, D., Belangero, William, D., Santana, Maria, H, S. (2014). Fibrin network architectures in pure platelet-rich plasma as characterized by fiber radius and correlated with clotting time. Journal of Materials Science- Materials in Medicine, Vol 25, issue 8. doi 10.1007%2Fs10856-014-5235-z.
[26] A. A. M. Shimojo, A. G. M. Perez, S. E. M. Galdames, I. C. D. S. Brissac, and M. H. A. Santana, "Performance of PRP associated with porous chitosan as a composite scaffold for regenerative medicine," Scientific World Journal, 2015, 1-13.
[27] Rowe, S. L., Lee, S. Y., and Stegemenn. J. P. (2006). Influence of Thrombin Concentration on Mechanical and Morphological Properties of Cell-Seeded Fibrin Hydrogel. Acta biomaterial, 3, pp. 59-67. Doi : 10.1016/j.actbio.2006.08.006.
[28] N. Nagasawa, T. Yagi, T. Kume, and F. Yoshii, "Radiation Crosslinking of Carboxymethyl Starch," Carbohydrate Polymers, 58, 2004, 109–113.
[29] Wu, L. dan Ding, J., 2004. In vitro degradation of three-dimensional porous poly (D,L-Lactide-co-Glycolide) Scaffold for tissue engineering. Biomaterials, Issue 25, pp.5821-5830.
[30] Mahanani, E. S. (2013). Perancah Hidogel untuk Aplikasi Rekayasa Jaringan Tulang. IDJ, 2(2), pp. 51–56.
[31] Nirmala, L., Oley, M. C., Prasetyo, E., Hatibie, M. dan Loho, L. L., 2016. Pengaruh Pemberian Plasma Kaya Trombosit dan Karbonat Hidroksiapatit Pada Proses Penutupan Defek Tulang Kepala Hewan Coba Tikus. Jurnal Biomedik (JBM), Vol.8 No.3, pp. 172-178.
[32] Muljanti, M., Hernaningsih, Y., Nugraha, H. K. dan Nugraha, J., 2014. Upaya Optimasi Pembuatan Plasma Kaya Trombosit Sebagai Pengobatan Sel Punca. Indonesian Journal of Clinical Pathology and Medical Laboratory, 20(3), pp. 196-200.
[33] Tello, M. S., Martorell, S., Roig, M. M., Milian, L., Gonzalez, MA. G., Ribelles, J. L. G. dan Carda, C., 2017. Human Platelet-Rich Plasma Improves The Nesting and Differentiation Of Human Chondrocytes Cultured In Stabilized Porous Chitosan Scaffolds. Journal of Tissue Engineering, Vol.8, pp. 1-6.

[34] Shalehah, A., Cahaya, N. dan Faradilaturahmah., 2015. Pengaruh Pemberian Ekstrak Etanol Daun Kajajahi (Leucosyke Capitellata Wedd.) Terhadap Efek Pembekuan Darah dan Penurunan Agregasi Platelet Pada Darah Manusia Sehat Secara In Vitro. Pharmacy, Vol.12 No.02, pp.140-152.
[35] Weisel, J. W. 2007. Structure of Fibrin: Impact on Clot Stability. Journal of Thrombosis and Haemostasis, 5 (1), pp.116-124.

Potential Analysis of Majapahit Fruit Powder (Crescentia cujete L) as Shrimp Immunostimulants using the in Silico Method

Sri Rahmaningsih[1], Riska Andriani[2]

[1]Universitas PGRI Ronggolawe, Tuban - Indonesia
[2]Universitas PGRI Ronggolawe, Tuban - Indonesia

[1]* rahmaningsihsri@unirow.ac.id, [2] andrianiriska@unirow.ac.id

Abstract. Shrimp production is currently constrained by many diseases. These diseases are conventionally prevented by administering chemicals as immunostimulants, with their attendant side effects. Alternative natural ingredients as immunostimulants in shrimp production are being sought, and Majapahit plant (*Crescentia cuje* L.) is one. *In vitro* and *in silico* studies on the plant's stems, leaves and fruits showed the plant's desirable effects on the immune system of Vannamei Shrimp, as the plant contains important bioactive compounds (e.g. carotenoids, phenolics, alkaloids, pectin, tannins, flavonoids and terpenoids). The present study determined the ability of Majapahit fruit powder mixed into shrimp feeds in enhancing the immune system of the seafood. Liquid chromatography–mass spectrometry (LC-MS) revealed quercetin flavonoids as the effective active ingredient in the powder, which *in silico* study proved to be an effective immunostimulant for shrimps.

1. Introduction

Vanname shrimp is a sea shrimp that can be found widely in Pacific ocean waters starting from Mexico, Central and South America. Life inhabits all water columns, from the bottom to the surface layers, for example in muddy sandy areas, estuaries, and the sea, with a depth of 0 to 70 m [1]. According to [2] Vanamei shrimp (*Litopenaeus vannamei*) is a type of penaeid shrimp that has a higher endurance compared to tiger species. Even so, in the cultivation of vannamei shrimp, there are still many obstacles that can hamper success, including disease attacks caused by, among others, bacteria and viruses that can cause great losses to farmers.

One effort that can be done to reduce disease is by providing immunostimulant materials that can improve the immune system. Increasing the body's defense or immune system against attacks by t disease can be done by feeding with balanced nutrient composition, or by providing immunostimulants in feed, so that it will be directly related to immune system cells that make these cells more active [3]. Majapahit plants are flowering dicot plants originating from South and Southeast Asia including Indonesia, commonly found in tropical and sub-tropical regions. In the leaves of the majapahit plant it contains active ingredients; flavonoids, saponins, tannins, steroids, and triterpenoids. While the stem contains saponins, tannins, steroids and teriterpenoids [4]. According to [5], based on phytochemical tests of Majapahit fruit powder containing saponins, triterpenoids and flavonoids. The results of the analysis using FTIR, UV-VIS Spectrophotometry and GCMS, Majapahit fruit extracts contain active ingredients such as furfural, 1,2,3 Benzentriol, 4H-Pyran-4-one and others which have potential as antibacterial. Previous research proves that majapahit leaf powder can increase the immune system of sangkuriang catfish (*Clarias gariepenus*) that are infected with aeromonas bacteria [6], insilico majapahit leaf extract has proven potential as an antibacterial for cannamei shrimp infected with Vibrio harveyi bacteria [7]. Use of Majapahit leaf extract also plays a role in reducing the activity of Aeromonas hydrophila bacteria in tilapia (Oreochromis niloticus) cultivation [8].

According [9] insilico method or bioinformatics test is a test using a computer to determine the potential of a compound. The stages in insilico analysis include; selection and characterization of target molecules, visualizing the structure of target molecules and designing the mechanism of relationships between drug molecules or chemical compounds. This research determines the ability of Majapahit fruit powder mixed into shrimp feed to improve the shrimp's immune system in an insilico manner.

2. Research Methods

2.1. Material
The material used in this study was Majapahit fruit powder (*Cresentia cujete* L.) which was analyzed using LCMS. The result is a flavonoid type Quercetin -3β-D-glucoside, Quercetin, Chlorogenic Acid and Taxifolin [10].

2.2. Modelling Protein
Modeling of the LGBP protein (Lipopolysaccharide Glucan Binding Protein), Toll like receptor, and Lectin from Penaeus vannamei was done using the Swiss Model webserver (https: // swissmodel. Expasy.org/). Protein modeling is done because LGBP and Toll like receptor proteins are not yet available in the database, so homology needs to be done with 3D protein models that are already available in the database. LGBP protein construction uses a template with PDB ID (3AZX) with a sequence identity of 33.3%. While the Toll like receptor protein construction uses the PDB ID (5ZSA) protein template with a sequence identity of 21.3%. Proteins with PDB ID 6HUG were used to make Lectin protein with sequence identity of 22.22%.

2.3. Molecular Docking
Docking is done by using Autodock Vina on the PyRx 9.5 Protein program which is used as a target and controls are attached to the Excel data. The PyMol 2.3.1 program is used to visualize the docking results, while the LigPlot 2.1 program is used to see the amino acid interactions that occur.

3. Results and Discussion

3.1. Molecular Docking Analysis
The molecular docking method is also used to explore possible ligand binding modes and the relationship between 3D binding modes and structural features of 2D-molecular ligands so that they can be explored further. Wang et al (2010) [11]. Docking conducted in this study is reverse docking by mimicking the interaction of target proteins and ligands that are considered as controls (peptidoglycan from bacterial walls). Docking is a method for predicting the strength of interactions between receptors and ligands, based on binding affinity values. The more negative the value, the stronger the interaction that occurs between receptors and ligands. Compared to other bioactives, Quercetin-3β-D-glucoside has a stronger interaction with TLR, whereas Taxifolin with GLBP protein.

Table 1. Results of Majapahit Bioactive Docking

Pubchem ID	Compound (as a ligand)	Binding Affinity (kkal/mol)		
		GLBP	TLR	LECTIN
5280804	Quercetin-3β-D-glucoside	-7.9	-8.1	-7,3

5280343	Quercetin	-7.8	-7.8	-6,6
1794427	Chlorogenic Acid	-6.9	-7.6	-7,2
439533	Taxifolin	-8	-7.7	-6,4
9816401	Peptidoglycan	-5.9	-5.4	-5,5

Table 1. describes the docking results of LCMS compounds from majapahit fruit powder (*Crescentia cujete* L.) namely Quercetin -3β-D-glucoside, Quercetin, Chlorogenic Acid and Taxifolin, as ligands with proteins that play a role in shrimp immune system, namely LGBP (Lipopolysaccharide Glucan Binding Protein), Toll like receptor, and Lectin as receptor. The value of binding affinity (kcal / mol) of receptors with ligands shows a higher value than control (peptidoglycan) both in GLBP, TLR and Lectin ligands, this shows that the compounds contained in Majapahit fruit powder have stronger affinity binding bonds than controls.

According to [12] Ligands are substances that form complexes with biomolecules to serve biological purposes. In protein-ligand binding, ligands are usually molecules that produce signals by binding to sites on the target protein (receptors). For example the binding of DNA-ligands, ligands can be small molecules, ions, or proteins. Besides the bonding relationship between ligands and receptors in the form of charge functions, hydrophobic, and molecular structure.

Ligands binding to receptor proteins can change conformations by affecting the orientation of three-dimensional shapes. The binding level is called affinity, (binding affinity) which indicates the tendency or strength of the effect. Binding / affinity bonding is actualized not only by host-guest interactions, but also by the effect of solvents which can play a dominant steric role that encourages non-covalent binding in solution. Besides the binding occurs by intermolecular forces, such as ionic bonds, hydrogen bonds and Van der Waals [13]. Figures 2, 3 and 4 show the visualization of docking results between ligands (Quercetin -3β-D-glucoside, Quercetin, Chlorogenic Acid and Taxifolin) and receptors (LGBP (Lipopolysaccharide Glucan Binding Protein), Toll-like receptors, and Lectin).

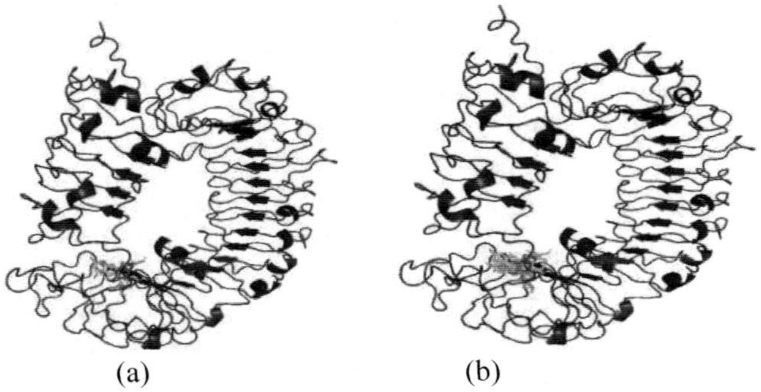

Figure 1. a) Visualization of TLR docking results. Red (Control), Green (Quercetin-3β-D-glucoside) b). Visualization of TLR docking results. Red (Control), Green (Quercetin-3β-D-glucoside), Cyan (Chlorogenic Acid), Pink (Taxifolin), Orange (Quercetin)

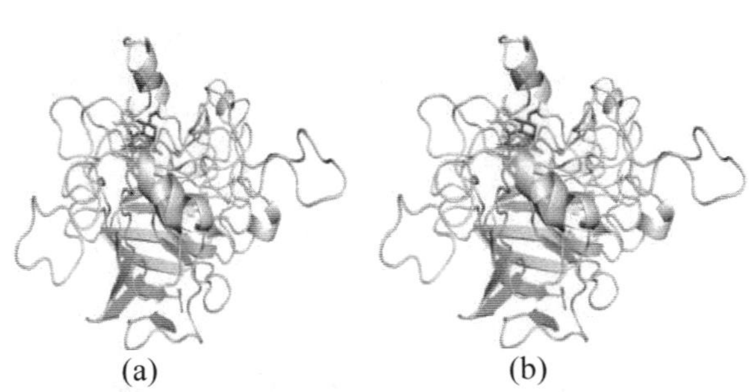

Figure 2. a) Visualization of GLBP docking results. Red (Control), Pink (Taxifolin), b) Visualization of GLBP docking results. Red (Control), Green (Quercetin-3β-D-glucoside), Cyan (Chlorogenic Acid), Pink (Taxifolin), Orange (Quercetin)

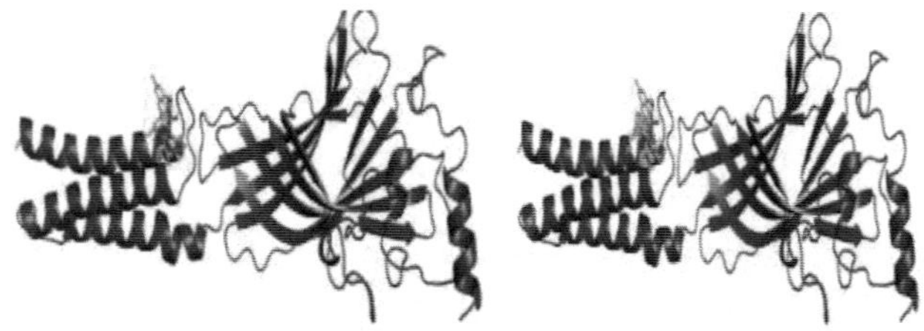

(a) (b)

Figure 3. a) Visualization of Lectin docking results. Red (Control), Green (Quercetin-3β-D-glucoside), b) Visualization of Lectin docking results. Red (Control), Green (Quercetin-3β-D-glucoside), Cyan (Chlorogenic Acid), Pink (Taxifolin), Orange (Quercetin)

After knowing the value of binding affinity, it is necessary to do further analysis, namely the interaction of amino acids between proteins and ligands, using the LigPlot program. Based on LigPlot analysis there are several amino acids that are the same between bioactive and ligand control, which means that predicted bioactive Majapahit can interact with the target protein similar to the interaction between the target protein and the peptidoglycan used as control. The following table shows a visualization of the results of the Lig Plot.

Table 2. LigPlot Visualization Results

		Hydrophobic Bonds	Hydrogen Bonding
TLR	Control	ILE725, LEU694, ARG681, LEU722, THR691	ARG727, ARG726, ARG690, ASP692
	Quercetin-3β-D-glucoside	THR691, ARG690, ILE725, LEU722, ARG681, ASP692, ARG678, LEU718, LEU694	ALA729, ARG726, SER724
	Quercetin	LEU718, LEU694, ARG678, LEU722, ILE725, ASP692	SER724, VAL689
	Chlorogenic Acid	ARG690, ASP692, ARG678, LEU718, ARG681, LEU694, LEU722, ILE725	ARG726, SER724, ARG727
	Taxifolin	LEU718, ARG678, LEU694, ASP692, LEU722, ILE725, THR691	ARG690, VAL689, SER724
GLBP	Control	SER188, GLY203, PHE307, VAL222, TYR201	THR220, HIS221, ARG189, VAL222
	Quercetin-3β-D-glucoside	GLU223, ARG189, SER188, PHE307, GLY203, VAL222	ASN191, TYR201, GLU187, ASN199, THR220, HIS221
	Quercetin	GLU223	ASN199, ARG189, TYR201, HIS221
	Chlorogenic Acid	TYR201, GLY203, PHE307, GLU223, HIS221	ARG189, ASN191, THR220, GLU187, THR205
	Taxifolin	SER188, GLU187, GLY203, HIS221, PHE307	TYR201, ARG189, THR205, VAL222, THR220
LECTIN	Control	LEU652, THR655, LYS654, ASP662, SER649, THR648, TYR657	PRO653, SER656, LYS659
	Quercetin-3β-D-glucoside	LEU645, ASP662, PHE658, LYS659, TYR657, THR648, LEU652	LYS654, SER649, PRO653, THR655
	Quercetin	SER649, LEU645, LEU665,	LYS654, THR648, LYS659

	Hydrophobic Bonds	Hydrogen Bonding
	LEU652, ASP662, SER656, VAL661	
Chlorogenic Acid	THR648, SER649, LEU645, VAL661, PHE666, PHE658, LEU652, LYS654	PRO653, THR655, LYS659, ASP662
Taxifolin	LYS659, PHE658, PHE666, LEU645, LYS654, SER649	LEU652, PRO653, ASP662

From table 2. it is known hydrophibic bonds and hydrogen bonds between ligands and receptors and bonds on control. Amino acids that bind to control are also present in ligand compounds, this shows that ligand compounds also work like controls that play a role in the shrimp immune system. Based on its affinity bond, the Taxifolin compound binds best with GLBP WITH VALUE -8 kcal / mol, cholargen acid compound binds best
with TLR at -8.1 kcal / mol, whereas quercetin compound binds best with TLR and GLBP at -7, 8 kcal / mol, while quercetin 3β-D-glucoside compounds are best bound to TLRs of -8.1 kcal / mol. Figure 5 is a visualization of the bonds that occur between compounds as ligands with amino acids both hydrophobic and hydrophilic bound.

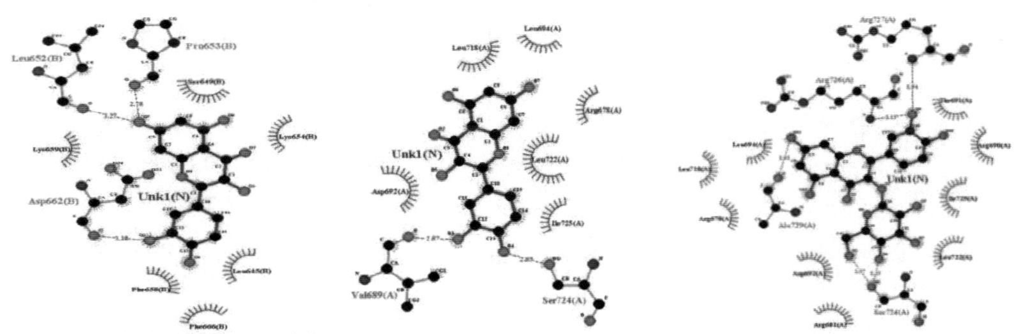

(a) (b) (c)

Figure 4. a) Visualization of 2D Lectin with taxifolin, b) TLR with Quercetin, c) TLR with quercetin 3β-D-glucoside

4. Conclusions

Quercetin-3β-D-glucoside is predicted to be bioactive which has the best potential to influence immunity in shrimp.

5. Acknowledgments

Thank you to RistekDikti for funding this research through Higher Education Applied Grants Contract Number 313/071073 / PGRI / KL / III / 2019.

References

[1] Rusmiyati S. Menjala Rupiah Vannamei Shrimp Cultivation New Varieties Featured. (New Library Press, Yogyakarta. 2012. pp. I62
[2] Zainuddin, Haryati, Aslamyah, S., Surianti, Effect of Carbohydrate Level and Feed Frequency on Feed Conversion Ratio and Juvenile Litopenaeus vannamei survival. J.Fish. Sci XVI (1) (2014), 29-34.

[3] Ekawati, A.W, Nursyam. H., Widjayanto, E., Marsoedi, Diatomae Chaetoceros ceratosporum in Feed Formula Increases Cellular Immune Response of Tiger Shrimp (Penaeus monodon Fab.) J. Exp. Life Sci Vol. 2 No. 1, (2012), 20-28.

[4] Rahmaningsih, S. Dan Jumiati, Study of the Utilization of Majapahit Plants (Crescentia cujete) for the Control of Vibrio Harveyi Bacteria In In vitro, Proceedings of the VI National Fisheries and Marine Seminar, Faculty of Fisheries and Maritime Sciences, Malang University, (2016) Hal 52-58

[5] Rahmaningsih, S., Prajitno A., Aulanniam and Maftuch, ioactive Compounds From Majapahit Fruit (Crescentia cujete) As a Potential Natural Antibacterial. International Journal of Chem Tech Research. Vo. 10. No.3 (2017) pp 90-99.

[6] Rahmaningsih S., Zaenudin, Sudianto, Overview of Blood Hematocrit of Sangkuriang Catfish (Clarias Gariepenus) which are fed majapahit leaf powder and infected with Aeromonas Hydrophila Bacteria. JKPT Vo. 1. No.2. (2018), 11-29.

[7] Rahmaningsih S. And Andrini R. Biological Activity of Majapahit Leaf Extract (Cresentia cujete) and Its Potential As Insilico Antibacterial Antibacterial. National Seminar Proceedings on Research and Community Service Results. (2017), pp 80 -87.

[8] M Luthfi, S Rahmaningsih, A Sudianto, Effect of Majapahit Leaf Extract (Crescentia cujete) with different Doses To reduce the activity of Aeromonas hydrophila bacteria in tilapia (Oreochromis niloticus) cultivation. Proceedings of SNasPPM, 2017, 209-214

[9] According to Nursamsiar, Alprida Tandi T., Akbar A. Study of Insilico Compounds of Chalone and Pyrimidine Analogs as Anti-Inflammation: Prediction of Absorption, Distribution and Toxicity. Pharmacy. Volume. 13.No. 01. (2016)

[10] Rahmaningsih S. and Andriani R. Use of Majapahit Fruit Powder in Feed to Improve Shrimp Immune System. Research Report on Applied Grants for Higher Education Research and Technology Higher Education (2019)

[11] Z. Wang, Y. Li ,, C. Ai, and Y. Wang, Silico Prediction of Estrogen Receptor Subtype Binding Affinity and Selectivity Using Statistical Methods and Molecular Docking with 2-Arylnaphthalenes and 2-Arylquinolines, Int. J. Mol. Sci 11, 3434-3458; doi: 10.3390 / ijms11093434 (2010).

[12] Teif VB. "Ligand-induced DNA condensation: choosing the model". Biophysical Journal. 89 (4): 2574–87. (2005).

[13] Baron R, Setny P, McCammon JA "Water in cavity-ligand recognition". Journal of the American Chemical Society. 132 (34): 12091-7. (2010).

IOP Conf. Series: Materials Science and Engineering **874** (2020) 012003 doi:10.1088/1757-899X/874/1/012003

The Effect of Giving Okra (Abelmoschus esculentus L.) Juice on HDL Levels in the Blood Serum of Wistar Strain Rats Fed High-Fat Feed

Rizqi Nur Alifah [1], Endah Peniati [1], Ely Rudyatmi [1], Ning Setiati [1]

[1] Universitas Negeri Semarang, Semarang – Indonesia

* rizkinuralifah@mail.unnes.ac.id

Abstract. Cardiovascular disease is the biggest cause of death in the world with the number of deaths reaching 17.7 million people (31% of the total number of deaths). The main cause of cardiovascular disease is a disturbance in the function of the heart and blood vessels caused by atherosclerosis due to the deposition of cholesterol in artery walls. High Density Lipoprotein (HDL) functions to transport free cholesterol contained in the endothelial peripheral tissue to HDL receptors located in the liver so that it can be processed to be used as bile. Therapy to increase levels of HDL in the blood using okra juice (Abelmoschus esculentus L.) can be done in rats that have been given high-fat feed. The purpose of this study was to determine the effect of okra juice (Abelmoschus esculentus L.) on HDL levels in rats fed high fat. The study was conducted with giving okra juice (Abelmoschus esculentus L) with dosage difference of 1500 mg/kg bw, 3000 mg/kg bw and 6000 mg/kg bw. The research design used was complete random design with post-test randomized group design with a number of replications 5. Testing of HDL levels was carried out by the CHOD-PAP (Cholesterol Oxidase Para Aminophenazone) method. The research data were analyzed using one-way Anova followed by LSD tests. The results of the study giving okra juice (Abelmoschus esculentus L.) to HDL levels by induction of high-fat feed at a dose of 6000 mg / kg body weight were significantly different from lower dosage. Conclusion, okra juice (Abelmoscus esculentus L.) has a positive effect on increasing HDL levels..

1. Introduction

Cardiovascular disease is one of the non-communicable diseases with the highest mortality rates in the world with total deaths globally by 17.7 million people (31% of total deaths) [1]. Cardiovascular disease is a disease caused by impaired heart and blood vessel function such as coronary heart disease, heart failure, hypertension and stroke. The main cause of cardiovascular disease is atherosclerosis, a progressive multifunctional disease of the arterial wall. The central pathogenesis of atherosclerosis is cholesterol deposits in arterial walls [2]. Factors that can cause cardiovascular disease are unhealthy lifestyles such as irregular exercise, smoking, alcohol consumption and unhealthy diets. Unhealthy diets such as consumption of fast food can cause various diseases due to the tendency of consuming foods that are high in carbohydrates and saturated fats [3]. Foods high in carbohydrates and saturated fats can increase intracellular cholesterol, and cholesterol is stored as cholesterol esters. High-cholesterol diet can also lead to a decrease in the LDL receptor gene transcription so synthesis LDL receptor, and LDL level in the circulation increases [4].

LDL levels continue to increase can suppress HDL and cannot get rid of excess cholesterol in the blood so that HDL levels decrease [5]. Decrease in cholesterol levels in the blood can be influenced by changes in a healthier lifestyle, one of which is by changing diet by consuming fruits and vegetables. One of the fruits or vegetables that can reduce cholesterol levels in the blood is okra (Abelmoschus esculentus L). Okra (Abelmoschus esculentus L) is a plant from the malvaceae with a life cycle length of 90-100 days [6]. Okra plants (Abelmoschus esculentus L) are widely cultivated in the tropics, subtropics and regions with warm temperatures that exist throughout the world [7] with India as the country with the highest production of okra (Abelmoschus esculentus L). Okra is widely used as an ethnic treatment [8]. Okra (Abelmoschus esculentus L) is reported to have antioxidant activity in the form of quercetin and phenolic flavonoids [9]. Used cooking oil contains many free radicals due to lipid peroxidation [10]. Repeated use of cooking oil can cause damage to body organs such as blood vessels, heart, liver and kidneys due to the accumulation of fat (cholesterol) [11]. Used cooking oil has undergone chemical and physical changes. Chemical changes in the form of hydrolysis, oxidation and polimeration processes. While physical changes are marked by an increase in oil viscosity, a change in color and foam formation during frying [12]. The purpose of this study was to determine the effect of okra juice on HDL levels in rats fed high fat.

2. Method

The rats used were 25 male Wistar strain aged 2-3 months with a body weight between 150200g. Rats were grouped into 5 treatment groups with a number of replications 5. Experimental research was carried out with a completely randomized design method with Post-test Randomized group design. Producing Used cooking oil Used cooking oil is made using packaged cooking oil. Frying is done with frying tofu (250 g) for every 1 liter of oil with 4 repetitions [13] Producing Okra Juice Okra used in this research is green okra with species of Abelmoschus esculentus L. Moench which is still young with fruit size between 7-12 cm. Okra that has been washed clean, weighed at 150 grams and then cut into small pieces using a knife then added with 150 ml of water and blended until smooth then put in a bottle to be stored in the refrigerator with a temperature of -4oC.

Treatment The treatment was done for 7 days with induction cooking oil as high-fat feed as much as 1.5ml. The treatment was followed by induction of okra juice at each dose of each group. Group 1 (positive control) was given treatment in the form of standard feeding and drinking. Group 2 (negative control) was given treatment in the form of high-fat feed feeding and drinking. Groups 3-5 (T1, T2 & T3) were given high-fat feed and okra juice with doses of 1500 mg/kgBW, 3000 mg/kg BW and 6000 mg/kg BW respectively. On day 8th rat blood drawn through plexus retro orbitalus as much as 1.5 ml. HDL levels in rat blood serum were tested using the CHOD-PAP (Cholesterol Oxidase Para Aminophenazone) method. The data obtained were analyzed using the Shapiro-Wilk test for the normality test followed by the One-way Anova test and the LSD Test..

3. Results and Discussion

The process of atherosclerosis begins with an increase in cholesterol levels, causing LDL levels to rise and HDL to fall. Increased LDL triggers LDL oxidation and causes inflammation of blood vessel walls. The results of the study giving okra juice to rats fed high-fat feed are presented in Table 1.

Table 1. Summary of LSD follow-up test results on the effect of HDL levels in rats fed high fat feed

Groups	Mean
C -	91 a
C +	78 a b
T1	69 a b
T2	87,4 a b

Groups	Mean
T3	68 b

Note:
C- is a negative control
C+ is a positive control
T1 is a treatment with induction of okra juice with a dose of 1500 mg / kg BW
P2 is a treatment with induction of okra juice at a dose of 3000 mg / kg BW
T3 is a treatment with a dose of 6000mg / kg BW.

The 5% LSD test results showed that the response of HDL levels in the T3 group was significantly different when compared to the positive control but not significantly different from the negative control, the T1 and T2 treatment groups. This happens because the dose of okra juice is too low (1500 and 3000 mg/kg body weight /day) so that okra juice cannot bind all cholesterol and fat in the intestine.

In the study that has been done, it was shown that the value of HDL levels in rat blood serum was 69.6, 87.4 and 68 mg/dL for the treatment groups respectively 1500, 3000 and 6000 mg/kg BW/day respectively compared to 78 mg/dL for negative control and 91 mg/dL for positive control. Induction of used cooking oil that acts as a high-fat feed for 7 days shows the HDL levels in the blood with the highest average. This result is inversely proportional to the treatment given okra juice with a dose of 6000 mg/kg BW/day as the highest dose which has the lowest average HDL levels of all treatments. [14] mentioned a significant increase in HDL levels in diabetic rats after being given okra juice. Muchilage produced by okra is also beneficial for reducing the level of hyperglycemia and decreasing dyslipidemia in streptozotocin-induced diabetic rats [15]. The secondary metabolite compounds found in okra (*Abelmoschus esculentus* L) include alkaloids, glycosides, tannins, steroids/terpenoids and flavonoids [16]. Alkaloid compounds can inhibit the performance of the lipase enzyme in the digestive tract so that fat absorption in the body is reduced [17]. Terpenoid compounds play a role in inhibiting cholesterol biosynthesis by regulating the degradation of the enzyme 3-hydroxy-3-methylglutaril (HMG-CoA) reductase [18]. [19] stated that the skin of okra fruit and okra seed powder (*Abelmoschus esculentus* L) have potential as anti-hyperlipidemic in diabetic rats. Further research has identified the presence of two major flavonoids of glucoside (isoquercetin and quercetin 3-O-betaglucopyranosyl-(1-6) glucoside) in okra seeds (*Abelmochus esculentus* L) showing inhibitory activity of α-glucosidase which is responsible for decreasing blood sugar levels in rat. Quercetin works in decreasing free radicals by transferring an electron to free radical compounds and forming complexes with metals and increasing the activity of antioxidant enzymes such as SOD, CAT and GSH-Px. Quercetin can work as a strong scavenger because it has an aromatic hydroxy group [20]. Quercetin can reduce de novo synthesis in fatty acids thereby affecting cholesterol biosynthesis and lipoprotein formation [21]. Fiber in the okra can affect hypolipidemic, stabilized blood sugar levels by determining the absorption of blood sugar levels by the intestine and decreasing sugar assimilation through the intestine [22].

4. Conclusion

The administration of okra (Abelmoschus esculentus L) juice to Wistar strain rats fed with high-fat feed influences HDL levels in blood serum at a dose of 6000 mg / kg BW.

Acknowledgments

We thank the supervisor from the Biology Department, Faculty of Mathematics and Natural Sciences, Universitas Negeri Semarang.

References

[1] World Health Organization (WHO). 2017. Global Health Observatory (GHO) data. World Health Organization. Accesed September 11st 2019.

[2] Kumar S, Negi S. 2014. Transformation of waste cooking oil into c-18 fatty acids using a novel lipase produced by Penicilium chrysogenum through solid state fermentation. 3 Biotech. 5(5): 847-51

[3] Fairudz, Alyssa & Khairun Nisa. 2015. *Pengaruh Serat Pangan terhadap Kadar Kolesterol Penderita Overweight*. Majority. 4(8)

[4] Anas E & Asterina. 2011. Efek Pemberian Jus Tomat (Lycopersicum pyporme) terhadap Spermatogenesis pada tikus putih (Rattus norvegicus) Jantan Dewasa Hiperkolesterolemia. *Majalah Kedokteran Andalas* 35(1): 27-36.

[5] Riesanti DG, Masdiana CP, Herawati. 2013. Kadar HDL, Kadar LDL dan Gambaran Histopatologi Aorta Pada Hewan Model Tikus (Rattus norvegicus) Hiperkolesterolemia Dengan Terapi Ekstrak Air Benalu Mangga (Dendrophthoe pentandra). Student Journal Universitas Brawijaya 2(3): 1-10.

[6] Centelles VM, V Martinez, E Carballido, DM Bou, M Enrich, MC Saiz, J Masats, Y Canas, MT Gavalda, M Castella, J de la Fuente, JC Fajas, P Barnola, A Carreras, P Alarcon, S Sinclair, V Fedor, KA Cogollo. 2019. *Okra characteristics, uses and composition.* Botanical Online. [https://botanical-online.com/en/botany/okra-abelmoschus-esculentus] accesed on September 2019

[7] Lamont W. 1999. Okra a Versatile Vegetable Crop. Holticulture Technology. 9: 179-184

[8] Kumar, D. S., D. E. Tony, A. P. Kumar, K. A. Kumar, D. B. Srinivasa Rao & R. Nadendla. 2013. A review on: Abelmoschus esculentus (Okra). *International Research Journal of Pharmaceutical and Applied Scienc*e (IRJPAS) 3(4): 129-132

[9] Nabila, M, E Damayanthi, and S A Marliyati. 2018. Extracts of Okra (Abelmoschus esculentus L.) improves dyslipidemia by amelioratinglipid profile while not affectinghs-CRP levelsin streptozotocin-induced rats. FORUM IPIMA IOP Conf. Series:Earth and Environmental Science 196

[10] Iqbal HM, A Ulilalbab, AP Dani, T Estiasih. 2014. Effervescent rosella ungu mencegah penurunan sod dan mencegah nekrosis hepar tikus Wistar yang diberi minyak jelantah. *Jurnal Kedokteran Brawijaya*. 28(2): 85-90

[11] Shastry CS, PN Ambalal, J Himanshu & BJ Aswathanarayana. 2011. Evaluation of effect reused edible oils on vital organs of wistar rats. Nitte University of Health Science. 1(4): 10-15

[12] Kumar, Sanjeet. Dagnoko S. Haougui A. Ratnadass A. Pasternak D. Couame C et al. 2010. Okra (Abelmoschus esculentus spp.) in West and Central Africa: Potential and Progress on its Improvement. African Journal of Agricultural Research. 5(25): 3290-3598

[13] Muhartono, MA Yudhistira, NT Putri, TN Sari & Oktafany. 2018. Minyak Jelantah Menyebabkan Kerusakan pada Arteri Koronaria, Miokardium dan Hepar Tikus Putih (Rattus norvegicus) Jantan Galur Sprague dawley. JK Unila. 2(2): 129-135

[14] Onuoha, Nnenna O, Iroegbu L U, Uwaezuoke N J. 2017. Anti-diabetic effect of okra (Abelmoschus esculentus (L.) Moench) fruits in alloxan-induced diabetic rats. Biokemistri. 29(2): 94-101

[15] Hajian S, Asgary S, Rafieian-Kopaei M, Sahebkar A, Goli-Malekabady N, Rashidi B. 2016. Hibiscus esculentus seed and muchilage beneficial on reducing complications of diabetes in streptozotocininduced diabetic rats. *Annals of Research in Antioxidants* 1(2): e23

[16] Marpaung, Nona. 2018. Uji aktivitas antioksidan ekstrak etanol buah okra hijau (Abelmoschus esculentus (L.) Moench) secara spektrofotometri UV-Vis. Skripsi. Sumatera Utara: Universitas Sumatera Utara

[17] Budiarto M A, Yuniwarti E Y W, Isroli. 2016. Pengaruh pemberian tepung daun jati belanda (Gauzuma ulmifolia L.) dalam pakan terhadap kadartrigliserida darah dan lemak abdominal ayam broiler. *Bulletin anatomi dan fisiologi*. 1(1): 43-47

[18] Warditiani N K, Indrani A A I S, Swasti N A P P, Dewi N P A K, Widjaya I N K, Wirasuta I M A G. 2015. Pengaruh pemberian fraksi terpenoid daun katuk (Sauropus androgynous (L) Merr) terhadap profil lipid tikus putih (Rattus norvegicus L.) jantan galur Wistar yang di induksi pakan kaya lemak. Jurnal Farmasi Udayana. 4(2): 66-71

[19] Sabitha V. Ramachandran S. Naveen KR. Panneerselvam K. 2011. Antidiabetic and antihyperlypidemic potential of Abelmoschus esculentus (L.) Moench in streptozotocin induced diabetic rats. Journal of Pharmacy and Bioallied Sciences. 3(3): 397- 402

[20] Du Thie G, Crozier A. 2000. *Plant derived phenolic antioxidant*. Curr. Opin. Clin. Nutr. Metab. 447-51

[21] Gnoni G V, Paglialonga G dan Siculella L. 2009. Resveratrol inhibits fatty acid and triacylglycerol synthesis in rat hepatocytes. *European Journal of Clinical Investigation*. 39(3) 761-8

[22] Ngoc T, Ngo N, Van T dan Phung V. 2008. Hypolipidemic effect of extract from Abelmoschus esculentus L. (Malvaceae) on tyloxapol-induced hyperlipidemia in mice. Mahidol University Journalof Pharmaceutical Sciences. 35(1-4): 42-46

IOP Conf. Series: Materials Science and Engineering **874** (2020) 012004 doi:10.1088/1757-899X/874/1/012004

Environmentally friendly production method of virgin coconut oil using enzymatic reaction

Sabtanti Harimurti*, Rizka Meilisa Rumagesan, and Susanawati

Universitas Muhammadiyah Yogyakarta, Yogyakarta, Indonesia

*sabtanti@umy.ac.id

Abstract. Virgin Coconut Oil (VCO) has been known in Indonesia since a decade ago. This oil is made from coconut milk using unheated method, such as enzymatic, physical, and chemical method. The VCO has been proofed to have many health benefits such as for treatment of hyper-cholesterol, diabetes, viral and bacterial infections. Also, the oil has been utilized in the cosmetics industry for beauty skin care. In the increase of VCO demand for pharmaceutical industry, this study was conducted to find the method for VCO production using an environmentaly friendly process. The environmentally friendly process was conducted to maintain the sustainable environment. This paper deals with the enzymatic method of VCO production. The enzyme was obtained from fresh pineapple juice that is proofed contain bromelin. The enzyme will cleave the polypeptide bond of emulsifying agent in the coconut milk. Further, the oil will be separated and then collected as the VCO. During this work, the amount of the fresh pineapple juice varied to measure the rate of VOC production. Based on the observation during the production, it is found that the increase of fresh pineapple juice added increases the VCO production.

1. Introduction

Coconut oil has been known for centuries in human life. The type of coconut oil is divided into two, namely ordinary coconut oil and virgin coconut oil (VCO). Coconut oil is usually obtained from heating copra with high temperatures while VCO is obtained without heating [1]. VCO has advantages in helping the treatment process of disease. Active substance that helps in the treatment process is lauric acid [2]. Lauric acid content in VCO is able to cure a patient with high cholesterol, diabetes, stroke, hypertension, hepatitis, cancer, antibacterial and antiviral, and maintain healthy skin [3 – 5] In addition, lauric acid is also used as raw material for the cosmetics and pharmaceutical industries. According to the previous researches, the process of VCO production is divided into three methods, namely physical, chemical and enzymatic. Physical method is production involving low temperature heating and centrifugation. The chemical method is production involving acidification and stimulating method. While enzymatic method is production involving protein degrading enzyme [6 – 14].

The enzymatic method in VCO production uses coconut milk as the main ingredient and protease enzyme as a supporting ingredient. The enzyme will break down the coconut emulsion, then the VCO can be collected. The enzymatic methods have many advantages, since the enzyme is a non-toxic ingredient, reactions accelerating agent, and active at low concentrations. The enzyme reaction speed is regulated by temperature, pH and enzyme concentration [15]. The reaction is also easily cut when the results has been desired. Another important thing using enzymes in the VOC production is enzymes categorized as a natural ingredients that is environmentally friendly [16]. One of the protease enzymes can be obtained from pineapple which according to the literature contains bromelin enzyme [17]. Based on the background mentioned, this study was conducted to determine the effectiveness of the enzymes contained in the fresh pineapple fruit juice in the production of VCO.

2. Method

2.1. Chemicals

The main ingredient is coconut and raw pineapple which were obtained from Purworejo, Central Java. Chemicals such as NaOH, KOH, oxalic acid, alcohol, HCl, phenolphthalein, and distillated water were obtained from Brataco.

2.2. VCO production

One kg of grated coconut was added with 3 liters of clean water and the coconut milk is taken, then let it settle for 6 hours until a layer of cream and water are formed. The top layer of cream was taken and measured using measuring cilinder for 100 mL then put into countainer. While preparing the coconut cream, the pineapple juice was prepared by mincing 500 grams of raw pineapple using blender then filtered. Pineapple juice was then put into each container containing coconut cream at different concentrations of 0 mL (control), 5 mL, 10 mL, 15 mL, 20 mL, 25 mL, and 30 mL, respectively (Table 1). After that, the cream and pineapple juice were homogenized and let sit for 24 hours. In the end of 24 hours, it formed 3 layers, namely "blondo", oil and water. The oil layer is in the middle layer. Separation of the oil was done by drawing one by one from the lowest layer. The volume of the oil the measured to study the effect of enzyme concentration in the VCO production. All the experiments were done in 3 times. Water content, free acid number, and saponification number were tested to analyze the quality of VCO. The interpretation of the measuring data was according to Codex Stand.

Table 1. Experiment setting of VCO production

Group	Coconut milk (mL)	Raw pineapple juice (mL)
1	100	0
2	100	5
3	100	10
4	100	15
5	100	20
6	100	25
7	100	30

3. Results and Discussion

In this study, the VCO has been produced by enzymatic method. The enzyme was obtained from fresh raw pineapple juice. Pineapple juice (Ananas comosus) contains a protease called bromelin [18]. Bromelin is able to hydrolyze peptide bonds in proteins or polypeptides into amino acid molecules [19]. Addition of solutions containing proteases into coconut milk cream will break down the proteins that act as emulsifiers in coconut milk. Thus, the VCO is able to be extracted from coconut milk cream [20]. Enzymatic hydrolysis of peptide is depicted in Figure 1.

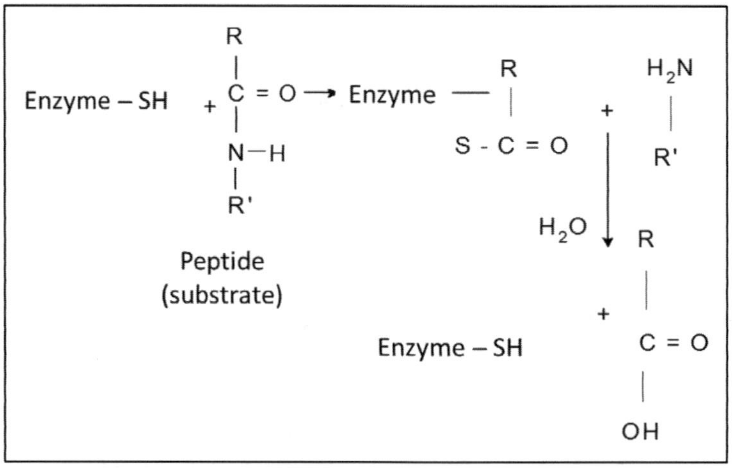

Figure 1. Enzymatic hydrolysis of peptide [21].

In the study of the effect of enzyme concentration, seven different experiments were done. All the experiments were conducted for 3 times to obtain the correct data. The experiments were started by preparing coconut milk than separation of coconut cream from the milk by settling the milk for 6 hours. Seven (7) clean containers were than filled with 100 mL of coconut cream and raw pineapple juice at 7 different volumes. The variation of the experiments and the VOC produced at 7 different experiments are explained in Table 2.

Table 2. VCO production in different concentration of raw pineapple juice

Group	Coconut cream (mL)	Raw pineapple juice (mL)	Volume of VCO			
			Data 1	Data 2	Data 3	Mean ± SD (%)
1	100	0	17.00	16.90	17.20	17.03 ± 0.15
2	100	5	17.50	17.50	17.60	17.53 ± 0.06
3	100	10	20.00	19.80	21.40	20.40 ± 0.87
4	100	15	22.50	22.30	22.50	22.43 ± 0.12
5	100	20	24.00	23.90	23.70	23.86 ± 0.15
6	100	25	24.20	24.10	23.90	24.06 ± 0.15
7	100	30	24.30	24.20	24.30	24.26 ± 0.06

Based on Table 2, in the 0 mL of raw pineapple juice added produce 17.03 ±0.15 mL of VCO. This may due to the production of VCO conducted by enzyme that come from the bacteria are available in the system. This can happen since the process was done in non-aseptic work and also was done without any sterilization process. In the increase of raw pineapple juice added, the VCO that is produced is increasing until in the certain level. It can be seen in the Figure 1, the VCO produced reach around 24 mL when 20 mL of raw pineapple juice was added. The higher concentration of pineapple juice added does not increase the VCO. A clear explanation can be seen also in Table 3. The production rate of VCO does not increase in the increase of raw pineapple juice added. This may due to all the substrate (coconut milk) have been reacted with the enzymes [22].

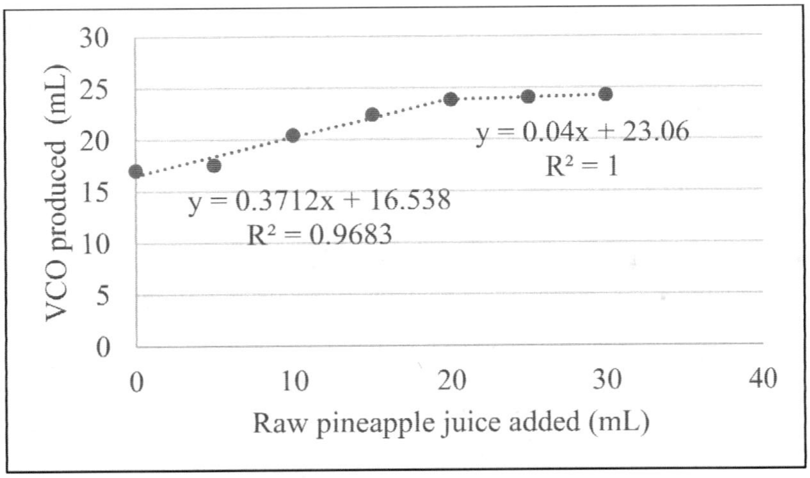

Figure 1. Profile of VCO production in the difference koncentration of pineapple juice

Table 3. Production rate of VCO in different concentration of raw pineapple juice.

Group	Raw pineapple juice (mL)	VCO produced (mL)	Fermentation period (minute)	Production rate (mL minute^{-1})
1	0	17.03 ± 0.15	1440	(1.10 ± 0.010) x 10^{-2}
2	5	17.53 ± 0.06	1440	(1.20 ± 0.004) x 10^{-2}
3	10	20.40 ± 0.87	1440	(1.40 ± 0.060) x 10^{-2}
4	15	22.43 ± 0.12	1440	(1.50 ± 0.008) x 10^{-2}
5	20	23.86 ± 0.15	1440	(1.60 ± 0.010) x 10^{-2}
6	25	24.06 ± 0.15	1440	(1.60 ± 0.010) x 10^{-2}
7	30	24.26 ± 0.06	1440	(1.60 ± 0.004) x 10^{-2}

The quality of VCO was evaluated by measuring water content, free acid number, and saponification number. While, the interpretation of the quality was according to Codex Stan. The tabulation of quality evaluation of VCO can be seen in Table 4.

Table 4. Quality evaluation of VCO.

Evaluation	Water content (%)	Free fatty acid (%)	Saponification number (mg KOH)
1	0.34	0.26	264.77
2	0.35	0.30	264.22
3	0.36	0.31	262.25
Mean ±SD	0.34 ±0.01	0.29±0.03	263.75±1.33
Standard (Codex Stan)	0.10 – 0.50	Less than 0.5	250-260

The water content is interpreting the amount of water contained in the VCO [23]. The measurement of water content was done by weighing as much as 5 grams of VCO in the evaporating disc and then placed into 105 ^{0}C oven until constant weight. The calculation of water content was based on Equation 1.

$$Water\ content = \frac{A-B}{A} x100\% \quad (1)$$

Where A is 5 grams of VCO and B is weight of dried VCO in the 105 ^{0}C oven until constant weight. In the 3 times measurement, the water content was found to be 0.34 ± 0.01 %. According to Codex Standard, the water content of VCO produced met to the standard, which is between 0.1-0.5% [24]. This water content is necessary to be evaluated since the presence of water in the VCO will initiate the hydrolysis reaction to produce fatty acid and glycerol. The hydrolysis reaction undergoes to rancidity process of VCO [21, 26, 27].

The number of free fatty acid measurement was done for 3 times measurement (Table 4). The average number of free fatty acid of VCO produced was found to be 0.29 ± 0.03 %. According to the Codex Stan, the number of free fatty acid of VCO produced met the standard, i.e.did not exceed than 0.5% [24]. There is a very close correlation between the number of free fatty acids and water content. According to Frega (1999), the presence of free fatty acids is the result of hydrolysis process of oil or fat. The more water contained in the oil leads the more free fatty acids found in the oil [26, 27].

The saponification number of VCO produced was found to be 263.75 ±1.33 mg KOH. This data is slightly higher than the Codex Stan, which mentioned 250-260 mg NaOH/KOH [24]. The larger saponification numbers of VCO produced may due to the VCO which contains high lauric acid i.e. a medium carbon chain (MCFA) in the large amount. The medium carbon chain takes the responsibility of the higher saponification number [21].

4. Conclusion

Based on the experimental data, the production rate of VCO increase in the presence of enzyme which is availble in the raw pineapple juice. The quality of VCO produced meet the standard available.

5. Acknowledgment

The facility and financial support from Universitas Muhammadiyah Yogyakarta to the authors are highly acknowledged.

6. References

[1] Silalahi J dan Nurbaya S 2011 Minyak Kelapa Sawit di dalam Makanan: Masalah dan Manfaatnya (Palm Oil in Food: Problems and Benefits) *Prosiding Seminar dan Workshop Pharmacy Update* **3** (Medan).

[2] Patil U Benjakul S Prodpran T Senphan T and Cheetangdee N 2016 Characteristics And Quality Of Virgin Coconut Oil As Influenced By Maturity Stages *Carpathian Journal of Food Science and Technology* **8(4)** 103.

[3] Chinwong S Chinwong D and Mangklabruks A 2017 Daily Consumption of Virgin Coconut Oil Increases High-Density Lipoprotein Cholesterol Levels in Healthy Volunteers: A Randomized Crossover Trial *Hindawi: Evidence-Based Complementary and Alternative Medicine* **2017** Article ID 7251562 (https://doi.org/10.1155/2017/7251562).

[4] Lima R and Block J M 2019 Review: Coconut oil: what do we really know about itso far? *Food Quality and Safety* **3** 61 (doi:10.1093/fqsafe/fyz004).

[5] Anzaku A A Akyala J I Juliet A and Obianuju E C 2017 Antibacterial Activity of Luaric Acid on Selected Clinical Isolates *Annals of Clinical and Laboratory Research* **5(2)** 170.

[6] Marina A M Che Man Y B and Amin I 2009 Virgin coconut oil:emerging functional food oil *Trends in Food Science & Technology* **20** 481.

[7] Soeka Y S Sulistyo J and Naiola E 2008 Analisis Biokimia Minyak Kelapa Hasil Ekstraksi Secara Fermentasi (Biochemical Analysis of Extracting Fermented Coconut Oil) *BIODIVERSITAS* **9(2)** 91.

[8] Raghavendra S N and Raghavarao K S M S 2010 Effect of different treatments for the destabilization of coconut milk emulsion *Journal of Food Engineering* **97** 341.

[9] Prapun R Cheetangdee N and Udomrati S 2016 Characterization of virgin coconut oil (VCO) recovered by different techniques and fruit maturities *International Food Research Journal* **23(5)** 2117.

[10] Hidayati N 2010 Modifikasi dan Aplikasi pengolahan Minyak Kelapa Dengan Pemanfaatan Limbah Buah Nanas (Modification and Application of Coconut Oil Processing by Utilizing Pineapple Fruit Waste) *Jurnal Kimia dan Teknologi* **6(1)** 27.

[11] Handayani R Sulistyo J Rahayu R D 2009 Extraction of Coconut Oil (Cocos nucifera L.) through Fermentation System *BIODIVERSITAS* **10(3)** 151.

[12] Ricochon G and Muniglia L 2010 Influence of enzymes on the oil extraction processes in aqueous media *Technologie – Innovation* **17** 356.

[13] Harni M and Putri S K 2014 Processing Methode Effect to Virgin Coconut Oil (VCO) Quality After Storaging *International Journal on Advanced Science Engineering Information Tecnology* **4(2)** 28.

[14] Rohyami Y Anjani R D and Purwanti N P 2017 The Influence of Saccharomyces cerevisiae Enzyme Ratio on Preparation Virgin Coconut Oil for Candidate in-House Reference Materials *International Conference on Chemistry Chemical Process and Engineering (IC3PE) 2017 AIP Conf. Proc. 1823 020086-1–020086-15* (doi: 10.1063/1.4978159).

[15] Kumaunang M and Kamu V 2011. Aktivitas Enzim Bromelin Dari Ekstrak Kulit Nenas (Anenas Comosus) (Activities of Bromelin Enzymes of Extracts Pineapple Skin (Anenas Comosus) *Jurnal Ilmiah Sains* **11(2)** 198.

[16] Gavrilescu M Chisti Y 2015 Biotechnology—a sustainable alternative for chemical industry *Biotechnology Advances* **23(7-8)** 471.

[17] Nadzirah K Z Zainal S Noriham A Normah I and Siti Roha A M 2012 Physico-Chemical Properties of Pineapple Crown Extract Variety N36 and Bromelain Activity in Different Forms *APCBEE Procedia* **4(2012)** 130 – 134.

[18] Hui Y H 1992 *Encyclopedia of Food Science and Technology vol 3* (New York: John Wiley and Sons Inc).

[19] Herdyastuti N 2006 Isolasi dan Karakterisasi Ekstrak Kasar Enzim Bromelin dari Batang Nanas (Ananas Comusus L.Merr) (Isolation and caracterization of Bromelin from pineaple stalk) *Berk. Penel. Hayati* **12** 75.

[20] Winarti S Jariyah dan Purnomo Y 2007 Proses Pembuatan VCO (Virgin Coconut Oil) Secara Enzimatis Menggunakan Papain Kasar (The Production Process of Enzymatic VCO (Virgin Coconut Oil) Using Raw Papain) *Jurnal Teknologi Pertanian* **8(2)** 136.

[21] Ketaren S 1986 *Pengantar Teknologi Minyak dan Lemak Pangan (Introduction of Technology of Food's Oil and Fat)* (Jakarta:Universitas Indonesia]

[22] Bennion 1988 *The Science of Food* (New York Jhon willey and Son).

[23] Sudarmadji S Haryono B dan Suhardi 1984 *Prosedur Analisa untuk Bahan Makanan dan Pertanian (Analysis Procedure for Food and Agriculture Materials)* (Yogyakarta: Liberty).

[24] Codex Standard 1999 *Codex Standard for Edible Fats and Oils not Covered by Individual Standards: Codex Stan 19-1981* (Rev.2-1999) FAO (London: United Nations).

[25] Sudarmadji 1997 *Prosedur Analisa Untuk Bahan Makan dan Pertanian (Analysis Procedure for Food and Agriculture Materials)* (Yogyakarta: Liberty).

[26] Fregaa N Mozzona M and Lerckerb G 1999 Effects of Free Fatty Acids on Oxidative Stability of Vegetable Oil *Journal of the American Oil Chemists' Society* **76(3)** 325.

[27] Almeida D T Viana T V Costa MM Silva CS Feitosa S 2019 Effects of different storage conditions on the oxidative stability of crude andrefned palm oil, olein and stearin (Elaeis guineensis) *Food Sci. Technol, Campinas*, **39(Suppl. 1)** 211-217.

IOP Conf. Series: Materials Science and Engineering **874** (2020) 012005 doi:10.1088/1757-899X/874/1/012005

Improving biodiesel properties by mixing jatropha oil and soybean oil

W Wahyudi, M Nadjib, D M Pahlevi, and B S N Rohman

Department of Mechanical Engineering, Faculty of Engineering, Universitas Muhammadiyah Yogyakarta, Jl. Brawijaya Tamantirto, Yogyakarta, Indonesia

Abstract. The need for biodiesel as an alternative fuel will increase along with the increasing energy consumptions and the depletion of fossil fuel reserves. Jatropha oil and soybean oil are potential raw materials for biodiesel. Each raw material gives different properties to biodiesel. Mixing the two types of oil makes it possible to improve the quality of biodiesel. The aim of this study was to determine the effect of variations in the composition of the oil mixture on the biodiesel properties. Jatropha oil was mixed with soybean oil in various compositions. The mixture was heated and stirred for 60 min at 90°C. The more the percentage of soybean oil biodiesel, the viscosity of the biodiesel mixture decreased. Mixing with 20% soybean oil has a significant reduction in viscosity. The increasing percentage of soybean oil also influences the higher heating value of the biodiesel mixture.

Keywords: heating value; viscosity; jatropha oil; soybean oil

1. Introduction

Energy needs will increase over time. So far, these energy needs have been supplied from fossil fuels. As a non-renewable fuel, fossil fuels will diminish over time (1). Biodiesel is an alternative fuel as a substitute for diesel fuels. The need for using biodiesel will increase. Biodiesel has many advantages, including renewable, non-toxic and biodegradable (2).

The main raw material for biodiesel is vegetable oil. Many vegetable oils, both edible and non-edible ingredients, can be processed into biodiesel. Biodiesel produced from edible feedstocks will compete with food needs. Among the potential non-edible raw materials is jatropha oil (3).

Vegetable oil is a triglyceride composed of glycerin and fatty acids. Each vegetable raw material has different fatty acids. The properties of fatty acids are influenced by the length of the chain and the number of double bonds in these fatty acids (4). The properties of fatty acids affect the properties of biodiesel such as viscosity, density, calorific value and flash point (5).

Jatropha oil has a relatively high viscosity and a relatively low calorific value. The use of jatropha oil biodiesel still requires a high percentage of diesel oil mixture. Efforts are needed to improve the properties of jatropha biodiesel, especially its viscosity and calorific value. One effort to improve the properties of jatropha oil is by mixing other vegetable oils such as coconut oil (6) and used cooking oil (7). Because each vegetable oils has a different fatty acid composition (5), research needs to be done for other feedstocks. One of the vegetable oils that are widely used as biodiesel feedstocks is soybean oil (8). Soybean oil has medium chain fatty acids, so this oil has the potential to be used in biodiesel mixtures with jatropha oil. Mixing jatropha oil and soybean oil was expected to improve the properties of the mixture as a biodiesel feedstock. The purpose of this study was to improve the biodiesel properties from jatropha oil by mixing the jatropha oil with soybean oil.

2. Material and methods

2.1. Materials

The raw materials used in this study were jatropha oil and soybean oil. Some properties of jatropha oil and soybean oil are presented in Table 1.

Table 1. Properties of jatropha oil and soybean oil

Properties	Jatropha Oil	Soybean Oil

Density, at 40 °C (kg/m^3)	936.93	881.88
Viscosity, at 40 °C (cSt)	198.7	26.9
Calorific Value (MJ/kg)	37.16	39.23
Flash Point (°C)	305	347

The fatty acid content in raw materials is presented in Table 2 below.

Table 2. The fatty acid content of jatropha and soybean oil

No	Fatty Acid	Abbreviation	Jatropha Oil	Soybean Oil
1	Methyl Butyrate	C4:0	-	9.37
2	Methyl Palmitate	C16:0	1.2	10.09
3	Methyl Octadecanoate	C18:0	-	2.7
4	Cis-9-Oleic Methyl ester	C18:1	9.82	20.66
5	Lenolelaidic Acid Methyl Ester	C18:2	1.42	-
6	Methyl Lenoleate	C18:2	-	50.82
7	Methyl Lenolenate	C18:3	-	0.21
8	Gamma-lenolenic Acid Methyl Ester	C18:3	-	0.26
9	Methyl Arachidate	C20:0	-	0.15
10	Methyl Cis-11-eicocenoate	C20:1	-	5.38
11	Cis-4-7-10-13-16-19-decosahexaenoate	C21:6	-	-
12	Methyl Docosanoate	C22:0	-	0.36
13	cis-11 14 17-eicosatrienoic acid methyl ester	C22:3	87.19	-

2.2. Experimental procedure

In the first method, mixing between pure jatropha oil and pure soybean oil was carried out on various compositions. The percentage of jatropha oil ranged from 0% and 100% with a 10% increase. Jatropha oil and soybean oil were mixed according to the ratio of the mixture that had been determined at a temperature of 90°C to be stirred for 60 minutes.

Then the esterification process was carried out on the mixture. The esterification process was done for 60 minutes at 60°C, using a base catalyst (H_2SO_4) which had been dissolved with methanol.

The transesterification process was conducted at 60°C within 60 minutes. The oil mixture was reacted with Methanol (15% of oil volume) using (KOH) as a catalyst. The settling process was carried out for 8 hours, then the separation between biodiesel and glycerol was carried out. Methyl esters which have been separated from glycerol are washed 3 times with water.

In the second method, each oil was processed into biodiesel by transesterification. The process of transesterification, separation and washing of this method was the same as the first method. Then mixing between the two types of biodiesel was carried out. Mixing was conducted at 90 °C and stirred for 60 minutes. The percentage of the mixture began from 0% to 100% with 10% increment. Then the density, viscosity, calorific value and flash point of each mixture were tested.

3. Results and discussion

From Table 2 it can be seen that jatropha oil was dominated by cis-11 14 17-eicosatrienoic acid methyl ester while soybean oil had a content of cis-9-Oleic Methyl ester as much as 20.66%, and Methyl Lenoleate as much as 50.82%.

Jatropha oil has many fatty acids that have double bonds. The structure of fatty acids has an effect on the oil properties, including viscosity, density and calorific value. Because jatropha has a longer carbon chain, its density, viscosity and flash point are relatively higher than that of soybean oil (5). Table 3 shows the properties of biodiesel from jatropha oil and biodiesel from soybean oil.

Table 3. Properties of biodiesel from jatropha oil and soybean oil

Properties	Jatropha Oil	Soybean Oil
Density, at 40 °C (kg/m^3)	907	862
Viscosity, at 40 °C (cSt)	19.8	7.7
Calorific Value (MJ/kg)	36.79	39.45
Flash Point (°C)	215	186

3.1. Density

Fuel is injected into the combustion chamber by volume, so the amount of mass of fuel injected depends on the density of the fuel. Fuel density affects the air fuel ratio.

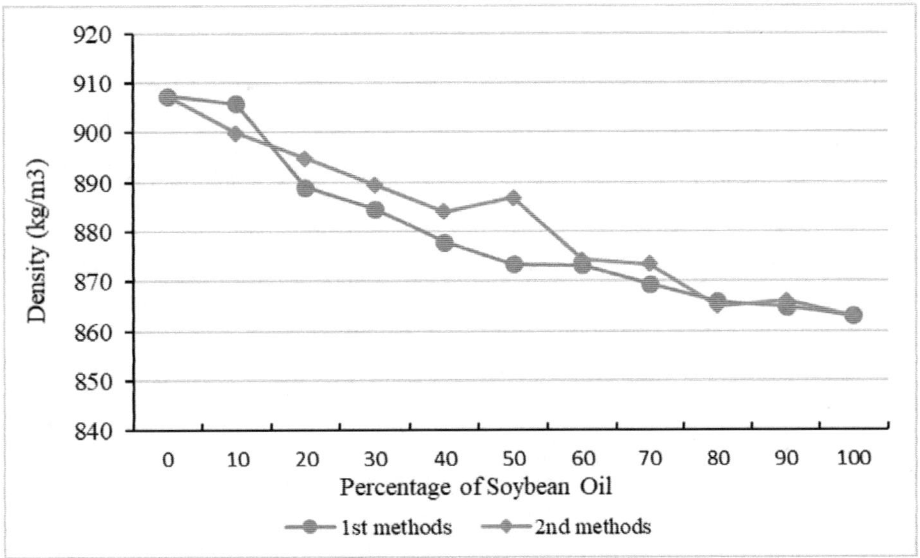

Figure 1. Density of biodiesel from a mixture of jatropha and soybean oil

Figure 1 shows the density change in a mixture of jatropha oil and soybean oil. The more volume fraction of soybean oil in the mixture, the density decreases. The density of biodiesel blends ranges from 850 to 900 kg/m^3. Density is influenced by the number of double bonds in fatty acids (9). Jatropha oil has more fatty acids with double bonds, so the density is relatively higher than soybean oil. It also affects the density of the mixture of both. Both types of mixing methods showed the same density results.

3.2. Viscosity

Viscosity is an important characteristic of liquid fuels, because it affects the quality of atomization when it is injected into the combustion chamber. High viscosity causes poor fuel atomization and incomplete combustion (10).

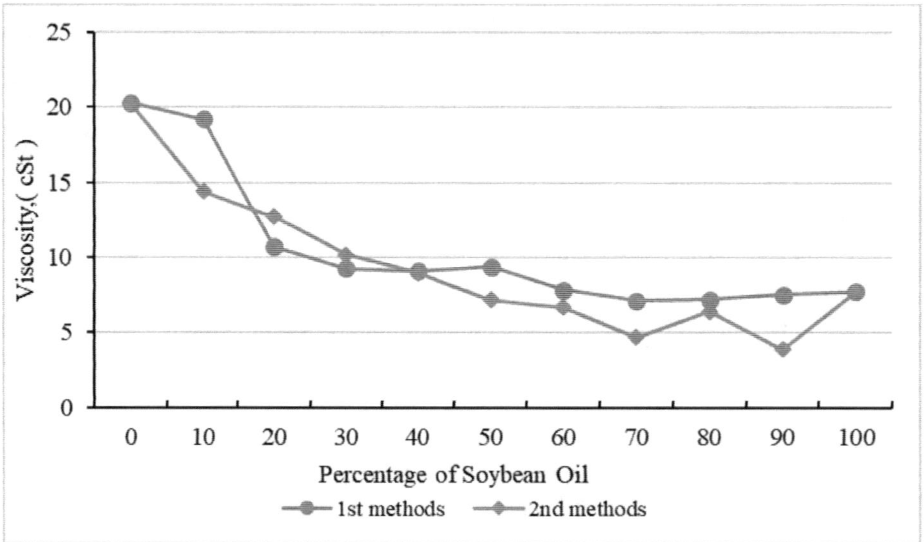

Figure 2. Viscosity of biodiesel from a mixture of jatropha and soybean oil

Figure 2 shows the change in viscosity of various compositions of a mixture of jatropha oil and soybean oil biodiesel. The higher the percentage of soybean oil biodiesel, the lower the viscosity of the biodiesel mixture. The length of the fatty acid carbon chain affects the viscosity. The longer the carbon chain of the fatty acid, the higher the viscosity (5). Jatropha oil has a relatively long fatty acid with a carbon chain. This causes the viscosity of jatropha oil biodiesel to be relatively higher than soybean oil biodiesel.

The decrease in viscosity from the mixture showed a non-linear relationship with the percentage of volume. Thus, mixing with 20% soybean oil has a significant reduction in viscosity. In the next volume fraction up to 100% soybean oil mixture, the biodiesel viscosity was relatively constant.

3.3. Flash point

The flash point is the properties of biodiesel which is related to the flammability properties (11). Mixing jatropha oil with soybean oil produces flash points in the range between 200 - 250°C. Figure 4 shows changes in flash point biodiesel from a mixture of jatropha oil and soybean oil.

The flash point value met the standard. The flash point decreased with increasing volume fraction of soybean oil in the mixture. This shows that the flash point is also influenced by the structure of fatty acids.

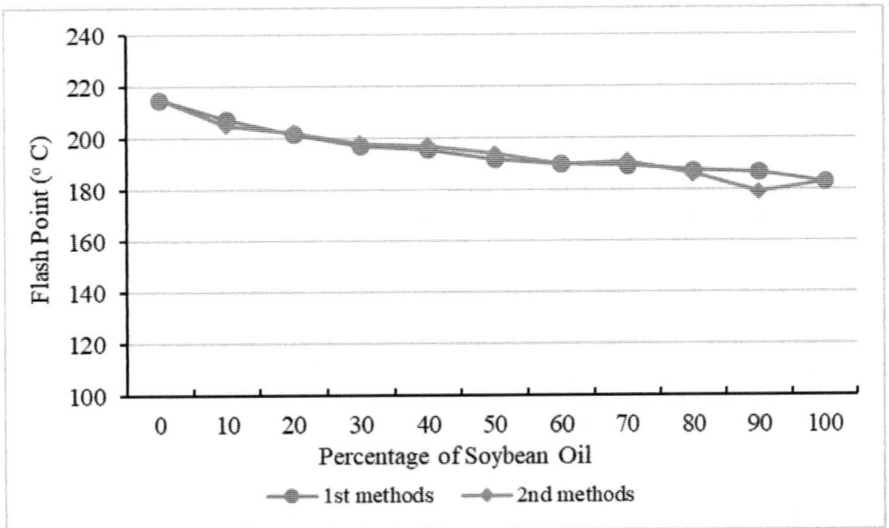

Figure 3. Flash point of biodiesel from a mixture of jatropha and soybean oil

3.4. Calorific value

Calorific value indicates the energy content of a fuel. Because of the presence of oxygen atoms in biodiesel, the calorific value is relatively lower than fossil fuels (2).

Figure 4 shows changes in the calorific value of various compositions of a mixture of jatropha oil and soybean oil biodiesel. Jatropha oil has more unsaturated fatty acids than soybean oil. The number of double bonds in fatty acids affects the calorific value of a fuel. The more fatty acid content that has a double bond in the carbon chain (C = C) in biodiesel, the lower the calorific value (5). The increasing percentage of soybean oil also influenced the higher heating value of the biodiesel mixture.

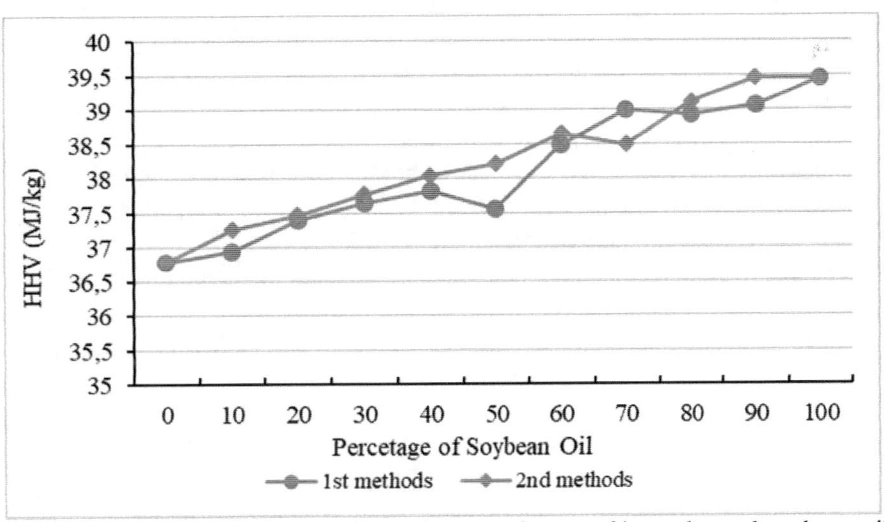

Figure 4. Calorific value of biodiesel from a mixture of jatropha and soybean oil

4. Conclusion

Mixing jatropha oil with soybean oil caused a change in the properties of the biodiesel produced. The density of biodiesel from mixing jatropha oil and soybean oil decreased with increasing volume fraction of soybean oil. This also affects the heating value of biodiesel from the mixture.

IOP Conf. Series: Materials Science and Engineering **874** (2020) 012005 doi:10.1088/1757-899X/874/1/012005

The viscosity of the mixed biodiesel decreased with the increasing amount of soybean oil mixture. The more the percentage of soybean oil biodiesel, the viscosity of the biodiesel mixture decreased. Mixing with 20% soybean oil has a significant reduction in viscosity.

Flash points are also influenced by changes in the structure of fatty acids. Flash points for all variations of the mixture above 100°C, meet ASTM standards. The increasing percentage of soybean oil also influenced the higher heating value of the biodiesel mixture.

Mixing jatropha oil with soybean oil provided a change in the composition of fatty acids which gives a positive change in the form of a decrease in viscosity and an increase in calorific value

References

1. Atabani AE, Mahlia TMI, Masjuki HH, Badruddin IA, Yussof HW, Chong WT, et al. A comparative evaluation of physical and chemical properties of biodiesel synthesized from edible and non-edible oils and study on the effect of biodiesel blending. Energy. 2013 Sep;58:296–304.
2. Demirbas A. Relationships derived from physical properties of vegetable oil and biodiesel fuels. Fuel. 2008 Jul;87(8–9):1743–8.
3. Silitonga AS, Masjuki HH, Mahlia TMI, Ong HC, Chong WT, Boosroh MH. Overview properties of biodiesel diesel blends from edible and non-edible feedstock. Renewable and Sustainable Energy Reviews. 2013 Jun;22:346–60.
4. Lanjekar RD, Deshmukh D. A review of the effect of the composition of biodiesel on NO x emission, oxidative stability and cold flow properties. Renewable and Sustainable Energy Reviews. 2016 Feb;54:1401–11.
5. Hoekman SK, Broch A, Robbins C, Ceniceros E, Natarajan M. Review of biodiesel composition, properties, and specifications. Renewable and sustainable energy reviews. 2012;16(1):143–169.
6. Wahyudi W, Wardana ING, Widodo A, Wijayanti W. Improving Vegetable Oil Properties by Transforming Fatty Acid Chain Length in Jatropha Oil and Coconut Oil Blends. Energies. 2018;11(2):394.
7. Wahyudi W, Nadjib M, Bari MF, Permana FW. Increasing of quality biodiesel of Jatropha seed oil with biodiesel mixture of waste cooking oil. Journal of Biotech Research. 2019 Jul 8;10:7.
8. Özener O, Yüksek L, Ergenç AT, Özkan M. Effects of soybean biodiesel on a DI diesel engine performance, emission and combustion characteristics. Fuel. 2014;115:875–883.
9. Refaat AA. Correlation between the chemical structure of biodiesel and its physical properties. International Journal of Environmental Science & Technology. 2009;6(4):677–694.
10. Knothe G, Steidley KR. Kinematic viscosity of biodiesel components (fatty acid alkyl esters) and related compounds at low temperatures. Fuel. 2007;86(16):2560–2567.
11. Acharya N, Nanda P, Panda S, Acharya S. A comparative study of stability characteristics of mahua and jatropha biodiesel and their blends. Journal of King Saud University - Engineering Sciences. 2019 Apr 1;31(2):184–90.

IOP Conf. Series: Materials Science and Engineering **874** (2020) 012006 doi:10.1088/1757-899X/874/1/012006

Effect of Aloe Vera Gel Extract to The Physicochemical Properties and Effectiveness of Burn Cream Formulation

Novianti*, T Y Hendrawati, R A Nugrahani

Chemical Engineering Department, Faculty of Engineering, Universitas Muhammadiyah Jakarta, Indonesia

*Novianti2302@gmail.com

Abstract. Aloe chinensis Baker is commonly cultivated in Indonesia especially in Pontianak and Bogor. Aloe vera gel contains various anti-inflammatory compounds. The objectives of this study are to determine the effects of Aloe vera gel extract additions to the formulations of burn cream and to obtain the best formulation in healing burn wounds on outer epidermis through testing on white mice. Aloe vera gel extract was prepared with a rotary evaporator for 90 minutes at 40 °C and 110 mBar. The basic formulation of the burn cream included oil phase and water phase. There were six concentrations of Aloe vera gel extract added to the above burn cream at 0%, 2.5%, 5%, 7.5%, 10% and 12.5%. The resulting formulations produced pHs that were within the acceptable range of skin physiology (4-8). The viscosities were in good consistencies that were easily applied on skin without drips. They also exhibited fineness values of 9-10 μ, refractive indices ranging from 51.5 to 78.3, and opaque color of milky white. Biological experiments showed that burn cream sample with Aloe vera gel extract at 5% concentration to be the most effective by healing 2nd degree burn wounds the fastest during 20-day trials.

1. Introduction

Aloe vera is one of medicinal herbs that were widely used in the western society in the middle ages, and the production of its extract become one of the largest botanical industries in the world [1]. This plant belongs to the *Liliaceae* family and is distributed throughout the world. There are 350 species of Aloe, and the one that is commonly cultivated in Asia, including in Indonesia, is *Aloe chinensis Baker*. This species is commercially cultivated in Kalimantan Barat (West Kalimantan) Province and locally known as Aloe vera Pontianak [2].

Aloe vera is used in cosmetic, food, and pharmaceutical industries. In cosmetic and toiletry industries, it is incorporated as the basic ingredient in skin moisturizer, soap, shampoo, sunscreen lotion, makeup cream, perfume, shaving cream, and many more. Aloe-based pharmaceuticals come in topical (gel and cream) and oral (tablet and capsule) forms [3].

Tri Yuni Hendrawati Studied (2018), Aloe vera gel extract is mucilaginous gel obtained from parenchymatous cells called aloe vera gel that is colorless and testeless. Aloe vera is Aloe vera gel contains the main ingredients of aloin, emodin, and sap, in addition to other compounds, such as essential oil. Aloin is an active compound that has antiseptic and antibiotic properties. The phenolic

IOP Conf. Series: Materials Science and Engineering **874** (2020) 012006 doi:10.1088/1757-899X/874/1/012006

compounds found in Aloe vera have been determined to include aloin A and B, aloenin (B), and chrysophanol [4].

Aloe vera is known to have healing properties including antioxidant, anti-inflammatory, anti-diabetes, burn reliever, anti-aging, and anti-cancer [5]. Aloe vera gel exhibits strong anti-inflammatory effects because of the presence of anthraquinones and kromon. Oral intake of aloe vera gel at 2% has been reported to effectively relieve the severity of pain and reduce the wound size in stomatitis aphtous patients. The anti-inflammatory effect of aloe vera has also been shown to reduce joint pain. Aloe vera gel is more effective in fighting inflammations caused by prostaglandin synthesis and leucocyte infiltration, but less effective against allergen-induced ones [6]

Burn wound is a damage on skin caused by exposure to fire or other causes such as hot water, high electrical current, chemicals, and radiation. The damage may reach other tissues under the skin. Burn wound is a type of trauma with high mortality rate that requires a specific management from the initial to the follow-up stages. Burn wound can also cause necrotic coagulations on skin and exposure of deeper tissues, as well as effects on other organs [7].

The severity of burn wound can be classified into three classes:

a. 1st degree

   Damages are limited to the epidermal layer indicated by skin redness. Pains disappear in 48 to 72 hours and wounds heal without scar.

b. 2nd degree

   Damages to the whole epidermis and some dermis include pain and skin redness, edema, and presence of bullae. 2nd degree burn is further classified into:

   1. Superficial, which includes skin redness, edema, and presence of bullae. Pain is due to a significant number of undamaged basal cells. Vasodilation exists and heals within two weeks without or with minimal scar.
   2. Deep, in which damages to the whole epidermis and some dermis are indicated by moist but pale wounds. The pain is less than superficial 2nd degree with healing in several weeks to months with scar.

c. 3rd degree

Damages to the whole dermis or deeper appear as peeling epithelium and white area caused by dermal protein coagulation. Narrowing area of dry burned dermis called eschar exists with no blood perfusion nor pain. Spontaneous healing may occur.

Appearance of a granular tissue after the second week requires closing with skin graft, but when left untreated causing the formation of a contracture (a thick and narrow scar tissue) (Sudjatmiko, 2007).

Cream is a form of traditional medicine in a semi-solid state made from extract dissolved or homogeneously dispersed in the basic form of balm or cream that is used for topical treatments [8]. Ashni Studied (2013), topical formulation for cream have three main function, such as to have hydrate skin because of their omellient properties, to protect from external environment or heal an intact or injured area of the skin, to deliver medication to the skin [9].

Mice are mammals that have an important role in human lives for their involvement in scientific endeavors because of their high adaptability. The type of mice that is commonly used as an animal model in laboratories, as well as pets, is the white mice (*Rattusnorvegicus*). The advantages these animals have over other species include ease of handling and care because of its small size, health and cleanliness, high reproductive capacity with short gestation period, and their production and reproductive characteristics are similar to other mammals [10].

The objectives of this study is to determine the effects of aloe vera gel extract concentrations in burn cream on the burn wound healing upon treatment with the cream, and to obtain the best formulation for burn cream.

2. Experiment Method

2.1 Materialsandequipment

The materials include gel extract from *Aloe Chinensis Baker* ([Semplak-Bogor]), Triethanolamine, Tween 80, liquid Paraffin, Methyl Paraben ([BP/USP], [Rasula Pharmaceutical & Fine Chemicals-India]), Propyl Paraben ([BP/USP], [Rasula Pharmaceutical & Fine Chemicals-India]), Propylene Glycol ([USP/EP], [SKC-Korea]), Tocopherol, Dimethyl SulfurOxyde ([Merck]), EDTA ([Merck]), NatriumDihydrogen Phosphate ([Merck]), distilled water, and laboratory white mice (*Rattusnorvegicus*) ([Sparague Dawley-IPB]).
The equipment used includes standard glass apparatus, scale, blender, knife, plastic tub, spoon, water bath, spatula, graduated cylinder, glass pipettes, mortar and pestle, pH meter, centrifuge, rotary vacuum evaporator, viscometer, chromometer, refractometer, and micrometer.

2.2 Method

The study took place in the Applied Chemical Engineering Laboratory, Engineering Faculty, UniversitasMuhammadiyah Jakarta.

The steps in this study went as follows: (1) Aloe vera gel extraction consisting of peeling the rind of clean Aloe vera leaves, blending the remaining gel, and filtering the resulting puree; (2) Purification of Aloe vera gel extract using a rotary vacuum evaporator at 110 mBar and 40°C for 90 minutes. This process started with 1100 mL of filtered puree and ended up with 600 mL of Aloe vera gel extract, which translated into a 54% yield; (3) Base cream and burn creampreparation. The base formula of the burn cream comprised of 14.5 grams stearic acid, 0.9 mL triethanolamine, 3 mL liquid paraffin, 4.9 grams Polysorbate 80 (Tween® 80), 0.1 gram Nipagin™, 0.05 gram Nipasol™, 0.5 mL tocopherol, 5 mL Propylene Glycol, 0.1 gram Dimethyl Sulfur Dioxide, 0.1 ml EDTA, and 8 ml distilled water; (4) Burn cream quality testing was performed by physical and chemical analyses for pH, color, viscosity, density, degree of fineness, and refractive indices; (5) In testing the effectiveness of the cream it is carried out by simple random sampling. The study was divided into seven groups, namely 7 groups treated with Aloe vera burn cream with variations in the concentration of Aloe vera gel which was added 0% (without Aloe vera gel), 2.5%, 5%, 7.5%, 10%, 12.5%, and 100% (pure Aloe vera gel without cream additives). Each group consisted of 2 white rats (Rattus norvegicus). (6). Making burns. Burns are made on the back of the rat's body. Then the area is cleaned and shaved with a distance of ± 5 cm. The area to be injured is sterilized by using alcohol and leave to dry. Then rats were anesthetized using ketamine as much as 0.3 cc. After the rats fainted to give burns on the back, metal iron with a size of 2 cm x 2 cm was attached which had been dipped in hot water ± 90-100 °C. Metal iron is attached to the wound area for 40 seconds, then washed with sterile water and dried (Samsudin et al., 2018); (7) Burn wound treatments by applying burn cream samples were conducted as described in step 5. Application was performed once a day for 20 days and effects were recorded. The parameter observed was the wound healing.

3. Result and discussion

Analyses of Aloe vera gel extract resulting from rotary vacuum evaporation process yielded pH of 4.7, color of 0.4708, viscosity of 3.47 cP, density of 0.9940 g/mL, and refractive index of 1.1. Visual observation resulted in white color of the burn cream. The results of physical and chemical analyses are presented in Table 1.

Table 1. Results of Physical and Chemical Characterization of Burn Cream with Aloe Vera

No	Parameter	Unit	Results at Various Concentration of Aloe Vera Extract (%)						
			0	2.5	5	7.5	10	12.5	100
1.	pH	-	7.25	7.83	7.85	7.89	7.85	7.8	4.7
2.	Colour		0.13069	0.59079	0.5591	0.66823	0.50224	0.74852	0.4708
3	Viscosity	cP	20,000	14,800	14,600	15,000	18,000	18,800	3,47

5.	Density	g/ml	0,7713	0.6463	0.6353	0.638	0.6887	0.6510	0.9940
6.	Fineness	µm	9	12	12	12	12	11	-
7.	Refractive Index		78.3	59.4	69.8	51.5	56.6	701	11

3.1 The effects of Aloe Chinensis Baker gel extract concentration on pH

The pH of the burn cream at 0% Aloe vera gel extract was 7.25, whereas that at 12.5% was 7.8. However, the highest pH of 7.89 occured at 7.5% concentration. The effect of Aloe vera gel extract addition at various concentrations on pH is described in Figure 1.

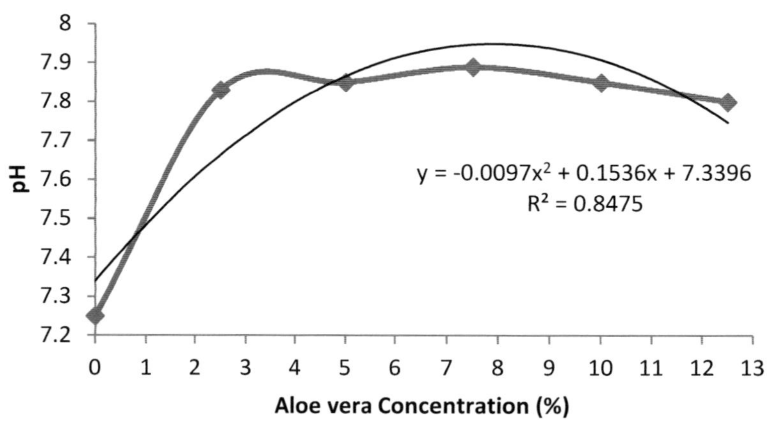

Figure 1. The effect of Aloe vera gel extract on the pH of burn cream samples

Further analyses with 2nd order polynomial regression yielded a correlation formula of $y = -0.0097x^2 + 0.1536x + 7,3396$ with $R^2 = 0.8475$, while y represents pH and x represents concentration of Aloe vera gel extract.
From the picture above shows the addition of aloe vera gel extract to the wound cream affects the pH value. If the pH of the cream is too acidic it can cause irritation, while the pH that is too alkaline causes scaly skin [11]. Based on the results of the pH analysis, it was found that the pH value ranged from 7.25-7.89 which meets the topical pH requirements, namely 4-8 [12]. According to Hasniar, et al. (2015), pH values can be affected due to chemical changes in active substances or additives, carrier or environmental influences and the effects of carbon dioxide (CO_2).

3.2 The effects of Aloe AloeChinensis Baker gel extract concentration on viscosity

Data analyses with 2nd order polynomial regression yielded a correlation between Aloe vera gel extract concentration and viscosity. The formula for this correlation is $y = 122.29x^2 - 1482.9X + 19129$ with $R^2 = 0.7946$, with y representing viscosity and x representing concentration of Aloe vera gel extract (Figure 2).

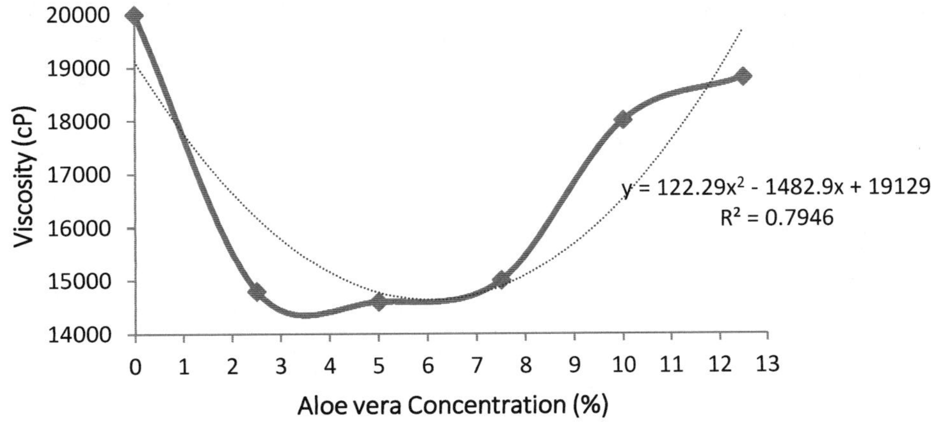

Figure 2. The effect of adding Aloe vera gel extract on the viscosity of burn cream samples.

Factors that determine the viscosity of creams are ingredients classified into the oil phase, especially cocoa fat, stearic acid, and cetyl alcohol [13]. The picture above shows a minimum viscosity value of 14600 cP and a maximum value of 18,800 cP.

3.3 The effects of Aloe veragel extract concentration on density

Data analyses with 2nd order polynomial regression yielded a correlation between Aloe vera gel extract concentration and density. The formula for this correlation is $y = 0.002x^2 - 0.0298x + 0.7461$ with $R^2 = 0.6356$, with y representing density and x representing concentration of Aloe vera gel extract (Figure 3). The picture above shows the minimum density value is 0.6353 g/ml and the maximum density 0.7713 g/ml. For cream density testing there's no specific fixed standard. That's why the validation using acceptable range. By visual observation observation does not visible the difference.

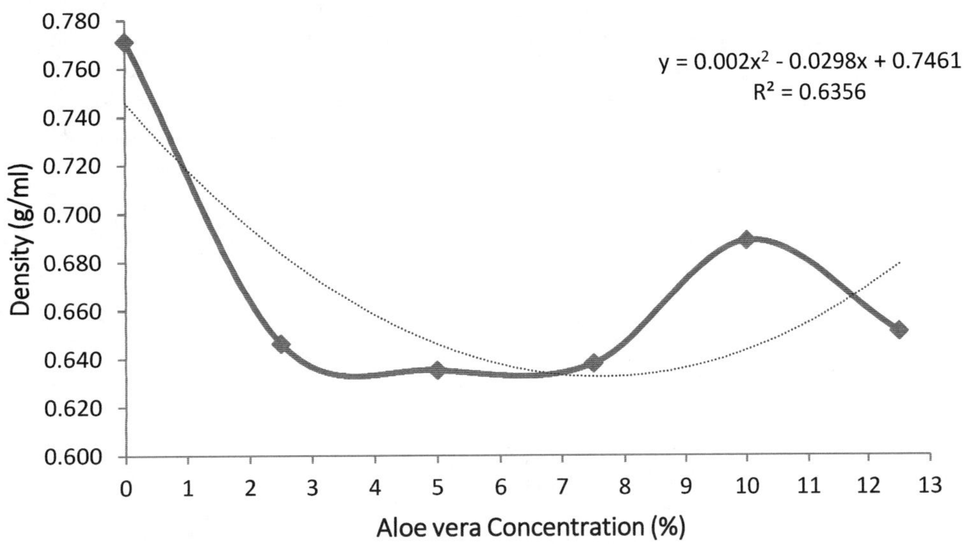

Figure 3. The effect of Aloe vera gel extract on the density of burn cream samples

3.4. The effects of Aloe vera gel extract concentration on fineness

Data analyses with 2nd order polynomial regression yielded a correlation between Aloe vera gel extract concentration and fineness. The formula for this correlation is y = -0.0571x+ 0.8286x +9.4286 with R^2 = 0.8442, with y representing fineness and x representing concentration of Aloe vera gel extract (Figure 4).The picture above shows the minimum fineness value is 9 μ and the maximum fineness 12 μ. For cream fineness testing there's no specific fixed standard. That's why the validation using acceptable range. By visual observation observation does not visible the difference.

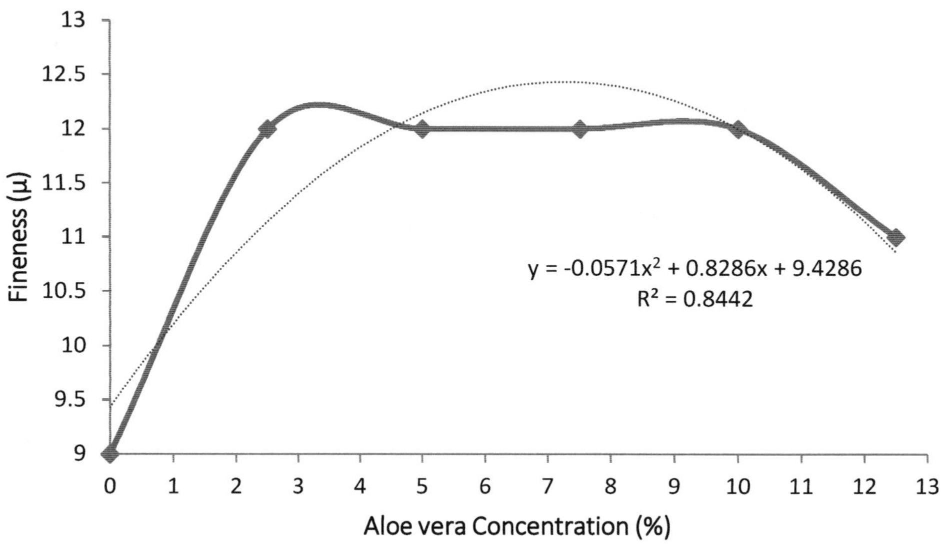

Figure 4. The effect of adding Aloe vera gel extract on the fineness of burn cream samples

3.5 The effects of Aloe vera gel extract concentration on refractive index

Data analyses with 2nd order polynomial regression yielded a correlation between Aloe vera gel extract concentration and refractive index. The formula for this correlation is $y = 0.4023x^2 - 5.8023x + 77.5$ with $R^2 = 0.5946$, with y representing refractive index and x representing concentration of Aloe vera gel extract (Figure 5). The picture above shows the minimum refractive index value is 51.5 and the maximum refractive index 78.3. For cream refractive index testing there's no specific fixed standard. That's why the validation using acceptable range. By visual observation observation does not visible the difference.

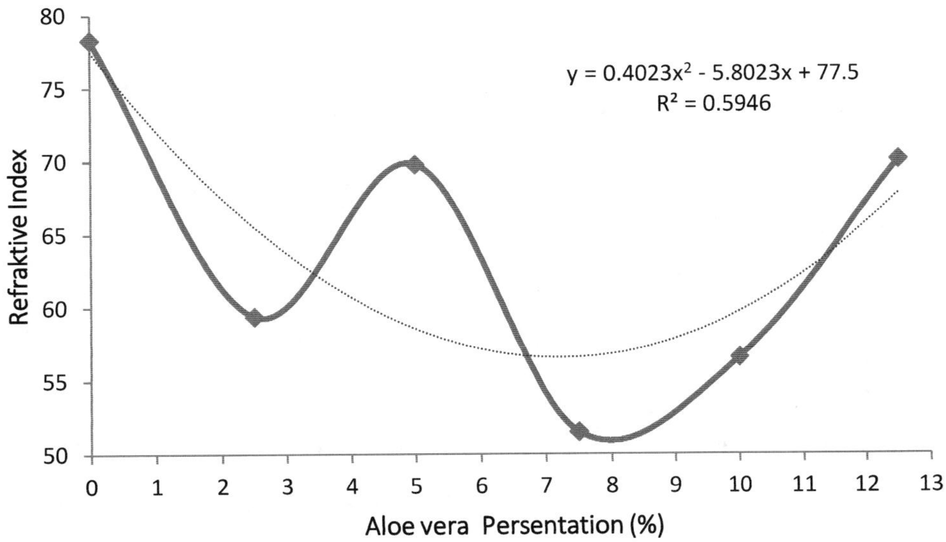

Figure 5. The effect of Aloe vera gel extract on the refractive index of burn cream samples

3.6 The effects of Aloe vera gel extract concentration on color

The results of data analyses on the effects of Aloe vera gel extract concentration on the color of the burn cream are presented in Figure 8. Data analyses with 2nd order polynomial regression yielded a correlation between Aloe vera gel extract concentration and refractive index. The formula for this correlation is $y = 0.019x^3 - 0.0409x^2 + 0.2568x + 0.1375$ with $R^2 = 0.9202$, with y representing the color value of burn cream and x representing the concentration of Aloe vera gel extract (Figure 6).The picture above shows the minimum color value is 0.13069 and the maximum color 0.74852. For cream color testing there's no specific fixed standard. That's why the validation using acceptable range. By visual observation observation does not visible the difference.

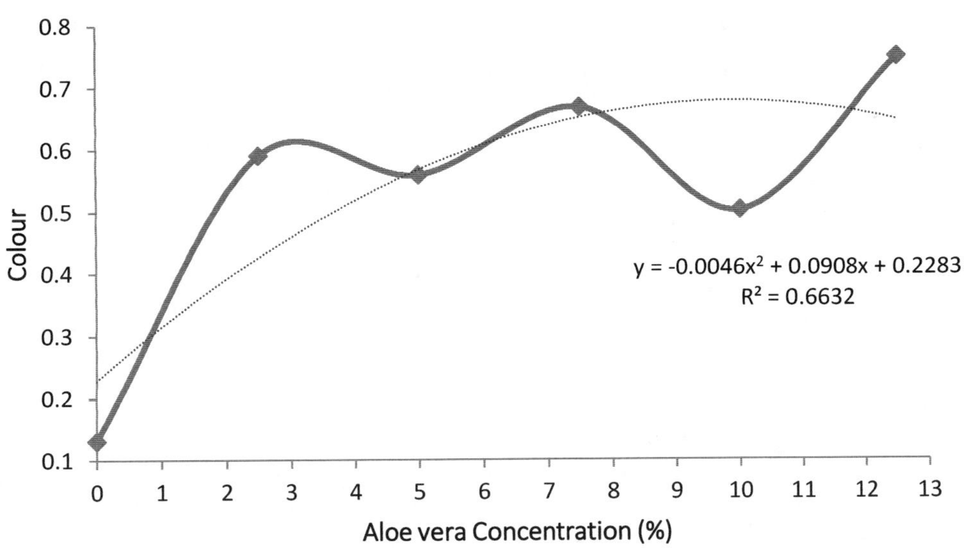

Figure 6. The effect of Aloe vera gel extract to the colour of burn cream samples

3.7 The effects of Aloe vera gel extract concentration on burn wounds

Based on the results of testing physical and chemical properties, the best burn cream is a cream with a 5% aloe vera gel content tested on white mice (Rattusnorvegicus) until healed and lost his burns for 20 days. Testing the effectiveness of second degree burn creams on white mice (*Rattusnorvegicus*) is characterized by skin damage in the epidermis. White mice that were declared cured were those whose burns had a change in the size of the wound. The wound healing process consists of 3 phases, namely the inflammatory phase, the proliferation phase, and the phase of exclusion. The inflammatory phase is characterized by swelling, the proliferation phase is characterized by the formation of exudates and fibrolas that look like crust at the top of the wound, and the phase of denial which is characterized by the formation of new tissue which means the wound has narrowed or healed [13]. In burn cream with 5% Aloe vera gel content, the inflammatory phase takes place on the first day to day 4, the profileration phase occurs on day 5 to day 12 and on day 20 the wound has healed.

4. Conclusion

The active ingredient of aloe vera extract (Aloe Vera) was successfully produced by means of a vacuum rotary evaporator at a pressure of 110 mBar temperature of 40°C for 90 minutes by producing a yield of 54.55%, the resulting cream preparation are cream with type M / A obtained from Aloe Vera gel extract with 0%, 2.5%, 5%, 7.5%, 10%, 12.5% enhancements where the cream has met the requirements based on organoleptic test parameters and pH with a tropical pH range of 4-8, based on the results of the analysis of pH 4-8, viscosity 14600 cP, density 0.6463 g/ml, fineness level of 12 micron, refractive index 69.8 colors 0.559104, the analyses of Aloe vera gel extract prepared in a rotary vacuum evaporator yielded pH of 4.7, color difference of 0.4608, viscosity of 3.47 cP, density of 0.97 gr/mL, and refractive index of 1.1. The burn cream formulation that yielded the fastest healing time of 2nd degree burn wounds in 14 days contained 5% Aloe vera gel extract. Physical and chemical analyses of this formulation yielded pH of 7.85, viscosity of 14,600 cPs, density of 0.6463 g/mL, fineness of 12 micron, refractive index of 69.8, and color difference of 0.559104.

5. Acknowledgment

We would like to express gratitudes to the Ministry of Research, Technology, and Higher Eduction of the Republic of Indonesia, Master in Chemical Engineering Program Faculty of Engineering Universitas Muhammadiyah Jakarta, LPPM Universitas Muhammadiyah Jakarta for the facilities and financial support to the implementation of this thesis research.

6. References

[1] T YHendrawati (2015, January). Aloe Vera Powder Properties Produced from *Aloe Chinensis Baker*, Pontianak, Indonesia. Paper presented at The Internasional Seminar on SOMCHE 2014 & RSCE 2014: Taylor's University, Selangor.

[2] T YHendrawati, R A Nugrahani., S Utomo., A I Ranadhan. (2018). Formulation Process Of Aloe Vera Mask With Variable Percentage Of Aloe Vera Gel Extract. Paper presented at The IOP Conference on Engineering and Applied Technology (ICEAT), 403 (2018) 012013, DOI: 10.1088/1757-899X/403/1/012013.

[3] Tri Yuni Hendrawati, Ratri Ariatmi Nugrahani, Suratmin Utomo, and Anwar Ilmar Ramadhan (2018). Effect of Evaporation Time on Physical and Chemical Properties of Aloe vera Gel Extract (Aloe Chinensis Baker). ARPN Journal of Engineering Applied Sciences, Vol 13, No. 24.

[4] R Rajeswariet,al. (2012).Aloe Vera The Miracle Plants Its Medicinal And Traditional Uses In India Journal of Pharmacognosy and Phytochemistry. Vol 1. ISSN 2278-4138

[5] Abid Aslam Maan, Akmal Nazir, Muhammad Kashif Iqbal Khan, Tahir Ahmad, 2018, The therapeutic properties and applications of aloe vera: A review. Journal of Medicine.
[6] Sudjatmiko, Gentur. (2007). Petunjuk Praktis Ilmu Bedah Plastik Rekonstruksi. ISBN: 978-979-16288-0-8.
[7] Badan Pengawas Obat Dan Makanan. 2014. Peraturan Kepala Badan Pengawa Obat dan Makanan, Peraturan Kepala Badan Pengawas Obat dan Makanan Indonesia, Nomor 12, tahun 2014. Tentang Persyaratan Mutu Obat Tradisional.
[8] Pribadi, Gutama Agus. (2008). Penggunaan Mencit dan Tikus Sebagai Hewan Model Penelitian Nikotin [skripsi]. Bogor: Fakultas Peternakan, Institut Pertanian Bogor.
[9] Ashni V, Sukhdev S, Rupinder K, Upendra K J, 2013, Topical Gels as Drug Delivery Systems: A Review, ISSN0976-044X, 23(2) .
[10] Rinza Rahmawati Samsudini. (2018). PotensiEkstrakKulitPacitan (*Citrus sinensi*) Sebagai StimulusRegenerasiSelPada Luka Bakar*Rattusnorvegicus*. JurnalLaboraMedika, Vol 2 No 2 Hal 19-23.
[11] Fitri Djumaati, dkk, (2018), Formulasi Sediaan Salep Ektrak Etanol Daun Kelor (Moringa oleifera Lamk) dan Uji Aktivitas Antibakterinya Terhadap Bakteri Staphylococcus, Pharmacon Jurnal Ilmiah Farmasi, Vol 7, No 1, ISSN 230202493.
[12] Adeltrudis Adelsa Danimayostu, Nilna Maya Shofiana, Dahlia Permatasari. (2017). PengaruhPenggunaanPatiKentang (*Solanumtuberosum*) TermodifikasiAsetilasi-Oksidasisebagai*Gelling agent* terhadapStabilitas Gel NatriumDiklofenak. Parmaceutical Journal Of Indonesia, Vol. 3 No 1 Hal 25-32.
[13] Hasniar, Yusriadi, Akhmad Khumaidi, (2015), Antioxidant Cream Formulation of *Gossypium sp.* Leaf Extract, Galenika Journal of Pharmacy, Vol 1 (1): 9-15, ISSN: 2442-8744.

IOP Conf. Series: Materials Science and Engineering **874** (2020) 012007 doi:10.1088/1757-899X/874/1/012007

Smart Laboratory Based On Internet Of Things In The Faculty Of Electrical Engineering, University Of Muhammadiyah Sidoarjo

Indah Sulistiyowati, Arif Rachman Sugiarto, Jamaaluddin Jamaaluddin

Universitas Muhammadiyah Sidoarjo, Sidoarjo, Indonesia

Abstract. Technology continues to grow from year to year that aims to facilitate human work or automation of a system. In this case, the concept of smart laboratory is applied in the laboratory room of Electrical Engineering Faculty of Muhammadiyah University of Sidoarjo by utilizing temperature sensor. This application is used to determine the room temperature as well as the proximity sensor to know the person entering and exiting the room. The sensor is controlled by the NodeMCU microcontroller board to make the system work automatically. For example if someone goes into the room the lights will light up and adjust the temperature with the number of people who are indoors. By applying the internet of things, the system can monitor and control via smartphone updated on realtime firebase databases so that it is not limited by distance because it uses the internet network.

keywords: smart laboratory, NodeMCU, internet of things, firebase,

1. Introduction

The presence of the internet for decades has really revolutionized the way the world works so fast. A new era is at the door, the Internet of Things (IoT) era. Smart devices that will be connected are predicted to reach 20 billion devices worldwide by 2020. Utilization of IoT is expected to be more massive, because IoT is not only talking about devices that are "smart", but also scanning and communication between machines.

The Internet of Things is a technology that innovates surrounding objects with the internet so that everyday activities become easier and more efficient. An easy example is how we can order all electronic equipment in our house to light up or die just by using our voice

In this paper internet of things (IoT) is used to design a smart laboratory. Smart laboratory is one of the technological advances that aims to facilitate human activities by utilizing several electronic devices, so that they can be controlled through smart phone applications [1].

As well as the use of electronic sensors to support an always updated information system and early warning system [2]. With this system, the use of electrical power can be controlled by turning off unused electronic equipment with actuators that are integrated with sensors and smartphone applications, so that it can reduce the electrical power load used [3]

At present, technological developments are also expanding in the field of communication, this can be seen from the number of internet usage from year to year [4]. One of them is Internet of Things technology. With the application of internet technology of things (IOT) where all electronic equipment or microcontrollers as a control of sensors and actuators can be connected to the internet so that they can be controlled or controlled from anywhere and anytime as long as they can connect to the internet. This is very useful for people who often leave the room for a long time or who are not able to indulge freely in the room [5]. The smart laboratory concept here is the efficiency of electricity use in air conditioner (AC) devices, by applying the working principle if the number of people in a room is increasing, the temperature of the air conditioner will be cooler and vice versa, if the number of people in the room is less then the temperature of the air conditioner the rising or normal temperature that the user wants. Ultrasound proximity sensor (HC-sr04) is used to calculate the number of people entering or exiting the room by utilizing the distance read by the sensor so that we can calculate the number of people in the room. Where the entire system will be connected to the internet that can be accessed with an android phone so that it can monitor the condition of the room or control the air conditioning temperature of the room via android.

2. Experimental Method

2.1. System Block Diagram

Sensors in this system will sense the conditions that occur on the sensor. Ultrasound HC-SR04 proximity sensor is used as sensing someone in or out of the room, while the temperature sensor is used as temperature sensing to determine the temperature value in the room.

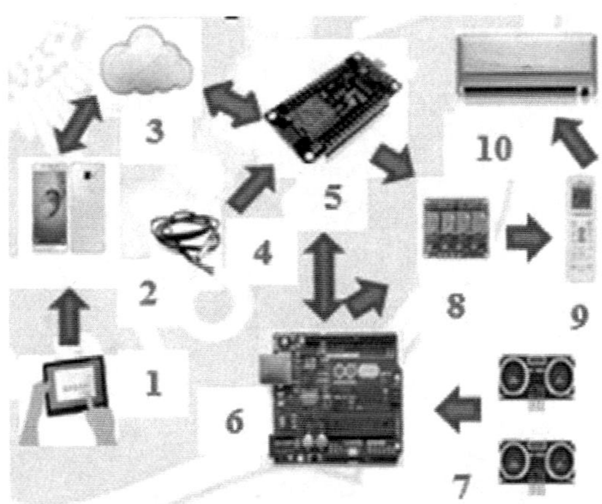

Figure 1. system block diagram

Description:
1. User.
2. Android smartphone.
3. Database (cloud).
4. Temperature sensor.
5. NodeMCU.
6. Arduino Uno.
7. Proximity sensor.
8. Relay module
9. Remote control
10. AC

The sensor will give a value to the Arduino microcontroller which will be processed based on the programmed instructions in it and will work automatically if there is someone entering the room then the lights will turn on and turn on the AC and adjust the AC temperature based on the number of people in the room .

IOP Conf. Series: Materials Science and Engineering **874** (2020) 012007 doi:10.1088/1757-899X/874/1/012007

The results of the microcontroller will provide output to the relay module and the remote as an actuator between the control device and the electronic device.

The following functions of each block block diagram above

1. Power supply is needed to provide voltage on the microcontroller for 12 volts DC while the sensor and actuator devices get a voltage of 5 volts DC to function properly
2. The microcontroller in this study uses Arduino Uno as a control center or processing data from sensors. While the nodeMCU microcontroller as access to connect to the wifi network.
3. HC-SR04 proximity sensor is used to detect the number of people entering or exiting by looking at the change in the detected distance value. While the DS18B20 temperature sensor is used to detect the temperature around the room
4. Actuators function as movers or connectors between the microcontroller and the load or equipment that is controlled, for example: remote AC
5. Android is used as an interface display with users so that it is easier and more efficient to perform system controls.

2.2 Flow Chart System

The data obtained by the micro-controller will be uploaded to the firebase database as an internet of thing system so that the value can be accessed again from other devices such as smartphones based on the values contained in the firebase database. Thus smartphone users can control and monitor the condition of the room based on the value and based on the instructions on the smartphone application mentioned. Like the flow chart below

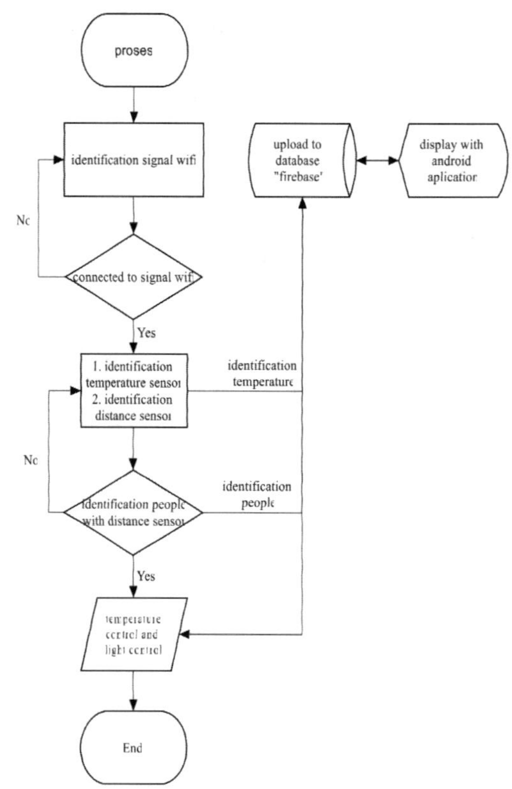

Figure 2. Flow Chart

The flow chart above explains that the system workflow starts by connecting the device to the power supply and the system is ready to process. NodeMCU is connected to the internet as a communication between software and hardware. If the two factors have been met, then the sensor sensor will start detecting and sensing so that the data will be processed immediately and sent to the server which will be displayed on a smartphone or 16x2 LCD. The next process is if the microcontroller gets the pairing code from android in accordance with the previous program it will activate or deactivate the actuator in the form of a relay, if there is no control from android the microcontroller will process based on the number of people who have been in the program.

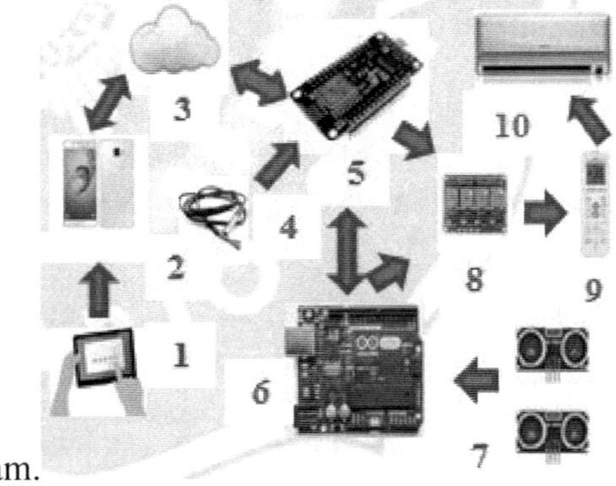

2.3. MIT APP INVENTOR

MIT App Inventor is an open source web application that was originally released or developed by Google, and is currently managed by an institution in the United States called the Massachusetts Institute of Technology. MIT App Inventor is used to create a Sketch program that will become an application for use on Android.

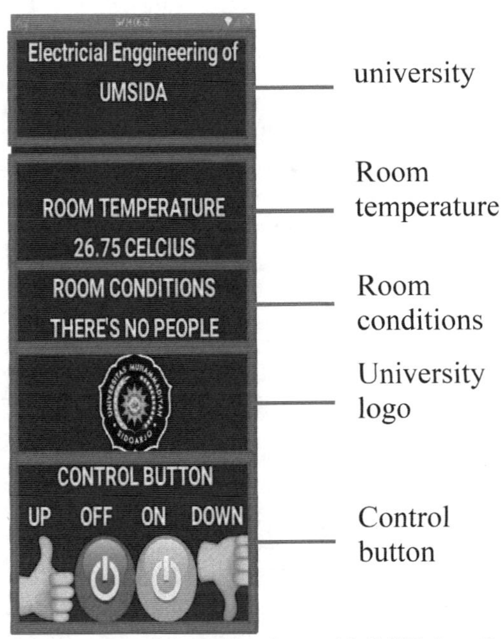

Figure 3. Design android application with MIT App Inventor.

3. Result and Discussion

3.1. Power Supply Testing

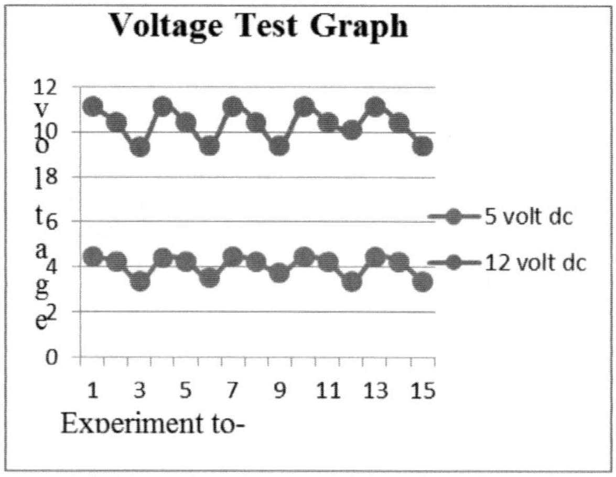

Figure 4. power supply testing chart

Table 1. Power Supply Test Results

Average			Standard Deviasi			Accuracy %		
No burden	Standby load	Peak load	No burden	Standby load	Peak load	No burden	Standby load	Peak load
4.43	4.25	3.47	0.0071	0.0045	0.1518	88.60	85.04	69.40
11.13	10.44	9.51	0.0045	0.0114	0.3253	92.77	86.97	79.25

The power supply voltage has decreased voltage starting from the test without load, standby load and arrived at the peak load caused by conditions where all the devices are working together to supply 2 arduino UNO microcontrollers and MCU nodes, I2C LCDs, relays, indicators and sensors. However, from the test value data has a very small standard deviation value of less than 1 (one), this indicates that the power supply has a very good voltage value stability for supply of equipment so that it can work well without any errors caused by the power supply voltage that is not good.

3.2. LCD TESTING

Table 2. Suitable LCD

Testing	LCD suitability testing with the program to						
	1	2	2	3	3	Average	Standart Deviasi
character	√	√	√	√	√	-	-

From the data value in table 3.2 we can conclude that: LCD can work well and is ready to be used in a system that functions to display information carried out by the microcontroller in the form of a temperature value detected by the DS18B20 sensor and display the number of people who are indoors.

3.3. Testing of HC-SR04 Ultrasonic Distance Sensor

This test is to determine the distance value that is read by the sensor as a scanner for people entering or leaving the room by utilizing the distance that is read then comparing the value of the sensor with a standard measuring instrument that has an accuracy of 1 cm.

The test steps are as follows:

1. Construct the sensor, microcontroller and power supply correctly.
2. Prepare the sensor program and upload it to the microcontroller.

IOP Conf. Series: Materials Science and Engineering **874** (2020) 012007 doi:10.1088/1757-899X/874/1/012007

3. Compare the measurement results of sensor values with standard tools.
4. Do it like the following picture and record the results in the table

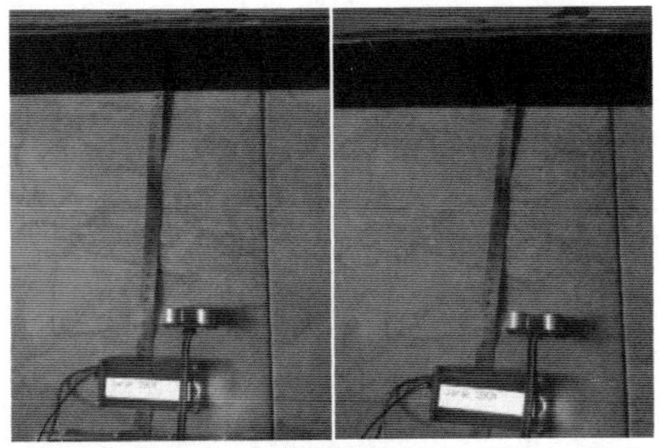

Figure 5. Examples of HC-SR04 ultrasonic testing

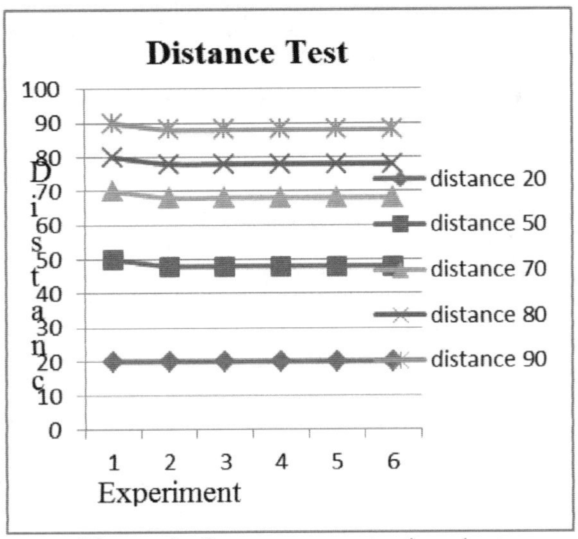

Figure 6. distance sensor testing chart

in testing the ultrasound sensor HC-SR04 has a very good measurement stability when viewed from the standard deviation value. But at a distance of 20cm and above, there is an increasingly large error. In this system the ultrasound sensor is used as an indication of the person who entered or the person who came out through the door with a difference in the measured distance from the ceiling of the door.

3.4. Testing of DS18B20 Sensor with Digital Termometer Standard Tools

DS18B20 sensor testing is carried out to determine whether the sensor is working properly and then compared to the standard Thermo ONE ALPHA 1 brand temperature sensor (MOH AKD 20901600231).

The DS18B20 sensor testing steps are as follows:

1. Connect the DS18B20 sensor to the NodeMCU Microcontroller pin with the cable to the specified pin.
2. Leave the computer and run the Arduino IDE program, then connect the USB cable to NodeMCU,
3. Then upload the program to measure room temperature

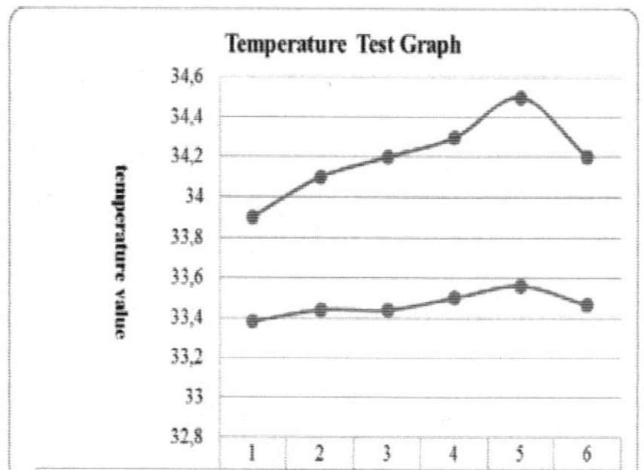

	1	2	3	4	5	6
temperature sensor	33,38	33,44	33,44	33,5	33,56	33,464
temperature actual	33,9	34,1	34,2	34,3	34,5	34,2

Figure 7. DS18B20 sensor testing chart

In testing the DS18B20 sensor with a standard tool ratio there is a difference in accuracy of 0.7 C with an accuracy of 97.85%. on the test value the DS18B20 sensor is good to use as a measurement even this sensor has a smaller standard deviation than the standard tool. This means that this sensor has better stability compared to standard tools

3.5 Overall System Testing

Overall testing of this system aims to be able to see whether the system is running as desired or not. To do the testing process are carried out several stages as follows:

1. Establish the system properly to avoid mechanical errors.
2. Prepare a wifi and internet network so that the microcontroller can be connected as the first condition.
3. Time the time according to BMKG
4. Record and observe the results of the tests that have been carried out

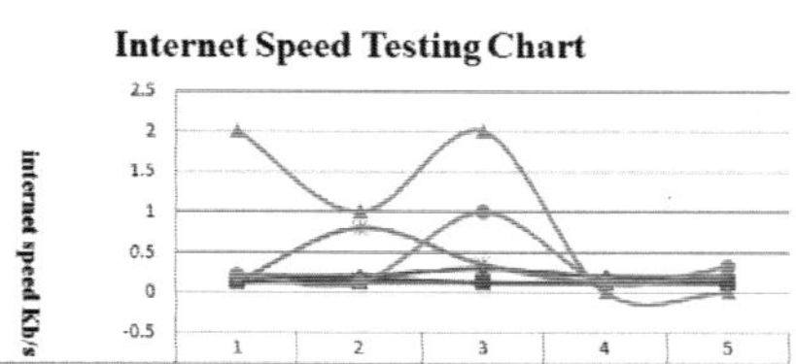

	1	2	3	4	5
internet speed 1	0.2	0.2	0.3	0.2	0.2
internet speed 2	0.12	0.12	0.12	0.12	0.12
internet speed 3	2	1	2	0	0
internet speed 4	0.12	0.17	0.12	0.12	0.12
internet speed 5	0.13	0.8	0.35	0.12	0.2
average internet speed	0.22	0.12	1	0.13	0.32

Figure 8. Results of internet speed testing graph

The graph above is a test of the internet speed needed to control and monitor the system

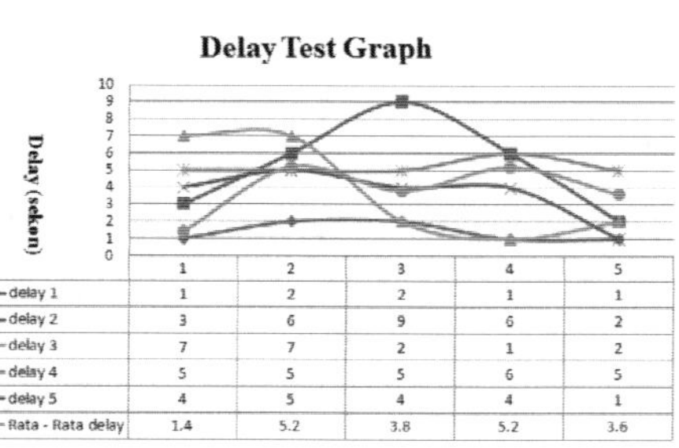

	1	2	3	4	5
delay 1	1	2	2	1	1
delay 2	3	6	9	6	2
delay 3	7	7	2	1	2
delay 4	5	5	5	6	5
delay 5	4	5	4	4	1
Rata - Rata delay	1.4	5.2	3.8	5.2	3.6

Figure 9. graph testing delay when using the application

The picture above is a graph of testing the delay to update data to the database from the microcontroller.

From the results of several tests that have been done above, the results of the overall tool system testing, starting from a room temperature monitoring system that uses an internet of things system can be monitored without any constraints at a certain distance. By using a temperature sensor DS18B20 which has an accuracy of up to 97% compared to a standard temperature measuring device so as to reduce the value error when monitoring. As well as monitoring the presence of someone or not in the room by using 2 ultrasound HC-SR04 proximity sensors, as well as controlling the temperature conditions in the room of the electrical engineering faculty laboratory so that the laboratory head can monitor the condition of the room.

To support the internet of things system we use the NodeMCU microntroller because it already includes a wifi module, so that it can connect to the internet network with the wirelees system and the android application using MIT APP INVENTOR as an interface to the user. The use of the internet system of things affects several factors, namely the distance and signal strength and network on the smartphone which results in a delay in the monitoring process and the process of controlling the temperature of approximately 3 to 4 seconds against the actual response of the device to the smartphone.

4. Conclusion

1. This system is expected to be able to use a microcontroller that is able to work in multitasking because this tool must be able to control and monitor at the same time.
2. The use of a better program, for example PID to be able to determine the desired temperature set point so that it can work automatically and smartly.
3. Use of internet networks that are good from network strength factors and data rates that are expected to be able to communicate over the network without any interference in the form of time delay.

References:

[1] F. Masykur and F. Prasetiyowati, "Aplikasi Rumah Pintar (Smart Home) Pengendali Peralatan Elektronik Rumah Tangga Berbasis Web," vol. 14, no. 1, pp. 93–100, 2016.
[2] B. Mikrokontroler, A. Dan, and R. Time, "VOL . 6 NO . 1 Maret 2013," vol. 6, no. 1, pp. 146–162, 2013.
[3] S. S. Kusumawardani, "PEMODELAN MONITORING PEMAKAIAN DAN PENGHEMATAN ENERGI," vol. 2014, no. Sentika, 2014.
[4] kominfo.go.id, "penggunaan internet." [Online]. Available: https://kominfo.go.id/index.php/content/detail/3415/kominfo+%3A+pengguna+internet+di+indonesia+63+juta+Orang/0/berita_satker2.
[5] A. Uno, "Fakultas Teknik – Universitas Muria Kudus 297," pp. 297–302, 2017.
[6] K. Nasional, I. Sosial, A. Dhiwantara, and B. K. Simpony, "Rumah pintar berbasis mikrokontroler dan android sebagai pengendali," 2016.
[7] P. Ilmiah, "PURWARUPA SISTEM MENYALAKAN DAN MEMATIKAN LAMPU RUANGAN BERBASIS ANDROID DENGAN WEMOS D1 MINI," 2017.
[8] Arduino, "Arduino UNO," 2015. [Online]. Available: http://www.arduino.org/products/boards/arduino-uno.
[9] Anonymous, "Handson Technology User Manual V1.2," *Hanson Technol.*, pp. 1–22, 2017.
[10] kominfo.go.id, "penggunaan internet." [Online]. Available: https://kominfo.go.id/content/detail/4286/pengguna-internet-indonesia-nomor-enam-dunia/0/sorotan_media.
[11] M. Irsan, P. Studi, T. Informatika, F. Teknik, U. Tanjungpura, and J. Prof, "RANCANG BANGUN APLIKASI MOBILE NOTIFIKASI BERBASIS ANDROID UNTUK MENDUKUNG KINERJA DI INSTANSI PEMERINTAHAN."

[12] appinventor.mit.edu, "mit app inventor." [Online]. Available: http://appinventor.mit.edu/explore/index-2.html.

[13] M. Hakimah, "Pembelajaran Embedded System Berbasis Arduino Mega 2560 dan MIT App Inventor," no. xx, pp. 2–6, 2016.

[14] E. B. Lewi, U. Sunarya, and D. N. Ramadan, "Sistem Monitoring Ketinggian Air Berbasis Internet of Things Menggunakan Google Firebase Water Level Monitoring System Based on Internet of Things Using Google Firebase," pp. 1–8, 2016.

[15] G. Magdalena, A. Aribowo, and F. Ati Halim, "Perancangan sistem akses pintu garasi otomatis menggunakan platform Android," *Pros. Conf. Smart-Green Technol. Electr. Inf. Syst.*, vol. D-025, no. November, pp. 301–306, 2013.

[16] Maxim Integrated, "Datasheet DS18B20," *Maxim Integr.*, vol. 92, p. 20, 2015.

IOP Conf. Series: Materials Science and Engineering **874** (2020) 012008 doi:10.1088/1757-899X/874/1/012008

Cyber Security System With SIEM And Honeypot In Higher Education

Jarot S. Suroso[1], Caesario Putra Prastya[2]

Bina Nusantara University, Jakarta, Indonesia

jsembodo@binus.edu[1], caesarioputraprastya@gmail.com[2]

Abstract—Communication science and information technology (ICT) has been progressing and contain complex implication for individual, groups and inter-country relationship. Along with ICT evolution, threats (including theft of information) increase and electronic systems are adopted. As a government agency in higher education with the country's strategic data, XYZ cannot escape hackers trying to access the assets. Strategic assets include system and company's data as well as systems and data which might be taken for granted by irresponsible parties. Therefore, it is necessary to secure information and data systems in XYZ. Moreover to test whether the implemented cyber security system can optimally standardize the company's information and data systems. The application of Security Information Event Management (SIEM) to cyber security systems to detect incoming threat alerts and block access to suspicious threats. While the honeypot is used to deceive the hacker into a fake system intentionally attacked and learning the tricks that intruders use when performing the action. The results of the test of the application of SIEM and Honeypot at XYZ can monitor the monitoring of information system and data-related threats in real-time, and deter and transmit cyber-attacks to adequately protect information security so that the use of cyber security systems with SIEM & Honeypot can secure the information and data systems owned by XYZ optimal.

Keywords— Information Technology, Threats, Cyber Security, SIEM, Honeypot, Realtime.

1. Introduction

The use of science and communication information technology (ICT) has progressed and has a complex impact on the lives of individuals, groups and inter-country relationships. With the advancement of ICT, it brings with it a new concept that connects all of them electronically to a zone called cyberspace, where cyberspace facilitates communication without the limitations of distance and time. Cyberspace is a zone created by electronic media that work with computer networks to communicate in one direction or both directions. Activities in cyberspace do not require direct physical relationships in traditional forms, but these activities are performed virtually. Cyberspace has formed a separate part of the world known as Maya World. The virtual world is today a necessity of human life, which is used for social media, commerce, health services, governance and the defense sector of a country.

The virtual world becomes part of the interface between the real world and the virtual world, with the side of the overlapping part having both positive and negative effects. In the virtual world, anyone can quickly change places while performing activities in different places at the same time, as well as being able to communicate and exchange information quickly in cyberspace, and many other activities that are possible in the world virtually. Like the two opposing sides of a currency, cyberspace also has a negative side, namely the number of threat activities targeting individuals, particular groups, authorities, and threats to a country. These threats include attempts to steal information, adopt electronic systems, and take various other types of action that violate the law, commonly referred to as cybercrime.

Cybercrime is considered an exceptional crime in the international community because many countries are involved. Cybercrime has many forms of performing cyberattacks (cyber attacks) such as phishing, malware propagation, hacking passwords, hacking systems / applications, malformations, backdoor, distributed denial-of-service (DDoS). *Cyber attack* techniques continue to evolve as information technology evolves. Therefore, a cyber threat information security system is needed to monitor cyberattacks in real time. Figure 1.1 shows that cyber threats increase each year.

Similarly, it happened with XYZ, where many hackers are trying to access strategic assets. Strategic assets include systems and data that are related to the enterprise and where the system and data are very dangerous when accessed by irresponsible parties. The following data shows the large number of attacks leading to multiple strategic assets of XYZ, the attack can reach the target if it can not be prevented by the firewall, so that hackers can access the system or data of XYZ following the above statement, the cyber world needs special attention to avoid potential threats, especially those that affect information security. Even the most dangerous threats are linked to cybersecurity today. Cybersecurity is a non-military defence action for a country to develop a sense of defence in cybersecurity in the cybersecurity sector. Cybersecurity targets all groups, individuals, groups, organizations and governmental interests. Priority is the protection of strategic assets of defence and security that are directly related to the public. If it comes to a disruption of the system that disturbs the stability of the nation, this is the reference for building cybersecurity to protect against cyber threats.

Based on background observations, this research study was conducted to find solutions to the problem :

1. How to design the information security system using SIEM and Honeypot in XYZ?
2. Can the use of information security systems with SIEM and Honeypot effectively secure information and data systems?

This template, modified in MS Word 2007 and saved as a "Word 97-2003 Document" for the PC, provides authors with most of the formatting specifications needed for preparing electronic versions of their papers. All standard paper components have been specified for three reasons: (1) ease of use when formatting individual papers, (2) automatic compliance to electronic requirements that facilitate the concurrent or later production of electronic products, and (3) conformity of style throughout a conference proceedings. Margins, column widths, line spacing, and type styles are built-in; examples of the type styles are provided throughout this document and are identified in italic type, within parentheses, following the example. Some components, such as multi-leveled equations, graphics, and tables are not prescribed, although the various table text styles are provided. The formatter will need to create these components, incorporating the applicable criteria that follow.

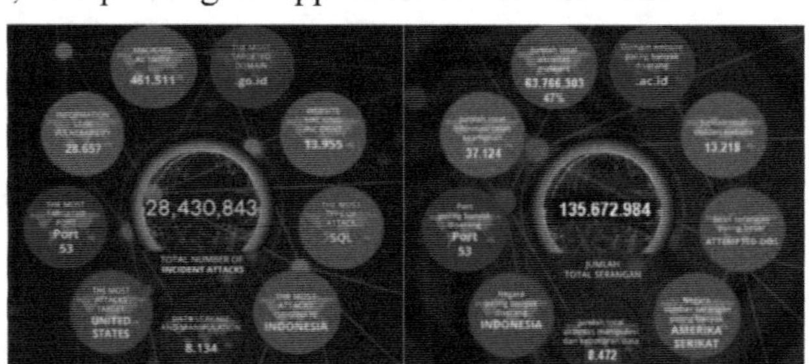

Figure 1. Indonesia threat 2015 & 2016

2. Literature Review

2.1 Information Security
The definition of information security has different opinions from several sources / experts. According to Sarno and Iffano, information security is an attempt to protect information assets from potential

threats. In the case of information security, the business can indirectly guarantee business continuity because information security can reduce the risks that occur in day-to-day business processes. In addition to information security, it can help optimize return on investment [1] More and more corporate information is being stored, managed and shared, the greater the risk of damaging, losing or disclosing data to unwanted external parties. [2] According to ISO / IEC 17799: 2005, an information security information security management system is an effort to protect against a variety of threats to ensure the sustainability of the business, minimize business risk, and increase investment in identifying opportunities. There are three key information security principles in maintaining information security from multiple sources of threat: confidentiality, availability, and integrity [2].

2.2 IDPS (Intrution Detection Prevention System)
IDS Intrusion Detection System is a system that monitors network traffic and suspicious activity in a network system [5]. If suspicious activity related to network traffic is detected, IDS warns network administrators through the IDS console system. In some cases, IDS can also respond to anomalous traffic anomalies by blocking actions against users or Internet Protocol (IP) addresses attempting to access the network. IPS (Intrusion Prevention System) is a system that combines firewall functions with IDS features. This IPS technology can be used to prevent cyber threats leading to the local network by checking in and out each packet (input and output), and the packet is detected by the sensor if an identified attack threatens the local network denies access (blocking) and logs (logs) all data packets identified as threats. In this way, IPS acts like a firewall that allows and blocks network access and, like IDS, combines with full packet detection. IPS uses signatures as packet filtering rules to detect traffic activity on the network or terminal. The detection of incoming and outbound (inbound-outbound) threats can be prevented as early as possible before they gain access to and damage or receive access to local area networks through the use of IPS. Thus, early detection and prevention can be promoted by using IPS [3]. IDS and IPS are commonly known as IDPS (Intrusion Detection and Prevention Systems). IDPS focuses on identifying possible events. For example, IDPS can detect when an attacker successfully exploited the system by exploiting security holes in the system. The IDPS then reports the incident to the security administrator, who can immediately perform an incident response to minimize the damage caused by the incident. IDPS can also record information that can be used by event handlers [4]. There are two types of IDS and IPS, NIDS (Network-based IDS) and HIDS (Host-based IDS). IPS also has two types, NIPS (Network based IPS) and HIPS (Host based IPS). There is a fundamental difference between IDS and IPS. The difference is that IDS does not work in-line on the network when placed on the network, or it can be said that IDS only monitors the network by "connecting" or "tapping" the network to be protected. Unlike IDS, IPS is in-line in the network. To make things suspicious, IPS can close the access immediately.

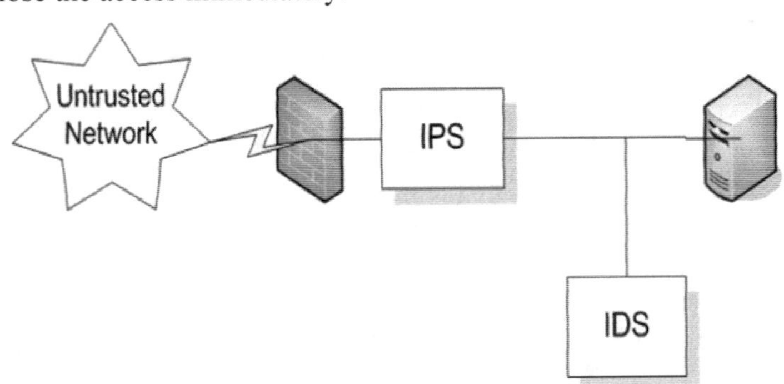

Figure 2. Different placement of IDS and IPS

Link : https://keamanan- informasi.stei.itb.ac.id/2013/10/30/menangani serangan-intrusi-menggunakan-ids-dan-ips/

1) *Network-based Intrusion Detection System (NIDS)*
NIDS is an intrusion detection system that collects data packets that travel on a network medium (wired, wireless) and matches the data packet with the signatures in the database. If the data

packet matches the intruder's signature, a warning is issued and the data packet is stored in a file or database.

2) *Host Intrusion Detection System (HIDS)*
Host-based Intrusion Detecton systems or HIDS are placed as agents of a host. Intrusion detection systems such as these can examine system and application log files to detect intrusion activity. Some of these systems are reactive, which means that HIDS only notifies users when an attack has occurred. But there are also some who are proactive who can track network traffic from specific hosts installed with HIDS and perform peringantan in real time.

2.3 SIEM (Security Information Event Management)

Security Information and Event Management (SIEM) is a system that can analyze security events in real-time while providing long-term log storage, historical reporting, and trend analysis. They serve as event alerts and correlate vulnerability data and threat intelligence to provide insights into risk predictions (and to prioritize those risks) and audit and audit reports for compliance purposes, such as: the PCI provisions (Payment Card Industry) to perform. SIEM is able to provide information security capabilities as a holistic view of security risks through historical data collections from multiple sources such as firewalls, switches, load balancers, weblog servers, and proxy servers. This logarithmic correlation is intended to extract the meaning and intelligence of this mass data; remember when certain event conditions are met; and a dashboard that visualizes events that previously required management of multiple dashboards and reporting tools. The use of SIEM technology is determined by the need for information security. Investigating incoming attacks requires intensive resources, data access, and monitoring. SIEM is a combination of Information Management Security (SIM) and Security Event Management (SEM) [6]. While SEM or Security Event Management consolidates and stores multiple log sources and security events, SIM or Security Information Management provides an analysis and analysis of long-term trends from this log.

- *Security Information Management (SIM) - log management and compliance reporting.*
- *Security Event Management (SEM) - Monitoring management and real-time events for security events from networks, security devices, systems, and applications.*

  SIEM technology is typically used for three primary use cases:
- Compliance - log management and compliance reporting.
- Threat Management - direct monitoring of user activity, data access and application activity and incident management.
- A deployment that offers a mix of compliance and threat management.

SIEM was developed to meet the governance compliance reporting requirements that have been developed, but companies are taking this as an opportunity to implement technologies that enhance their threat management and incident response capabilities.

2.4 *HONEYPOT*

The honeypot is a computer system and the directory is like a real computer. However, the purpose of this computer is to make hackers fall into it when we as users observe and follow their behavior. So we can define it as a wrong system that looks like a real system. They differ from other security systems in that they not only find a solution to a particular problem, but are also capable of implementing various security issues and finding multiple approaches to them [7]. For example, they can be used to record malicious activity in compromised systems, they can also be used to learn new threats to users and to develop ideas on how to resolve these issues. In general, the main goal in designing a honey pot is to attack and examine the system. Needless to say that behind the honey pot no valuable data should be.

Unlike other security tools, honeypots do not provide a stand-alone solution to specific problems [8]. For example, the firewall analyzes incoming and outgoing traffic and then determines whether the packet is allowed or not. They basically control the traffic in the network. Intrusion detection systems monitor networks or systems for unauthorized activity. Instead, honeypots are more flexible and apply to many security situations. You do not monitor or control the traffic. Instead, they receive the attack and collect data.

IOP Conf. Series: Materials Science and Engineering **874** (2020) 012008 doi:10.1088/1757-899X/874/1/012008

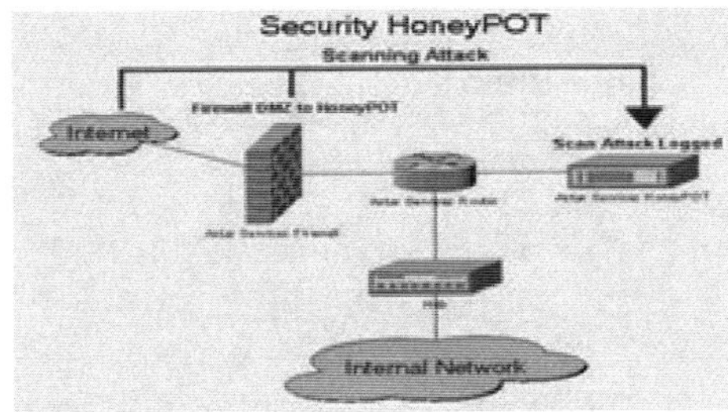

Figure 3. Topology Honeypot.

3. Research Methodology

The method used is the Network Development Life Cycle (NDLC), namely, analysis, design, simulation prototype, implementation, monitoring, and management. This method will be used to analyze and design the phases of research performed in this study, as shown in Figure 3.1. and research framework is shown in Figure 3.2.

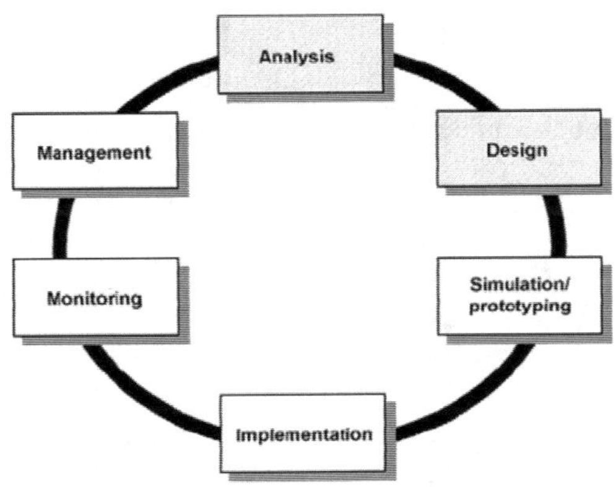

Figure 4. NDLC Method.

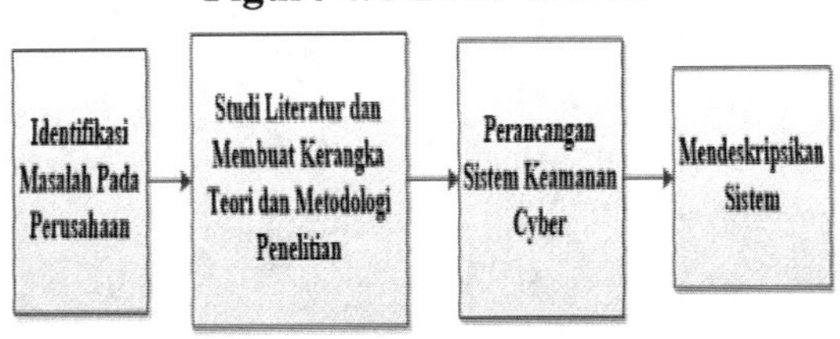

Figure 5. Research Framework.

Integrate and implement a security system designed with the actual system, both in monitoring cyber threats and automatically defending interference activity and redirecting inbound attacks to the shadow server to secure real data. Layered security is designed to ensure that data remains secure in accordance with CIA principles (confidentiality, integrity and availability).

4. Result and Discussion

SIEM is here an application of software or appliance hardware devices in which there are applications that monitor suspicious activity in a network or system. SIEM checks the incoming and outgoing data on the inbound and outbound pages of a network and the system it owns. Honeypot here is a wrong system that uses the actual system to stop the shadow where here Honeypot hackers want to access the data from the server, in addition to the Honeypot here, as hackers are tempted to start their action at the Backup data against cyber threats intended for administrators. The software used to create this information security system is software that is open source, meaning that the software is used freely without having to pay the license. Open source software is also easy to get and can be freely downloaded from the Internet. The hardware needed to create an information security system is a server that has enough RAM and has been installed with the software needed to build an information security system.

The steps of the discovery system and the SIEM server are as follows:
1. The sensor as a collector in SIEM receives the protocol and returns it with various types of communication protocols.
2. The collector stores data in a SIEM database.
3. Subsequently, an analysis or risk assessment of the incoming and outgoing events is carried out by the network.
4. If the incident is considered a threat, IPS sends a command to the firewall to block access and provide information from the control panel.
5. The SIEM system displays all alarms and events detected by the sensor, the information coming from an information security system that is synchronized.
6. The user interface of the SIEM system allows the administrator to monitor events in real time.

4.1 SIEM and Honeypot General Architecture

The cyber security system consists of two main parts, the section having different tasks and functions. For the first part as a collector in information related to Internet security, system cyber security implemented by IDS sensors and other devices that send their logs to the SIEM. While the second part is the part that processes information, and collects information obtained from collectors. Information - The information is collected in a cyber security analysis database. Below you will find an overview of the architecture of the SIEM sensor, the database and the server.

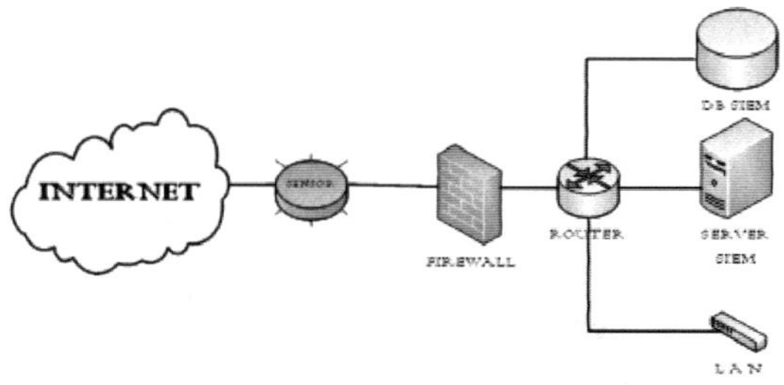

Figure 6. Overview of the general SIEM architecture.

Honeypot system is placed in the company's network, aligned with the server from system to divert threats by hackers.

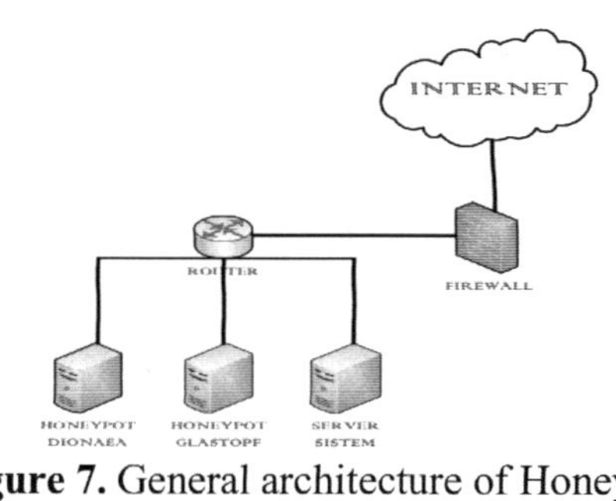

Figure 7. General architecture of Honeypot

4.2 Topology

Topology proposed, the system SIEM and Honeypot in the implementation can be applied as shown in Figure 3, where the system SIEM and the Honeypot All these devices are connected to the Internet network, where devices are used between LAN sensor, firewall, router , Switch, SIEM Server, SIEM DB Server, System Server, Honeypot Server, and Switch are connected for employee PC distribution. Sentry leader in the system Cyber security Inbound sensor installs events capture from the outside, the sensor is connected to the firewall to ward off attacks on cyber security systems. Firewall is with a router to set the routing on cyber security system and the needs of the corporate network. Once it is connected to the sensor outgoing to capture the event from the inside, then connected to 3-switch, SIEM in connection with the system SIEM, switching system connected to the system by a company honeypot from the server, and the last one Router connected to the distribution switch was overshadowed employee PC needs.

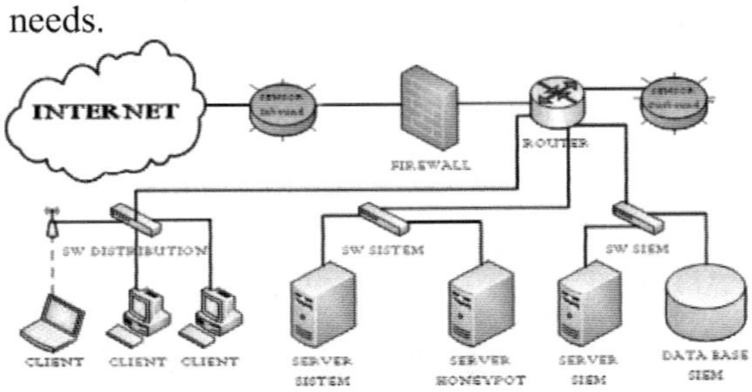

Figure 8. Suggested topology.

4.3 Attack Test System Test Results

If the server suffers a hacker attack in the form of sql injection in a network of companies that have implemented the cyber security system using SIEM and honeypot. The attack on the server of the company was detected by the sensor.

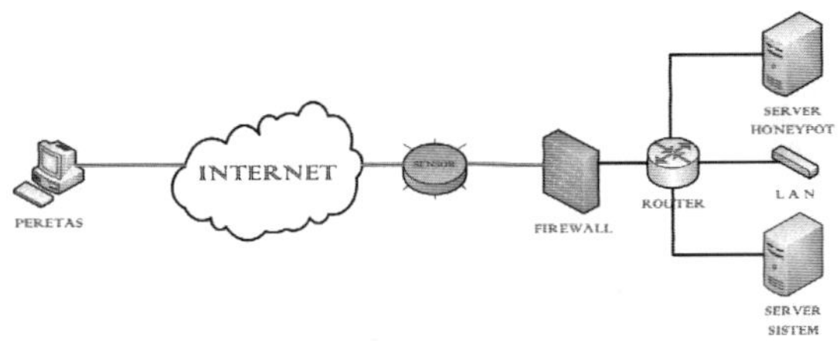

Figure 9. Attack block topology.

4.4 Results of Attack Attack System Tests

Attacked by DDOS hackers on the network of companies implementing the cyber security system with SIEM and Honeypot. The attack on the server of the company was detected by the sensor. The event, captured by the sensor, is forwarded to Iptables and forwarded to the Honeypot server to redirect the load on the server. Can see in the image revenue attack, which had been in a state of full memory to be normal, so the attack is not on another server, but on the honeypot.

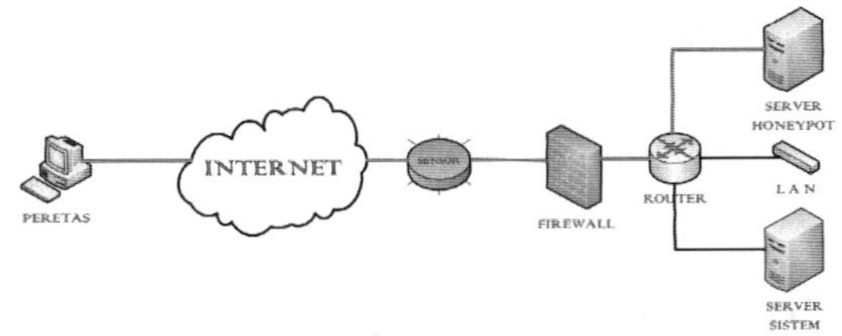

Figure 10. Attack transfer topology.

B. Test Results of The Target Attack Transmision System

If the server receives a hacker attack in the form of port scanning in the network of companies that have implemented the cyber security system using SIEM and Honeypot. The attack on the server of the company was detected by the sensor. From events intercepted by the sensor, iptables are routed and redirected to the honeypot server to distract hackers. so that the scanning process does not point back to the server, but to the honey pot.

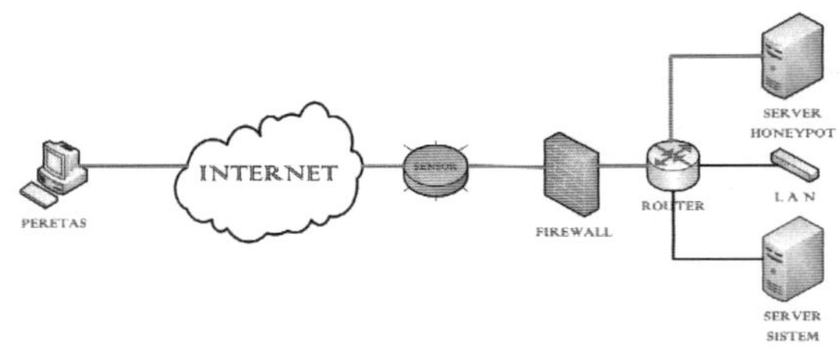

Figure 11. Topology of the transmission of the target attack.

The template is designed for, but not limited to, six authors. A minimum of one author is required for all conference articles. Author names should be listed starting from left to right and then moving down to the next line. This is the author sequence that will be used in future citations and by indexing services. Names should not be listed in columns nor group by affiliation. Please keep your affiliations as succinct as possible (for example, do not differentiate among departments of the same organization).

1) *For papers with more than six authors:* Add author names horizontally, moving to a third row if needed for more than 8 authors.

2) *For papers with less than six authors:* To change the default, adjust the template as follows.

  a) Selection: Highlight all author and affiliation lines.

  b) Change number of columns: Select the Columns icon from the MS Word Standard toolbar and then select the correct number of columns from the selection palette.

  c) Deletion: Delete the author and affiliation lines for the extra authors.

5. Conclusion

Based on the analysis of the application of Cyber Security System using SIEM and Honeypot at XYZ can close as follows:

1. The Cyber Security System was developed using a dual security system between Security Information Event Management (SIEM) and the Honeypot to effectively secure critical assets of the company. successfully developed by using company equipment to protect the system and data at XYZ.
2. The cyber security system is developed using enterprise devices and networks implemented through the Network Base IDS (NIDS) method.
3. Based on the results of the tests, the cyber security system with SIEM and Honeypot has the reliability and ability to maintain cyber security. A proven cyber security system with SIEM and Honeypot can overcome attacks on SQL Injection, DDOS, and Port Scanning testing.

References

[1] Indrajit, R. E. (2011). Pengantar Konsep Keamanan Informasi di Dunia Cyber. Jakarta: APTIKOM..

[2] Sarno, Riyanarto., Iffano, Irsyat. (2009). Sistem Manajemen Keamanan Informasi. ITSPress.

[3] Deris Stiawan, Intrusion Prevention System (IPS) dan Tantangan dalam pengembanganya, Jurusan Sistem Komputer, FASILKOM, Universitas Sriwijaya.

[4] Mokube,I. & Adams M.,2007.Honeypots:Concepts,Approaches,and Challenges.ACMSE 2007, March 23-24, 2007, Winston-Salem, North Carolina, USA ,pp.321-325.

[5] O'Brien, James A. 2005. Introduction to Information System. Mc Graw Hill.

[6] Research, G. (2011). How to deploy SIEM technology. Technical report..

[7] Mokube,I. & Adams M.,2007.Honeypots:Concepts,Approaches,and Challenges.ACMSE 2007, March 23-24, 2007, Winston-Salem, North Carolina, USA ,pp.321-325..

[8] Spitzner, Lance. (2012), Honeypots: tracking hackers. Addison-Wesley Longman Publishing Co., Inc..

IoT-Based Smart Home Controller Using NodeMCU Lua V3 Microcontroller and Telegram Chat Application

Yulian Findawati, Achmad Idris, Suprianto, Yunianita Rachmawati, Eko Agus Suprayitno

Universitas Muhammadiyah Sidoarjo, Sidoarjo, Indonesia

Abstract. The house is a valuable asset that needs to be guarded, not infrequently some people experience undesirable things happening in the home environment, these things include fires and theft. Anticipation is not just enough by manually such as locking the door, or preparing fire extinguishers, in addition to the manual way of anticipation can be added by automation, namely by installing devices that can monitor the condition of the house and also can control electrical equipment in the house. The automation tool intended in this study is an IoT-Based Smart Home Controller Using the NodeMCU Lua V3 Microcontroller and Telegram Chatting Application. The use of automation systems like this is an option to be implemented because it saves power and the price is relatively cheap. This device / system uses 2 NodeMCU Lua V3 microcontrollers and several sensors (fire sensors, motion sensors, temperature and humidity sensors) and also this role utilizes the 4 Chanel Relay Module which functions as a switch to turn off or turn on electrical equipment in the home. The control function of the smart home controller device also utilizes the Telegram chat application bot that is planted on an IoT-based system device with the function of receiving commands from users through the Telegram chat application to monitor the condition of the house and also controls the relay module used as a switch of equipment.

***Keywords**:* IoT, Internet of Things, Monitoring, Controller, NodeMCU, Bot, Telegram, Smart Home, Arduino

1. Introduction

The development of technology is now increasingly fast so that the emergence of a new technology spoil people. Even the phrase automation is often heard in the ears of many people, this indicates that the decrease in human intervention in terms of doing routine activities in daily activities. Automation of the system has now penetrated into the house, and is known as the 'Smart Home' system. System automation that will be discussed here is the installation and configuration of electronic equipment that is integrated with the Telegram chat application that serves as a tool to manage / control all the needs of an electrical installation and control of devices that use electrical energy (such as TV, AC, Refrigerator , etc.) in the home environment. For example: if a homeowner in general has a house where the method usually turns on the lamp using the help of conventional / general switch, and usually turns on or turn off the lights manually, then if the homeowner uses system automation, the lights can be turned on automatically with a timer or use a motion sensor (the lamp will automatically turn on itself if the sensor detects the presence of humans or dies when there are more people) or the homeowner can turn on the lamp remotely by using the remote control via a smartphone or even through internet site services. This can also apply to all electrical appliances at home. Even though you are already using system automation, electrical equipment can still be activated or deactivated manually. Not only that, homeowners can also make systems in a room by designing memory programs or electrical installation habits that work, so just by pressing one button then the lights / electrical equipment will be active in accordance with the program that has been made. This is usually used for rooms that have lots of lights but homeowners do not want to have a lot of switches so this system is very flexible following the wishes of the homeowner.

Various studies have been carried out to monitor the condition of the house [1] and control of home electrical appliances [2]. Furthermore [3] has conducted research on IoT-based temperature monitoring monitoring systems, but the microcontroller used is not NodeMCU, the Arduino UNO microcontroller and its web-based control system are used, not using the telegram chat application, there are also previous research studies on the making of electrical equipment control systems. home, but the user interface and user experience that is used uses the Blink application to control the device. [4]. There is also previous research on iot smart fish farm design with Raspberry Pi and Webcam controls. This study discusses the monitoring system of fish farms where the system was built using the Telegram chat application as a control tool for monitoring the system, the system built using Raspbery Pi as a microcontroller, is different from the one used in this study [5]. There is also previous research on the design of home security systems using relays. This research focuses on controlling electrical home appliances. However, in this study the system built is an automatic control system (no need to do manual control of the tool) and also the system built does not use a NodeMCU Lua V3 microcontroller [6]. There is also previous research on fire spot detectors for early detection of fires. This study discusses the fire detection system. The system is automatic (not controlled manually) and only uses Buzzer as the output media when a fire occurs. The system built uses the Arduino Mega microcontroller, different from the one used in this study [7]. There is also previous research on the internet of things on room security systems, case studies of raharja college server rooms. The study discusses the server room monitoring system that uses motion sensors and camera modules as input. The system built uses Arduino Mega as a microcontroller and uses Web media as the controller of the built system. Web media is a media controller that is quite fast and effective. However, this media is considered ineffective because the user cannot know if the sensor in the server room detects movement because the user must first open the browser to see the detected movement [8]. Based on the background above, the author wishes to create an IoT-based Smart Home Controller by using the NodeMCU Lua V3 Microcontroller and Telegram Chat Application. This is an effort to facilitate users in monitoring the condition of the house and facilitate users in saving electricity as well as facilitate users in controlling electronic home appliances so that their use is more effective.

2. Design

2.1 Block Diagram System

In making tools / systems to monitor and control home electrical appliances, a tool / system design will be built to make it easier for writers to build the system / tool. Making tools are divided into 2 function blocks. General description of the work system and the division of system blocks from this study is shown in the following figure.

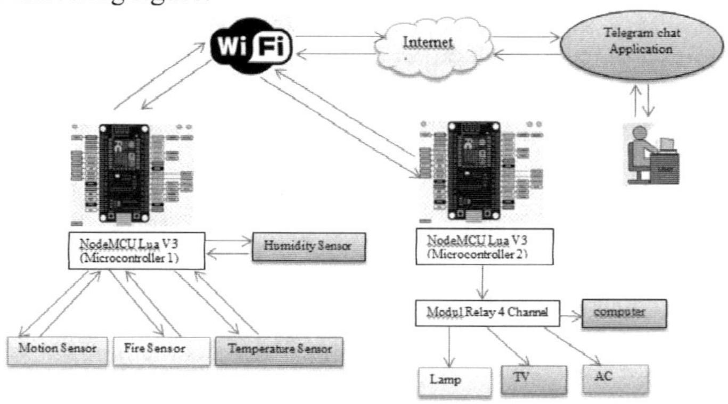

Figure 1. Block Diagram System

The smart home controller system that was built in this study is a system used to control household electronic equipment and also to monitor the condition of the house. This control and monitoring can be run by the user through the Telegram chat application media as HMI (Human Machin Interface) media. The system was built using a NodeMCU Lua V3 microcontroller that is connected to the Telegram chat application server via a WiFi network that is connected to the internet. Smart home controller that is built is divided into 2 parts, the first part is in charge of monitoring the condition of the house, and the second part is in charge of the system to control electrical home appliances. Home condition monitoring system uses 3 sensors as input media, including sensors fire sensor (Flame Sensor), motion sensors, and also temperature and humidity sensors. Whereas in the electrical control system home appliances the controlled electrical appliances will be simulated by using 2 lamps as output .

2.2 Design of NodeMCU with Relay Module and several sensors (Electrical Equipment Controller Circuit)

This circuit is a prototype of the electrical equipment controller circuit and the condition monitoring circuit of the house, the two series are combined into one then the writer adds a battery / power supply with a voltage of 5V 1.5A and also 2 light bulbs that are powered by 220V AC voltage as indicators to find out whether the home electrical control system that has been made works well or not. The series can be seen in the following figure 2.

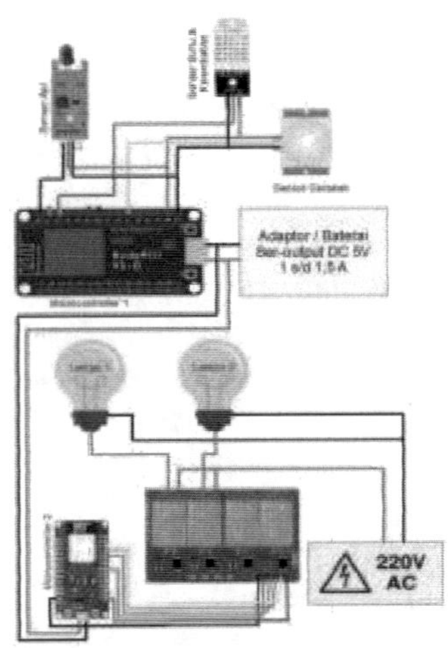

Figure 2. Combination of a home electrical equipment control circuit and a house condition monitoring circuit

3. Results

3.1 Assemble the System

This stage is to assemble the tools / systems that are tasked with monitoring the condition of the house and controlling the house electricity.

a. Assembling House Condition Monitoring System

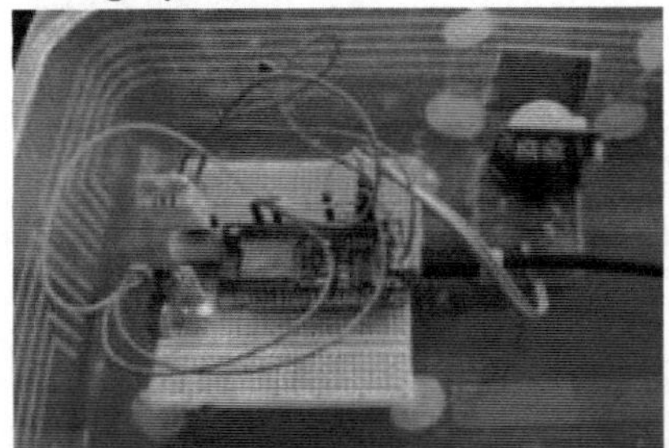

Figure 3. House Condition Monitoring System Assembly Results

In the prototype monitoring system above, the three sensors (Flame Sensor, PIR, and DHT22) are integrated with the microcontroller 1. Microcontroller 1 is powered by an adapter / battery with a 5V 1 - 1.5 amphere voltage and is connected via a micro USB cable.

b.Assembling a Home Electric Appliance Control System

Figure 4.Results of Assembling a Home Electric Control System

In the prototype picture of the electrical control system of the house above, two lights are seen powered by a parallel 220V AC current, each lamp does not use a conventional switch but directly uses a relay module as the main switch where the relay module is controlled via a microcontroller

3.2 System Testing
In this section, trials are divided into 2 types, the first is a trial of a home condition monitoring system and the second is the home electrical equipment control system.

3.3 Prototype Testing of House Condition Monitoring System
In the prototype testing phase, testing is done in different ways on each sensor. In testing the fire sensor, a match is used as a tool to create a small fire that will be used as a trigger for the fire sensor

Table 1. Prototype Testing of House Condition Monitoring System

NO	Sensor Tested	Testing Method	Results	Response Time (from the start of the test until the system sends a notification to the Bot chat)
1.	Flame Sensor	Conduct a test by igniting a flame using a match that is brought close to the flame sensor	Success	37 second
2.	Sensor Suhu Dan Kelembababan	Type the "/ sensor" command in the Telegram Bot chat which functions as a monitoring Bot.	Success	55 second
3.	Motion Sensor	Move the palm to the front of the motion sensor (PIR PIR HC-SR501)	Success	5 Minutes

3.4 Testing of Prototype Home Electrical Equipment Control Systems
In testing temperature and humidity sensors, testing is done by typing the command "/ sensor" in the chat bot telegram to trigger the microcontroller to read the temperature and humidity through the dht22 sensor connected to the microcontroller. Whereas in motion sensor testing, testing is done by moving the palm of the hand right in front of the motion sensor (PIR) so that the sensor can read hand movements easily, the motion sensor manages to detect movement, but the system's response time in sending notifications of the movement takes considerable time long, which is approximately 5 minutes

Tabel 2. Testing of Prototype Home Electrical Equipment Control Systems

No	Initial Condition (Tool)	Testing Method	Results	Response Time (from the start of testing until the electrical equipment turns on / off))
1.	Lamp 1 Off	Type the command "/ lampu1on" in the Bot Telegram chat that serves as a bot controller.	Successful,Light 1 Turns on 	35 Second
2.	Lamp 1 on And Lamp 2 Off	Type the command "/ lampu2on" in the Bot Telegram chat that serves as a bot controller.	Successful,Light 2 Turns on 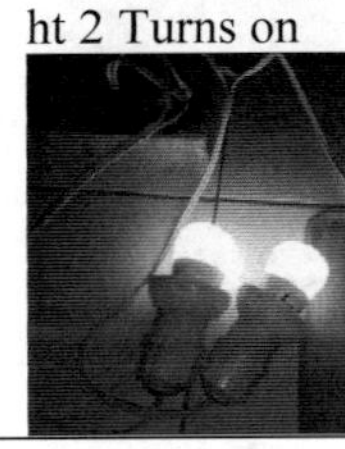	51 Second

4. Conclusion

The conclusion of the discussion of the implementation of the IoT-based Smart Home Controller system by using the NodeMCU Lua V3 Microcontroller and Telegram Chat Applications include:

1. The system built can facilitate the user in monitoring the condition of the house in real time (24 hours) anywhere and anytime.
2. The system built can make it easier for users to control home electrical appliances (be it lights, air conditioners, TVs, etc.) in real time (24 hours) anywhere and anytime.
3. This research is the result of the development of previous research in which connections using Bluetooth have not been able to control home electrical equipment or monitor the condition of the house in real time anywhere and anytime.
4. Using Bot telegram chat application as a media access smartphone application has advantages compared to using android applications or using web pages because it can be accessed with multi platforms both Android, iOS, Windows, and Linux, and also this telegram chat application bot can send notifications via directly when the sensor in the monitoring system detects fire or movement.

References

[1] Amaro, Najib. 2017. Sistem Monitoring Besaran Listrik Dengan Teknologi IoT (Internet of Things), Bandar Lampung.
[2] Afilusuf, Ryan, & Marisa, Fitri. 2016. Smarthome Automatic Lighting Berbasis Web, Malang, Informatika.
[3] Andika, Ichsan A. B. 2017. Monitoring Suhu Pemanas Portable Berbasis Arduino Yang Terintegrasi Dengan Android.
[4] Lusita Dewi, Nurul Hidayati, Rohmah, Mimin F., & Zahara Soffa. 2018. Prototype Smart Home Dengan Modul NodeMCU ESP8266 Berbasis Internet of Things (IoT), Mojokerto
[5] Nashrullah, Khabib Y, & Setyawan, M. Bhanu. 2019, Rancang Bangun IoT Smart Fish Farm Dengan Kendali Raspberry Pi Dan Webcam, Ponorogo.
[6] Shaleh, Muhammad, & Haryanti, Munnik. 2017. Rancang Bangun Sistem Keamanan Rumah Menggunakan Relay, Jakarta.
[7] Subekti, Arief, & Mudjiono, Urip. 2017. Fire Spot Detector Untuk Deteksi Dini Terjadinya Kebakaran Di Kapal, Surabaya. Mudjiono, Urip, & Subekti, Arief. - .Fire Spot Detector Untuk Deteksi Dini Terjadinya Kebakaran Di Kapal, Surabaya.
[8] Susanto, Fredy, & Rifai, M. Nur. 2017. Internet of Things Pada Sistem Keamanan Ruangan, Studi Kasus Ruang Server Perguruan Tinggi Raharja.

IOP Conf. Series: Materials Science and Engineering **874** (2020) 012010 doi:10.1088/1757-899X/874/1/012010

Smart Home Integrated With Internet Of Things (Iot) In The Digital Era Of Industry 4.0

Eko Agus Suprayitno[1*], Izza Anshory[2], Jamaaluddin[3]

Department of Electrical Engineering [1,2,3], Faculty of Science and Technology Universitas Muhammadiyah Sidoarjo

*eko.agus@umsida.ac.id

Abstract. Smart Home is one form of technological progress in the Industrial Revolution Era 4.0 that facilitates human activities in controlling and monitoring the condition of the home that is not limited to distance by utilizing electronic equipment, the internet, and smart phones[1]. In this research, a prototype of Smart Home with Internet of Things (IoT) which is equipped with a proximity sensor (HC-SRF04) to open and close the door of the house automatically, a temperature sensor (LM35) in detecting the integrated room temperature on and off based on the fan room temperature, and lights and fans that can be controlled remotely by using the internet and smartphones. Android application to control and monitor homes in Create by using the Blynk application. The Temperature Sensor and Proximity Sensor on Smart Home are tested using Industry standard distance and temperature measuring devices (digital thermometers). The accuracy value of the device reaches 98.22% for distance measurement, and 98.50% for room temperature when compared with industry standard tools. Android applications have been successfully tested in several types of smartphones with different brands and specifications. The distance of the smartphone to the Smart Home location in monitoring and control was successfully tested and worked well at the closest distance of 5 meters to the farthest distance of 17,466 km (State of Mexico), all of which use the Internet.

1. Introduction

Smart Home is one part of technological advances in the field of automation, especially in terms of controlling household electronic devices that are integrated with smartphones[1]. One example of its products is the use of sensors for several household electronic devices that can be controlled and monitored at any time with a smartphone connected to the internet. So that the use of electrical power for electronic devices and lights can be controlled and monitored by homeowners, in their efforts to save the maximum use of electric power load[2]. Aside from being a control and monitoring, smarthome in other studies can also be used for home security, one of the examples of its application is using an RFID Card (changing conventional keys using ID Card) for a key security system on a house door. This is done in an effort to reduce theft [3]. Smart home is a place to stay with automated systems that are sophisticated in controlling and monitoring lighting and room temperature, household appliances, multi-media equipment, and security systems and many other functions. Internet Of Things (IoT) [4] plays an important role in building smart homes. Through IoT almost every object of our daily life at home can be connected to the Internet. IoT allows monitoring and control of all these connected objects regardless of time and location [2]. In addition, the system of Smart Home can also be applied in the medical field, especially in patients who enter the recovery period after being hospitalized, where some physiological parameters need to be measured continuously and monitored by the hospital. Therefore telemedicine and remote monitoring of patients using Inernet of Things are increasingly important and urgent [5][6]. In this research, Prototype Smart Home was created to monitor Room Temperature, Room Fan, Room Light, and Open or close the door of the room with Internet of Things communication integrated with Android Smartphone. The working principle of Smarthome in this study starts from the room temperature which is read by a temperature sensor (LM35) which exceeds 28 oC will have an impact on the condition of the fan will be active. this is done to reduce the room temperature. Room temperature that can be read will also be monitored on an android smartphone. Proximity sensor (HC-SR04) is used to detect obstruction of objects in front of it, or human distance that is read on the sensor. If the read obstacle is less than 15 cm then the motor will be active to open the door of the house, but if the read distance is less than 15 cm then the door will close. For lights on this prototype smarthome can be turned on and off with an Android smartphone. The Smarthome prototype can be connected to the Internet, and the Android application is created using Blynk software or MIT App Inventor.

2. Research Methods

The Smart Home System diagram in this study (Figure 1) was made using NodeMCU ESP8266 as the main microcontroller which functions as a hardware liaison with the Smartphone. The working principle of the internet connected NodeMCU microcontroller will send data to the webserver from Blynk, then forwarded to an android smartphone so that the hardware can be controlled and monitored even over very long distances. The working principle of the Smarthome Prototype uses proximity sensor (HC-SRF04) which is placed in front of the door to detect the presence of humans or objects approaching. Distance information that is read will provide information to the microcontroller to enable the door to open or close automatically. The temperature sensor (LM35) placed on the Smarthome protopyte is used to detect room temperature, and activate the Fan when the room temperature is more than 28 oC and turn off the Fan when the room temperature is less than 28 oC. Room temperature can be monitored with an Android smartphone, to turn off and turn on the Fan and the Lamp can also use an android smartphone. The Android application that is used to control and monitor the Smarthome is created using the Blynk software application and the MIT App Inventor software. Distance is not an obstacle to control and monitor, as long as the Hardware and Smartphone are connected to the internet.

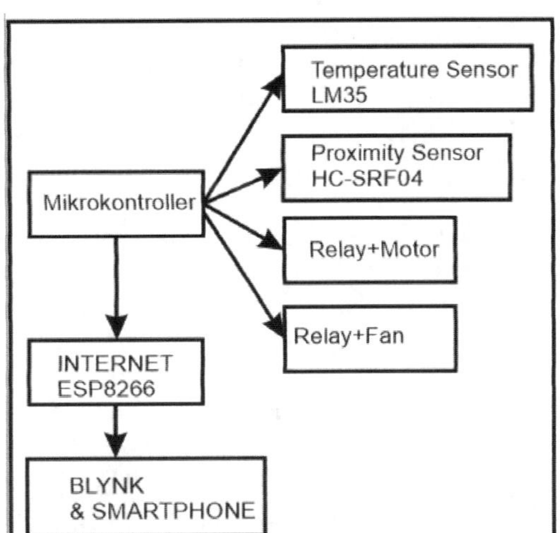

Figure 1. Android Smartphone Block diagram of Muhammadiyah University of Sidoarjo

IOP Conf. Series: Materials Science and Engineering **874** (2020) 012010 doi:10.1088/1757-899X/874/1/012010

3. Results And Analysis

The appearance of the Smarthome Prototype along with the Smartphone Application that has been made is shown in Figure 2 below. For Smartphone Applications, the authors use the Blynk Software and MIT App Inventor software in its manufacture. For MIT App Inventor, the writer uses ThingSpeak as a webserver. For Blynk there is automatically a web server. The HC-SR04 ultrasonic sensor test is performed to determine its level of accuracy when compared to distance measuring devices that are in accordance with industry standards. Tests in this study were carried out by giving different distances and were repeated 5 times.

Figure 2. Smarthome Prototype

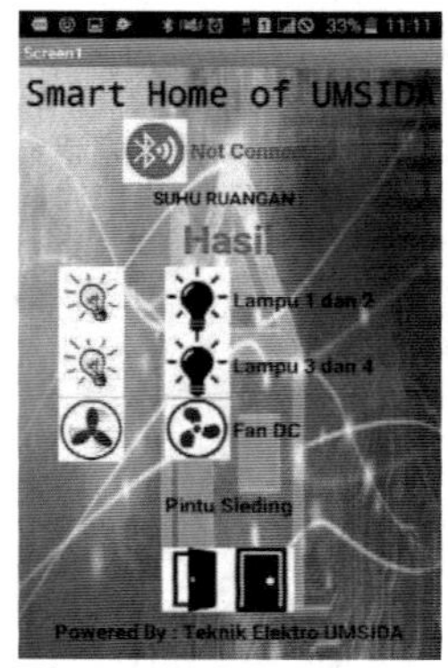

Figure 3. Smarthome Android Application with MIT App Inventor

Figure 4. Smarthome Distance control and monitoring testing

Table 1. Distance sensor Accuracy Testing (HC-SR04)

No.	Electronic Distance Measurement	Proximity Sensor test (HC-SR04) which is to number-					Average	Standart Deviation	Deviation	Accuracy
		1	2	3	4	5				Tools
	cm	cm	cm	cm	cm	cm	cm		cm	%
1	6	6	6	6	5.9	6	5.98	0.04	0.02	99.67
2	12	10	12.1	9	11.9	11.4	10.88	0.133	1.12	89.71
3	18	18.3	18.3	18.3	18.3	18.3	18.30	0.00	0.30	98.36
4	24	23.9	23.8	23.9	23.9	23.9	23.88	0.04	0.12	99.50
5	30	29.6	29.7	29.9	29.9	29.9	29.80	0.14	0.20	99.33
6	35	35	35.2	35.9	36	34.8	35.38	0.54	0.38	98.93
7	40	40.2	40.5	39.7	41	42	40.68	0.88	0.68	98.33
8	45	45	44	45.5	45.9	46	45.28	0.82	0.28	99.38
9	50	50.6	50.7	50.9	50	49.6	50.36	0.54	0.36	99.29
10	55	55	54.4	55.4	54.8	56	55.12	0.61	0.12	99.78
			Average					0.49	0.36	98.22

Based on table 1 for testing HC-SR04 sensors conducted with 10 variations of distance and repetition of measurements carried out 5 times, obtained information on the average accuracy of distance sensor (HC-SR04) is 98.22% of the Industrial standard Distance Measuring Instrument . The difference in the measurement results of the proximity sensor (HC-SR04) to the Industry standard Distance Measuring Instrument ranges from 0 - 1.12 cm, with an average standard deviation of 0.49. The resulting standard deviation is very small, the resulting small value implies that the proximity sensor (HC-SR04) works well and is stable even though it is tested with 10 different distance variables and the measurement of each distance is repeated five times. Proximity sensor (HC-SR04) in detecting distance or humans who are in front of it works well at a distance of 6-55cm. Besides being able to inform of obstacles and inform the distance value in the Smarthome case, the proximity sensor (HC-SR04) can also be applied to other objects such as "Smart blind stick". "Smart blind sticks" that use sensors (HC-SR04) can be implemented successfully. This can be used as an effective navigation tool for the blind. On the detection of obstacles in the path of the person (blind person), the "Smart blind stick" can ring the bell to make a warning. This system can detect any obstacles in the range of 5–35cm [7]. Besides the application of a proximity sensor (HC-SR04) can also be used to detect water level [8].

Table 2. Testing the accuracy of the temperature sensor (LM35).

No.	Electronic Temperature Measurement	Temperature Sensor test (LM35) which is to number-					Average	Standart Deviation	Deviation	Accuracy Tools	Error Tools
		1	2	3	4	5					
	cm	cm	cm	cm	cm	cm	cm		cm	%	%
room 1	27	27	26	26	27.8	26.8	26.72	0.76	0.28	98.95	1.05
room 2	26	26	26.4	26.3	26.2	25.8	26.14	0.24	0.14	99.46	0.54
room 3	27.5	27.8	27.3	26.8	26.2	27	27.02	0.59	0.48	98.22	1.78
room 4	28	27	27	26.3	27.8	29	27.42	1.03	0.58	97.88	2.12
room 5	25	25.4	25	25.4	25.8	26	25.52	0.39	0.52	97.98	2.04
		Average						0.60	0.40	98.50	1.50

Based on the Temperature sensor (LM35) testing table which is done with 5 variations of room temperature and 5 times the measurement repetition, it is found that the average accuracy of the Temperature sensor (LM35) is 98.50% against the Industrial standard Temperature Measuring Instrument. The difference in the measurement results of the Temperature sensor (LM35) against an Industrial standard Temperature Measuring Instrument ranges from 0.14 - 0.58 oC, with an average standard deviation of 0.60. The resulting standard deviation is very small, the resulting small value implies that the Temperature sensor (LM35) works well and is stable even though it is tested with 5 different room variables with temperature measurements in each room repeated five times. The average percentage error of the LM35 sensor temperature in detecting room temperature is 1.50% with a range of error percentage values of 0.54% to 2.12%. this is different when compared to the LM35 sensor temperature in detecting the temperature of burned peatlands in the disaster mitigation system of burning peatlands. In the case of detection of burning peatland temperatures, the temperature read by the temperature sensor (LM35) has an average percentage of temperature sensor error (LM35) of 0.22% with a range of error percentage values of 0% to 0.62% [9]. So it can be concluded that there is an error difference when the temperature sensor (LM35) is used to read the room temperature when compared to reading the temperature of burning peatlands (1.50% - 0.22%), the resulting error difference is 1.28%. This is due to the accuracy of each LM35 sensor and the way the temperature sensor (LM35) sensor is taken for each object and is different from each other. Apart from being read information on the room temperature in the Smarthome case and the hot temperature of the burning Peatland, the temperature sensor (LM35) can also be applied as another project, such as used to measure human body temperature [10].

IOP Conf. Series: Materials Science and Engineering **874** (2020) 012010 doi:10.1088/1757-899X/874/1/012010

Table 3. Testing distance control and monitoring of Smarthome prototypes with smartphones utilizing the Internet of Things

No	Distance (Km)	type of android smartphone	Button for			Information
			Lamp	Fan	Temperature	
1	0,005	Oppo F1s *4G *RAM 3GB Android 5.1 (Lollipop) *Wi-Fi 802.11 a / b / g / n, dual-band,	1	1	1	at Gempol - 5 meter
2	3,3	Asus z live *4G *Ram 2gb *Wifi 802.11b/g/n	1	1	1	at Gempol - Kejapanan
3	45	xiomi redmi note 5 *4G *Ram 3gb *Wifi Wifi802.11 A/B/G/N Wifi Diretct, Hotspot	1	1	1	at Gempol - Mojoagung
4	64	huawei y5 *3G *Ram 1gb *Os Android (5.1) Emotion 3.1Lite UI	1	1	1	at Gempol - Lowakwaru (Malang)
5	131	Oppo a37 *4G *Ram 2gb *Os ColorOs 3.0 based on Android 5.1	1	1	1	at Gempo - Kaliwates (Jember)
6	625	iphone 5s *Ram 1gb *Os ios7 *Wifi Hotspot wifi, Dua band	1	1	1	at Gempol - Cengkareng (Jakarta)
7	17466	xiomi 4x *4G *Ram 2gb *Os android 6.0.1 (marshmallow)	1	1	1	at Gempol - Cozumel (Mexico)
	Rata rata		1	1	1	
	Standart deviasi		0	0	0	

Information :

a. Condition 1 states that the communication conditions has been linked with a Smartphone.

b. Condition 0 indicates that the communication conditions are not connected to the Smartphone.

In table 3 about testing Android smartphones based on internet of things for 7 different distances can be sent properly when connected to a WiFi network or the internet. In this study, testing the control and monitoring of the prototype Smarthome with a smartphone for the closest distance is at a distance of 5 meters and the farthest distance is at a distance of 17,466 km (Country of Mexico). Testing on the Xiaomi Mi 4x Android smartphone that has an internet connection with a 4G network (faster internet connection speed) runs well. From the results of tests that have been done, the network and internet connection speed greatly affect the sending of data between the Smartphone and the Smarthome Hardware, and vice versa between the Smarthome Hardware and the Smartphone.

4. Conclusion

Prototype Smart Home with Internet Of Things (IoT) equipped with proximity sensor (HC-SRF04) and temperature sensor (LM35) in detecting room temperature, as well as monitoring and controlling to turn on and turn off lights with an android smartphone can work well. The accuracy value of the device reaches 98.22% for the measurement of distance, and its accuracy reaches 98.50% for the measurement of the room temperature if its value is compared with an industry standard tool. The Android application was successfully tested on several types of smartphones with different brands and specifications. The smartphone's range of distance to the Smart Home Prototype location to monitor and control the SMarthome Hardware was successfully tested and worked well at the closest distance (5 meters) to the farthest distance (17,466 km (Country of Mexico))..

Acknowledgments

This research work is supported by Universitas Muhammadiyah Sidoarjo and successfully assisted by R&D team from the Electrical Engineering Laboratory

References

[1] F. Masykur and F. Prasetiyowati, "Smart Home Application for Web-Based Household Electronic Equipment Controllers," *J. Teknol. Inf. dan Ilmu Komput.*, vol. 3, no. 1, p. 51, Mar. 2016.

[2] S. S. Kusumawardani, "MODELING OF USE AND ENERGY SAVING MONITORING," vol. 2014, no. Sentika, 2014.

[3] A. Mubarok, I. Sofyan, A. A. Rismayadi, and I. Najiyah, "Home Security System Using RFID, PIR Sensor and GSM Module Based on Microcontrollers," *J. Inform.*, vol. 5, no. 1, pp. 137–144, Apr. 2018.

[4] J. Gubbi, R. Buyya, S. Marusic, and M. Palaniswami, "Internet of Things (IoT): A vision, architectural elements, and future directions," *Futur. Gener. Comput. Syst.*, vol. 29, no. 7, pp. 1645–1660, 2013.

[5] B. Davidović and A. Labus, "A SMART HOME SYSTEM BASED ON SENSOR TECHNOLOGY," vol. 29, no. 3, pp. 451–460, 2016.

[6] J. Bangali and A. Shaligram, "Energy efficient Smart home based on Wireless Sensor Network using LabVIEW," *Am. J. Eng. Res.*, vol. 02, pp. 409–413, 2013.

[7] N. Dey, A. Paul, P. Ghosh, C. Mukherjee, R. De, and S. Dey, "Ultrasonic Sensor Based Smart Blind Stick," in *Proceedings of the 2018 International Conference on Current Trends towards Converging Technologies, ICCTCT 2018*, 2018.

[8] A. Prafanto and E. Budiman, "A Water Level Detection: IoT Platform Based on Wireless Sensor Network," in *2018 2nd East Indonesia Conference on Computer and Information Technology (EIConCIT)*, 2018, pp. 46–49.

[9] R. Amri, N. L. Marpaung, E. Ervianto, and Nurhalim, "Design of firing detector system by peat land with woody peat types using it's heat characteristics," in *Proceedings of the 2017 5th International Conference on Instrumentation, Control, and Automation, ICA 2017*, 2017, pp. 130–134.

[10] S. S. Thomas, A. Saraswat, A. Shashwat, and V. Bharti, "Sensing heart beat and body temperature digitally using Arduino," in *International Conference on Signal Processing, Communication, Power and Embedded System, SCOPES 2016 - Proceedings*, 2017, pp. 1721–1724.

Time Difference Calculation Settings for Very Short Term Electric Load Forecasting Using Interval Type-1 Fuzzy Inference System (IT-1 FIS)

Jamaaluddin[1,2], Imam Robandi[2], Izza Anshory[1,2]

[1]Program Studi Teknik Elektro, Universitas Muhammadiyah Sidoarjo, Jl. Raya Gelam No. 250, Candi, Jawa Timur, Indonesia.
[2]Departemen Teknik Elektro Institut Teknologi Sepuluh Nopember (ITS), Surabaya 60111, Jawa Timur, Indonesia.

jamaaluddin@umsida.ac.id

Abstract. Very short term electricity load forecasting (VSTLF) is done the day before forecasting every 30 minutes. This forecasting is imperative about the planning of operating plants, which are carried out in maintenance and related to economic considerations. This forecasting refers to the loading of electricity in a particular area and time. Very short-term electricity loading in the past few years has been used to predict the electricity load shortly. One method used for forecasting very short loads is using IT-1 FIS. In IT-1-FIS, the dataset is made by looking at the imposition of electricity on certain days at certain hours of the previous year. The data that calculate time difference forecast Friday 4th used Friday 3rd, 2nd, and 1st at the same hour and month of the past year. In this study, it proposed that the calculation of Time difference forecasting the 4th Friday be used data Thursday 4th, 4th Wednesday, 4th Tuesday. After entering into IT-1-FIS, the forecasting error result is 0,6123 %. This value is higher than if you use the time difference on Friday 3rd, 2nd, and 1st, which has a forecast error value of 0,4695 %. So it can be concluded that the time difference setting is very influential in the process of calculating forecasts, and the use of time difference data with the use of Friday data 3rd, 2nd, and 1st is better.

Keywords: Interval Type-1 Fuzzy Inference System (IT-1 FIS); Very Short Term Electric Load Forecasting; Time Difference

1. Introduction

Utilization of electrical energy should be set as well. This setting starts from generation planning and load sharing, operation, setup and other matters [1][2][3][4]. Planning is done on the load side for generation planning. This plan is also known as forecasting. The electrical load forecasting is divided into three, i.e. long-term load forecasting, short-term load forecasting and extremely short-term load forecasting[5][6]. The very short-term electrical load forecasting is required large-scale electrical systems. Forecasting of very short-term electrical burden is forecasting one day before the day of operation [7][8][9]. Forecasting is done every 30 minutes. The forecasting of this very short-term load is done by various methods. Among the methods used for forecasting a very short-term electrical load is using the Fuzzy[10][11]. The Fuzzy process required input, rule based and output. In general the Fuzzy process can be seen in the Figure 1.

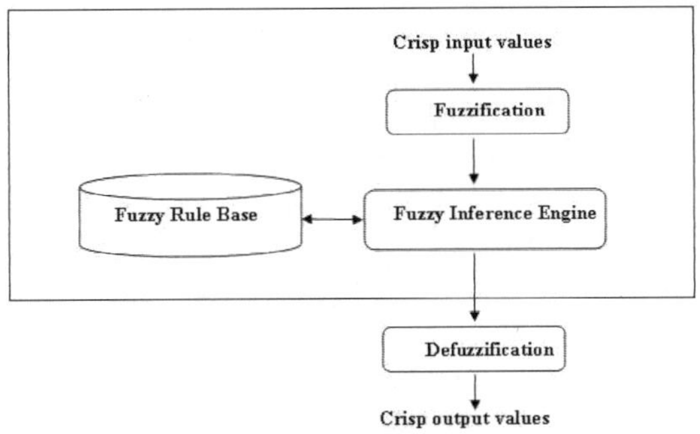

Fig.1. Fuzzy Process

Very short-term forecasting with this Fuzzy use, using data-electricity-loading data in 3 years before the year of forecast at a certain time [12][13]. In this study the analysed time was in October the 4th Friday. While the time of its ramation at out of peak load time, namely at 13.00 – 15.30 Wib. This time was chosen because it had a pretty dynamic electric load movement[14][15][16].

Inside the Fuzzy process, to define variables X, Y and Z then analysis is done – such as determinative analysis: Load Difference, Time Load Difference and Variation Load Difference [12]. In the process of determining the Load Difference, it takes time of electrical load data of the past. Data used is data 3 years ago with a time interval of 30 minutes. The analyzed time selection greatly affects the process and the results of analysis. In the previous research conducted analysis of Load Difference by using a certain month with the selection of certain days such as Friday, then the time data used is the time of the day Friday data of the week before. If the time to be in the search for an error is on Friday to 4, then the data in the analysis on Load Difference is data Friday to 3, Friday to 2 and Friday to one at the same hour[12]. It is currently proposed that the analysis is not on Friday the previous week, but on the previous day. If Friday to 4 will be searched for the value of the error, then the time data done is Thursday 4th, 4th Wednesday, 4th Tuesday.

By analyzing the time different from the previous research, then it will be known to use the analysis time which has a better error value. So this result can be used for data analysis process by using Interval Type-1 Fuzzy Inference System in future research research.

2. Material And Method

The stages of this research consist of pre-processing, processing and post-processing. The stages of the research will be explained as follows [17][12].

2.1 Pre Processing

The pre-processing stage is the preparation of daily load data every 30 minutes for 24 hours in a day, on the working day represented by Friday by classifying the out of peak load time which is at 13.00-15.30 to find the actual variation load difference, as the block diagram in Fig.2

Collecting daily electrical load data on the fourth Friday, Thursday, Wednesday and Tuesday in October from 2013 to 2017.

↓

Calculating the Difference Load at the same time on the fourth Friday (Time predicted). These results can be obtained using the formula below:

$$LD_{(i)} = \frac{SD_{(i)} - TD_{(i)}}{TD_{(i)}} x\ 100 \quad \ldots.(1)$$

$SD_{(i)}$ is the load at the predicted time.

↓

After that, the $TLD_{(i)}$ (The typical Time Load Difference) is calculated by calculating the average load of $LD_{(i)}$ at each time predicted by the same time last year and before.

$$TLD_{(i)} = \frac{LD_{(i-1)} - LD_{(i-2)}}{2} \quad \text{... (2)}$$

↓

After that, VLD (Variation Load Difference at Time) can be calculated at the predicted time.

$$VLD_{(i)} = LD_{(i)} - TLD_{(i)} \quad \text{......(3)}$$

Figure 2 Pre processing stage

2.2 Processing

In processing, the VLD value obtained from the above calculation will be entered into the IT-1 FIS, at the membership function with the following steps as Fig 3. The following input variables (X, Y) and output variables (Z) consist of 11 Fuzzy as follows :

1) Negative Very Big (NVB): with a range of values(ROV) of -6 s/d -4
2) Negative Big (NB): with (ROV)of -5 s/d -3
3) Negative Medium (NM): with (ROV) of -4 s/d -2
4) Negative Small (NS): with (ROV)of -3 s/d -1
5) Negative Very Small (NVS): with (ROV)of -2 s/d 0
6) Zero (ZE): with (ROV)of -1 s/d 1
7) Positive Very Small (PVS): with (ROV)of 0 s/d 2
8) Positive Small (PS): with (ROV)of 1 s/d 3
9) Positive Medium (PM): with (ROV) of 2 s/d 4
10) Positive Big (PB): with (ROV)of 3 s/d 5
11) Positive Very Big (PVB): with (ROV) of 4 s/d 6

Creating a membership function with inputs X and Y and output Z for the forecasted time

↓

Calculating the new candidates heading to the *Center Of Mass* by adding or multiplying the scrambled number, the number or value will decrease each time iterates

↓

Making IT-1 FIS rules as follows: IF X is A_i AND Y is B_i THEN Z is C_i

↓

Calculating the Defuzzyfication value until the REDUCER TYPE uses KERNIK MENDEL ALGHORITM to get the forecast value of VLD_{max}.

↓

Collecting load out of peak load, namely data load at 18.00; 18.30; 19.00; 19.30; 20.00; 20.30; 21.00 (Western Indonesia Time)

↓

Load identification is sought $TD_{(i)}$ on Friday before the fourth Friday when the load is analyzed.

$$TD_{(i)} = \frac{TD_{(i)D-3} + TD_{(i)D-2} + TD_{(i)D-1}}{3} \quad \text{... (4)}$$

Time Difference is the average load at the same time as different Fridays, namely Thursday*(D-1)*, Wednesday*(D-2)*, Tuesday*(D-3)*

Figure 3 Processing stage

2.3 Post Processing

In processing, the VLD value obtained from the above calculation will be entered into the IT-1 FIS, at the membership function.. The following input variables (X, Y) and output variables (Z) consist of the membership function. After the above process, the next step is to find the error value of the VLD_{max}

IOP Conf. Series: Materials Science and Engineering **874** (2020) 012011 doi:10.1088/1757-899X/874/1/012011

forecast, done in the following ways:
Calculating the forecast value of load difference for the predicted time using Equation (5):

$$F'cast\ LD_{MAX}(i) = F'cast\ VLD_{MAX}(i) + TLD_{MAX}(i) \ldots\ldots \quad (5)$$

The next step is to calculate the error value, using Equation (6):

$$Error\ (\%) = \frac{P_{forecast} - P_{actual}}{P_{actual}} \times 100 \qquad \ldots..(6)$$

Where:
$P_{forecast}$ = Power predicted at a certain time (MW).
P_{actual} = Actual power at a certain time (MW).

3. Result And Discussion

The next step is to group the load in peak load conditions at each time starting in 2013, 2014, 2015, 2016 and 2017. The LD values are calculated in 2013 as in Equation (1) and (2). In the next discussion for analyzing using Third Friday, Second Friday and First Friday called by **("a" analysis)**. While analyzing it using Thursday; Wedneday and Tuesday call by **("b" analysis)**. The result can seen in Fig. 4.a. (for "a" analysis) and Fig. 4.b. (For "b" analysis).

Furthermore, for the 2015 loading year, the TD value and LD were obtained. From the electricity load data obtained in the 2015 TLD calculation value, using the Equation as shown in Equation no. (3), the 2015 TDL value is the one-year LD average value and LD value two years earlier. Whereas to get the VLD value in 2015, it is obtained by calculation as shown in Equation (4) that is from the 2015 LD value reduced by the 2015 TLD value. The results of the analysis in 2015 with can be seen in Fig.5.

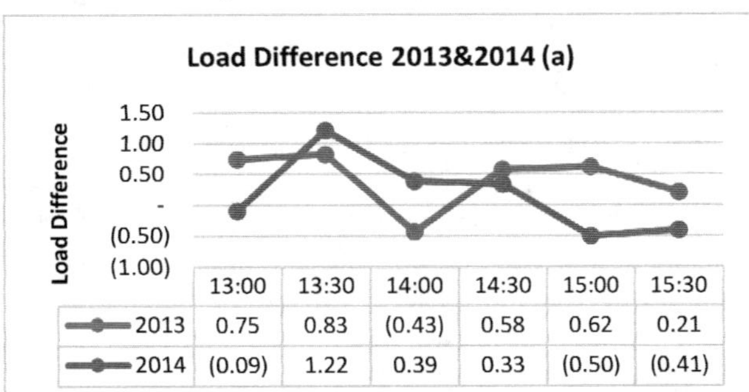

Figure 4.a. Load Difference of 2013 and 2014 (a)

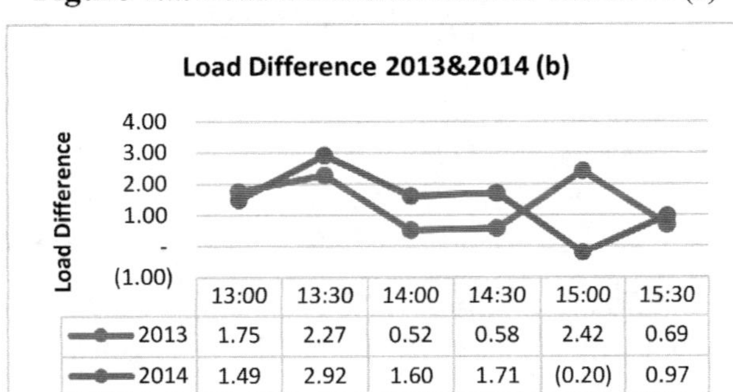

Figure 4.b. Load Difference of 2013 and 2014 (b)

For the TLD value in 2016, obtained the average value of LD 2015 and LD 2014. Whereas to get the 2016 VLD value, obtained from the 2016 LD value minus the 2016 TLD value. For the year of load calculation, 2017 TD, LD, TLD and VLD can be seen as in Fig. 5.a and 5.b.

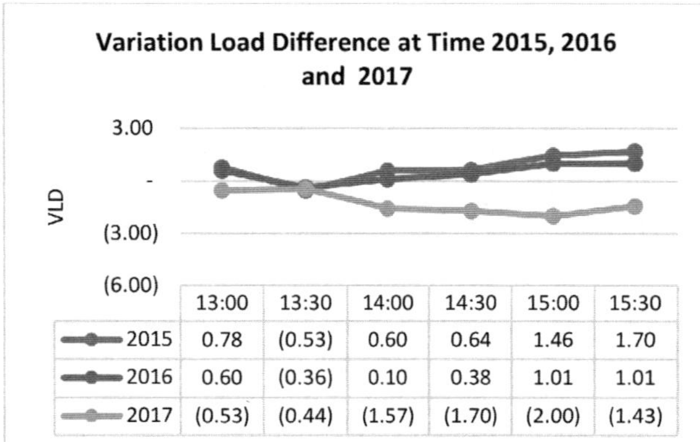

Figure 5.(a) VLD at 2015, 2016 and 2017

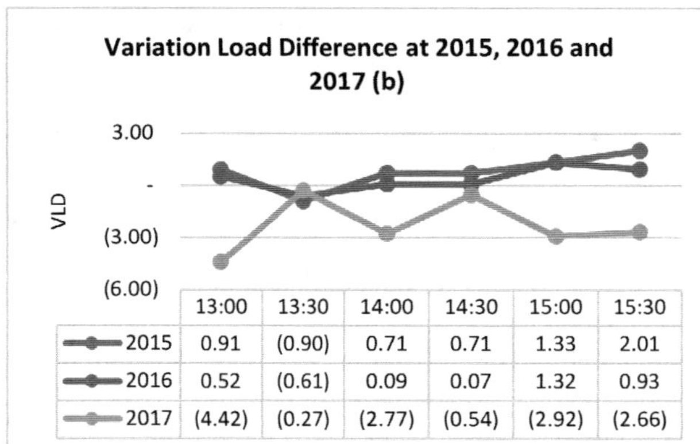

Figure 5(b) VLD at 2015, 2016 and 2017

The TLD value in 2017, obtained from the average LD 2016 and LD 2015 values. Meanwhile, the VLD 2017 value, obtained from the 2017 LD value minus the 2017 TLD value. The VLD values for 2016 and 2017, the X value of the 2016 VLD value is obtained and the output (Z) of the year VLD value is forecast in 2017, while the Y value is taken from the VLD value in 2017 for the adjacent time [18]. The result can seen in Table. 1.a and 1.b. After getting the X, Y and Z values, the next step is to enter the values into the membership function group.

Table 1.a. Calculation of determining the values of X, Y and Z

Time	VLD 2016	VLD 2017	X	Y	Z
13:00	0.6027	(0.5295)	0.6027	(1.4308)	(0.5295)
13:30	(0.3606)	(0.4432)	(0.3606)	(0.5295)	(0.4432)
14:00	0.0953	(1.5724)	0.0953	(0.4432)	(1.5724)
14:30	0.3826	(1.7022)	0.3826	(1.5724)	(1.7022)
15:00	1.0095	(1.9952)	1.0095	(1.7022)	(1.9952)
15:30	1.0052	(1.4308)	1.0052	(1.9952)	(1.4308)

Table 1.b. Calculation of determining the values of X, Y and Z

	VLD 2016	VLD 2017	X	Y	Z
13:00	0,522	(4,421)	0,522	(2,665)	(4,421)
13:30	(0,612)	(0,272)	(0,612)	(4,421)	(0,272)
14:00	0,086	(2,773)	0,086	(0,272)	(2,773)
14:30	0,073	(0,536)	0,073	(2,773)	(0,536)
15:00	1,319	(2,917)	1,319	(0,536)	(2,917)
15:30	0,933	(2,665)	0,933	(2,917)	(2,665)

Table 2.a. Based rule input X forecasting 2017 (a)

Time	Value	Membership Function (μ)											Sets
	X	NVB	NB	NM	NS	NVS	ZE	PVS	PS	PM	PB	PVB	X
		(5)	(4)	(3)	(2)	(1)	-	1	2	3	4	5	
13:00	0.6027						0.3973	0.6027					PVS
13:30	(0.3606)					0.3606	0.6394						ZE
14:00	0.0953						0.9047	0.0953					ZE
14:30	0.3826							0.6174	0.3826				PVS
15:00	1.0095							0.9905	0.0095				PVS
15:30	1.0052							0.9948	0.0052				PVS

Table 2.a. shows that the value of X at 13.00 is 0.6027 having a positive very small (PVS) membership degree of 0.6027, while the degree of membership value of zero (ZE) is 0.3973. Once the island for another X value is treated the same as the above calculation. That calculation is the same for the Y and Z. Also for "b" analysis treated the same calculation so that the result as table 2.b.

Table 2.b. Based rule input X forecasting 2017(b)

Time	Value	Membership Function (μ)											Sets
	X	NVB	NB	NM	NS	NVS	ZE	PVS	PS	PM	PB	PVB	X
		(5)	(4)	(3)	(2)	(1)	-	1	2	3	4	5	
13:00	0,5216						0,4784	0,5216					PVS
13:30	(0,6124)					0,6124	0,3876						NVS
14:00	0,0857						0,9143	0,0857					ZE
14:30	0,0726						0,9274	0,0726					ZE
15:00	1,3187							0,6813	0,3187				PVS
15:30	0,9328							0,9328	0,0672				PVS

From the data in Table 2.a. and 2.b., it is obtained the basic rules that are used to make the program in matlab, the rules are still in alphabetical form and will be changed to a number in order as in the order of the membership function above. The data can be seen in Table 3.a. and 3.b. With the conversion as in Table 3.a and 3.b., the matlab programming process can be done. Processing on matlab will generate value for VLD forecasting for 2017. After the process is done using matlab and found VLD forecasting in 2017 with the results as in Table 4.a. and 4.b. It shows very short term load forecasting in 2017 (out of peak load) fourth Friday October using IT-1 FIS.

Table 3.a. Conversion of 2017 Basic Forecasting Rules For The Matlab Software Code ("a" analysis)

No of	Antc		Cons	No of	Antc		Cons
rules	X	Y	Z	Rules	X	Y	Z
1	PVS	NVS	NVS	1	7	5	5
2	ZE	NVS	ZE	2	6	5	6
3	ZE	ZE	NS	3	6	6	4
4	PVS	NS	NS	4	7	4	4
5	PVS	NS	NS	5	7	4	4
6	PVS	NS	NVS	6	7	4	5

Table 3.b. Conversion of 2017 Basic Forecasting Rules For The Matlab Software Code ("b" analysis)

No of	Antc		Cons	No of	Antc		Cons
rules	X	Y	Z	Rules	X	Y	Z
1	PVS	NM	NM	1	7	3	3
2	NVS	NB	ZE	2	5	2	6
3	ZE	ZE	NM	3	6	6	3
4	ZE	NM	ZE	4	6	3	6
5	PVS	NVS	NM	5	7	5	3
6	PVS	NM	NM	6	7	3	3

Table 4.a. and 4.b. Results of VLD calculation in 2017 using IT-1 FIS (output matlab)

Actual VLD	Forecast VLD	Error VLD	Actual VLD	Forecast VLD	Error VLD
0,5295	(1,0185)	0,4890	(4,4214)	(1,5462)	(2,8752)
(0,4432)	(0,9582)	0,5150	(0,2723)	-	(0,2723)
(1,5724)	(1,0740)	(0,4984)	(2,7729)	(3,0000)	0,2271
(1,7022)	(0,9608)	(0,7414)	(0,5361)	(0,3859)	(0,1502)
(1,9952)	(1,5000)	(0,4952)	(2,9165)	(3,0000)	0,0835
(1,4308)	(1,5000)	0,0692	(2,6645)	(2,6515)	(0,0130)

Table 5.a. Value of 2017 forecasting and actual load expenses "a" analysis

Forecast P'(MW)	*Actual (MW)*	*Error (%)*
23.577,09	23.692,27	0,4861
24.618,75	24.746,13	0,5147
24.634,81	24.511,41	0,5035
24.612,72	24.430,49	0,7459
24.427,18	24.306,23	0,4976
24.316,84	24.333,66	0,0691
	Sum	2,8170
	Average	0,4695

Table 5.b. Value of 2017 forecasting and actual load expenses "b" analysis

Forecast P'(MW)	*Actual (MW)*	*Error (%)*
24.386,59	23.692,27	2,9306
24.812,57	24.746,13	0,2685
24.455,18	24.511,41	0,2294
24.466,71	24.430,49	0,1483
24.285,85	24.306,23	0,0839
24.336,83	24.333,66	0,0130
	Sum	3,6736
	Average	0,6123

Figure 6.a. Comparison between forecasting and actual ("a" analysis)

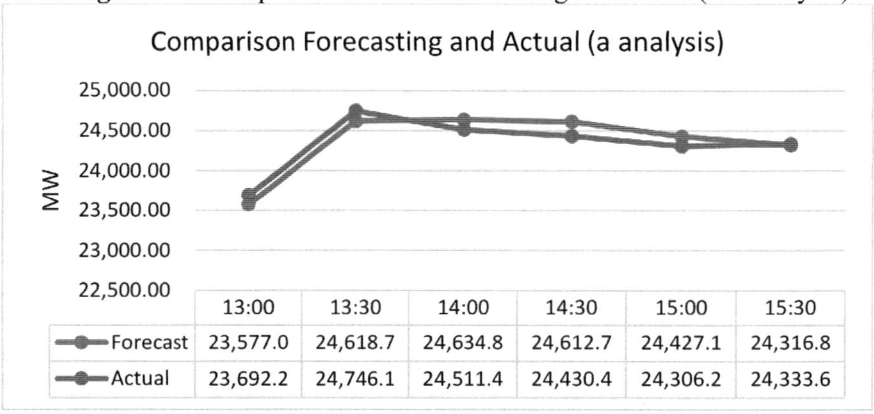

Figure 6.b. Comparison between forecasting and actual ("b" analysis)

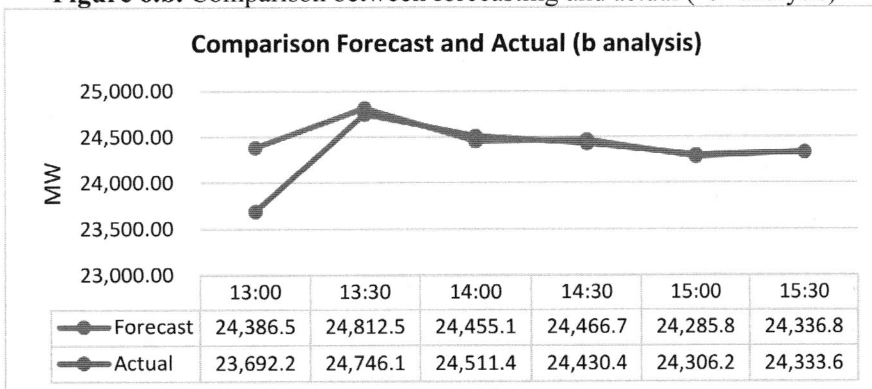

The value of Forecast entry to VLD electrical power calculation. The result of calculation above is show at the table 5.a and 5.b. In graphical form, it can be seen the comparison between forecasting and its behavior as in Figure 6.a. and 6.b. That's Table also shows the comparison of the value of power (P) forecasting results with the actual power (P) value. Therefore, the error value can be calculated. In Fig 7, it can be seen that the average error forecasting value using IT-1 FIS with "a" and "b" analysis.

Fig. 7 shows that comparison IT-1 FIS using ("a" analysis) and ("b" analysis). "a" analysis (Load Difference Analysis using Friday, 3rd Friday, 2nd Friday dan 1 st Friday have a small error value (0,4695). That the result show "a" analysis is better than "b" analysis for the electric load control agency in very short term load Forecasting.

Figure 7 Error forecasting 2017 using FT-2-BBBC

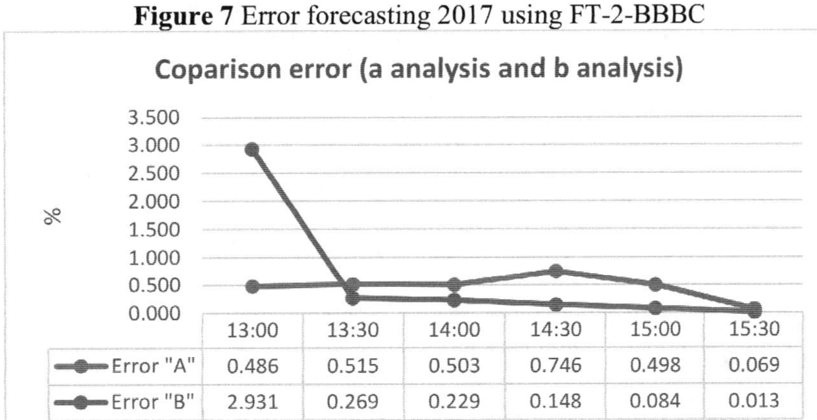

4. Conclusions

The findings of the research are as follows: IT-1 FIS there is a formula for managing Load Difference, the first is to use the day analyzed on the fourth Friday with data on the third Friday, the second Friday and the first Friday or called a "a" analysis. Whereas the second using the previous day, Thursday, Wednesday and Tuesday ("b" analysis). Then it can be concluded that:

1. The smaller error result is obtained from using the formula with data from the previous Friday with an error value: 0.4695%.
2. For analyzing data using IT-1-FIS several data processing formulas can be used. This can be studied further to get smaller errors.

Acknowledgements

The authors would like to thanks the Universitas Muhammadiyah Sidoarjo for funding this research.

References

[1] M. F. I. Khamis, Z. Baharudin, N. Hamid, M. Abdullah, and F. Nordin, "Short term load forecasting for small scale power system using fuzzy logic," *2011 Fourth Int. Conf. Model. Simul. Appl. Optim.*, 2011.

[2] F. Golestaneh, P. Pinson, and H. B. Gooi, "Very Short-Term Nonparametric Probabilistic Forecasting of Renewable Energy Generation; With Application to Solar Energy," *Power Syst. IEEE Trans.*, vol. PP, no. 99, pp. 1–14, 2016.

[3] I. Robandi, *Modern Power System Control*, 1st ed. Yogjakarta: Andi, 2009.

[4] X. Xia and A. M. Elaiw, "Optimal dynamic economic dispatch of generation: A review," *Electric Power Systems Research*. 2010.

[5] G. F. Casagrande, O. Hideo, A. Junior, M. O. Oliveira, O. E. Perrone, and H. Reversat, "Very Short-Term Electric Load Forecasting Considering Climate and Temporal Variable," *Int. J. Autom. Power Eng.*, vol. 3, no. 1, 2014.

[6] Y. Loewenstern, L. Katzir, and D. Shmilovitz, "The effect of system characteristics on very-short-term load forecasting," *12th Conf. Int. Sch. Nonsinusoidal Curr. Compens. ISNCC 2015 - Conf. Proc.*, 2015.

[7] S. A. Eroshenko, K. V Vinokurov, and A. Y. Smolina, "Electrical load forecasting," *1st Int. Conf. Energy Prod. Manag. 21st Century Quest Sustain. Energy*, vol. 190 VOLUME, pp. 299–305, 2014.

[8] Q. Liang and J. M. Mendel, "Interval type-2 fuzzy logic systems: Theory and design," *IEEE Trans. Fuzzy Syst.*, vol. 8, no. 5, pp. 535–550, 2000.

[9] M. Biglarbegian, W. W. Melek, and J. M. Mendel, "On the stability of interval type-2 tsk fuzzy logic control systems," *IEEE Trans. Syst. Man, Cybern. Part B Cybern.*, vol. 40, no. 3, pp. 798–818, 2010.

[10] C. Guan, S. Member, P. B. Luh, L. D. Michel, and Y. Bar-shalom, "Interacting Multiple Model Approach for Very Short-Term Load Forecasting and Confidence Interval Estimation," pp. 2680–2685, 2010.

[11] J. M. Mendel, R. I. John, and F. Liu, "Interval Type-2 Fuzzy Logic Systems Made Simple," *Fuzzy Syst. IEEE Trans.*, vol. 14, no. 6, pp. 808–821, 2006.

[12] Jamaaluddin, I. Robandi, and I. Anshory, "A very short-term load forecasting in time of peak loads using interval type-2 fuzzy inference system: A case study on java bali electrical system," *J. Eng. Sci. Technol.*, vol. 14, no. 1, pp. 464–478, 2019.

[13] J. Jamaaluddin, D. Hadidjaja, I. Sulistiyowati, E. A. Suprayitno, I. Anshory, and S. Syahrorini, "Very short term load forecasting peak load time using fuzzy logic," in *IOP Conference Series: Materials Science and Engineering*, 2018, vol. 403, no. 1.

[14] P. P. (PERSERO) P. J. B. – B. O. Sistem, *EVALUASI OPERASI SISTEM TENAGA LISTRIK JAWA BALI 2014*, 1st ed. Jakarta, 2014.

[15] P. P. (PERSERO) P. – B. Perencanaan, *EVALUASI OPERASI SISTEM TENAGA LISTRIK JAWA BALI 2015*, 01 ed. Jakarta: PT PLN (PERSERO) P2B – Bidang Perencanaan, 2015.

[16] P. P. (PERSERO) P. J. B. – B. O. Sistem, *EVALUASI OPERASI SISTEM TENAGA LISTRIK JAWA BALI 2013*, 1st ed. Jakarta: PT PLN (PERSERO) P3B Jawa Bali – Bidang Operasi Sistem, 2013.

[17] A. Ramadhani, I. Robandi, J. T. Elektro, and F. T. Industri, "Optimisasi Peramalan Beban Jangka Pendek untuk Hari Libur Nasional Menggunakan Interval Type-2 Fuzzy Inference System-Big Bang-Big Crunch Algorithm (Studi Kasus : Sistem Kelistrikan Kalimantan Selatan dan Tengah)," pp. 1–8, 2015.

[18] K. Liu *et al.*, "Comparison of very short-term load forecasting techniques," *IEEE Trans. Power Syst.*, vol. 11, no. 2, pp. 877–882, 1996.

IOP Conf. Series: Materials Science and Engineering **874** (2020) 012012 doi:10.1088/1757-899X/874/1/012012

Design of Pothole Detector Using Gray Level Co-occurrence Matrix (GLCM) And Neural Network (NN)

Imam Sutrisno[1], Albiyan Wanda Syauqi[2], Muhammad Khoirul Hasin[3], Iskandar[4], Mohammad Basuki Rahmat[5], I Putu Sindhu Asmara[6], I'ie Suwondo[7], Ardiansyah[8], Daviq Wiratno[9], Edy Setiawan[10]

[1,2,3]Shipbuilding Institute Polytechnic of Surabaya, Indonesia
[4,7,8,9,10]Politeknik Pelayanan Surabaya, Indonesia
[5,6] Shipbuilding Institute Polytechnic of Surabaya, Indonesia
imamsby01@yahoo.com[1],albiyan12@gmail.com[2],khoirul.hasin@ppns.ac.id[3],iskandarbp2ipsby@gmail.com[4], mbasuki.rahmat@gmail.com[5],putusindhu@ppns.ac.id[6],iiesuwondo@gmail.com[7], ardi9415096@gmail.com[8], daviq_34@yahoo.com [9], edy_setiawan@ppns.ac.id [10]

Abstract. Roads are land transportation infrastructure that covers all parts of the road. Roads with bad conditions will interfere with the achievement of activities to a destination. The situation also includes damage to the road surface in the form of holes. To overcome this, in this Final Project a hole detector was detected in the road using the Gray Level Co-occurrence Matrix (GLCM) and Neural Network (NN). The tool detects holes in the surface of the road using a camera by walking along the road being examined. The camera is used instead of the eye to detect road surface damage. The method used to detect holes is the GLCM. The GLCM method produces several features, namely entropy, contrast, energy, homogeneity, and correlation which will then be processed using a NN to produce a decision whether there is a hole or not. In addition to knowing where the location of the damage is equipped with GPS (Global Positioning System). The results of image feature extraction using the GLCM and road classification using NN can be used in the hole detection process. Testing is done using a car prototype that is monitored through the computer. The percentage of successful hole detection is 86.6% using 10 hidden. When a hole is detected the device manages to take a picture, then sends the hole coordinates to the server.

1. Introduction

Roads are land transportation infrastructure that covers all parts of the road, including auxiliary buildings and traffic equipment. To support safety, safety, comfort, and shorten travel time, good quality roads are needed. Roads with poor conditions will interfere with the achievement of activities to a destination. According to data from the Ministry of Public Works and Public Housing noted that the national road was in a slightly damaged state by 6.25% and in a heavily damaged state by 4.37%. The situation also includes damage to the road surface in the form of holes. Reported in the Depok Police online news portal from the end of January to the end of February 2017, there were 240 accidents caused by potholes.

To determine the condition of the road carried out the activities of road surveillance in the form of implementation activities, observation, utilization of the road and road quality observation reports. With this report, it can be proposed to related parties or agencies to improve or improve the quality of roads. To carry out road quality surveillance activities requires a road surface detection method with a faster, safer, and cost effective process.

Seeing these problems, a study was carried out to create a classification system design as well as monitoring the detection of potholes on the road. Detection is done by utilizing information from the results of video processing. The method used to detect holes in the road is the GLCM method in which the results of the resulting processing are entropy, correlation, contrast, energy and homogeneity of the detected image. The image processing results will then be classified using NN to determine whether the road has holes or not.

2. Methodology

Flowchart of this research is focused on getting classification of holes in the road with input data in the form of images. For image classification we use the GLCM and NN methods. In this study the so-called hole is a hole that has a diameter of about 10 cm.

In Figure 1. explain the work flow of the existing system in this study. In the initial condition the whole system is turned off. When the condition is on camera and the GPS sensor is on. Then do the selection of the route from the beginning of the departure point to the destination point you want to choose. After selecting the route the tool will walk down the road and the camera starts video processing using feature extraction using the GLCM method.

The results of video processing are then trained using NN. If it detects a hole then the tool will stop and take pictures of the hole. After that the tool takes a picture then the image and coordinates of the hole will be sent to the server.

IOP Conf. Series: Materials Science and Engineering **874** (2020) 012012 doi:10.1088/1757-899X/874/1/012012

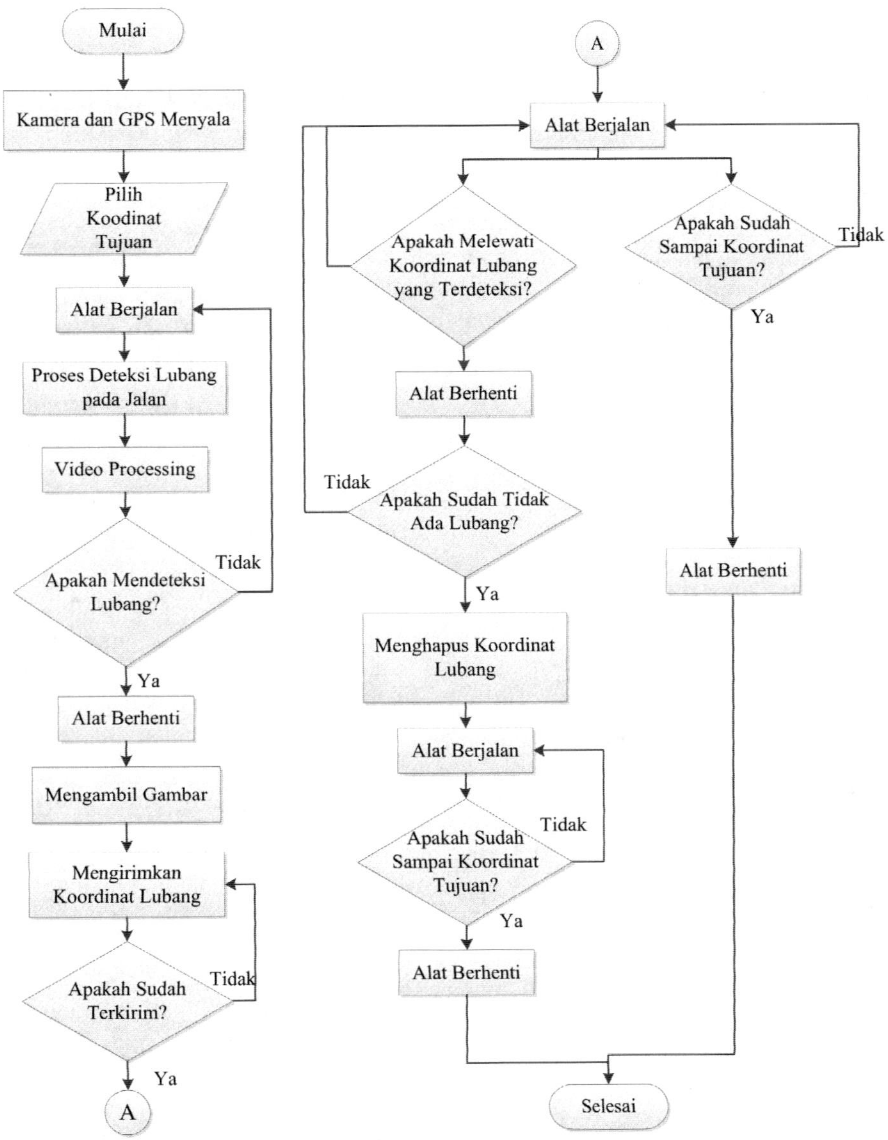

Figure 1. System Design's Flowchart

If it does not detect the device will still run. When the tool goes past the coordinates of the hole that has been detected it will stop and delete the coordinates of the hole that has been detected. When it has finished removing the coordinates from the hole, the tool will run again. When the GPS sensor detects that the coordinates of the device are at the coordinates of the destination point, the tool will stop and the inspection process is complete.

2.1. Image Preprocessing

The first step taken is taking images using the camera. In this process, the hole in the road as the main object is captured by the camera in the form of a frame. Then the preprocessing process is done by changing the color space of the RGB image to grayscale. This gray scaling image is used as input. The resolution of the image in this system is displayed with a size of 640 x 480 pixels with consideration that the computing that runs can work faster. This resolution setting will also reduce the use of memory used. 640 x 480 pixel resolution is an ideal size, it is not too small or large. This is because at this resolution the observations made are actually quite clear.

2.2. Gray Level Co-occurrence Matrix (GLCM)
The GLCM method is included in the statistical method used to obtain the texture of an image using a gray degree distribution (histogram) by measuring the degree of contrast, granularity, and roughness of an area of neighboring relations between pixels in the image. Statistical methods consist of first order feature extraction and second order feature extraction. The first order is done through image histograms or using image pixel values, while the second order statistical feature extraction is done by calculating the relationship between two pixel image pairs

2.2.1. 8 Degrees Quantitation
At this stage the aim is to get the value of pixel intensity or brightness by grouping the pixel intensity values into several levels such as Table 1. In Figure 2. this process produces a quantization matrix which will then be continued by noting the relationship between pixels in the form of a co-occurence matrix.

Table 1. Eight Levels of Gray

Level	Value
0	0 – 31
1	32 – 63
2	64 – 95
3	96 – 127
4	128 – 159
5	160 – 191
6	192 – 233
7	224 – 256

$$\begin{bmatrix} 17 & 10 & 60 & 0 \\ 43 & 35 & 0 & 20 \\ 80 & 28 & 0 & 50 \\ 130 & 15 & 10 & 73 \end{bmatrix} \longrightarrow \begin{bmatrix} 0 & 0 & 1 & 0 \\ 1 & 1 & 0 & 0 \\ 2 & 0 & 0 & 1 \\ 3 & 0 & 0 & 2 \end{bmatrix}$$

Figure 2. Eight Degrees Quantization

2.2.2. Make a Co-occurrence Matrix
The Co-occurrence Matrix is a matrix whose elements contain relationships between pixels such as pixel 0 with 0, pixel 0 with 1, etc. The relationship of the neighborliness can be seen from the results of the 8 gradation quantization.

IOP Conf. Series: Materials Science and Engineering **874** (2020) 012012 doi:10.1088/1757-899X/874/1/012012

$$\begin{bmatrix} 0 & 0 & 1 & 0 \\ 1 & 1 & 0 & 0 \\ 2 & 0 & 0 & 1 \\ 3 & 0 & 0 & 2 \end{bmatrix} \rightarrow \begin{array}{c|cccc} & 0 & 1 & 2 & 3 \\ \hline 0 & 4 & 2 & 1 & 0 \\ 1 & 2 & 1 & 0 & 0 \\ 2 & 1 & 0 & 0 & 0 \\ 3 & 1 & 0 & 0 & 0 \end{array} \rightarrow \begin{bmatrix} 4 & 2 & 1 & 0 \\ 2 & 1 & 0 & 0 \\ 1 & 0 & 0 & 0 \\ 1 & 0 & 0 & 0 \end{bmatrix}$$

Figure 3. Make a Co-occurrence Matrix

2.2.3. Addition of Co-occurrence Matrix with Transpose Matrix

After the co-occurrence matrix is formed, the sum of the co-occurrence matrix and the transpose matrix is done. The transpose matrix is formed by transposing the co-occurrence matrix.

$$\begin{bmatrix} 4 & 2 & 1 & 0 \\ 2 & 1 & 0 & 0 \\ 1 & 0 & 0 & 0 \\ 1 & 0 & 0 & 0 \end{bmatrix} + \begin{bmatrix} 4 & 2 & 1 & 1 \\ 2 & 1 & 0 & 0 \\ 1 & 0 & 0 & 0 \\ 0 & 0 & 0 & 0 \end{bmatrix} = \begin{bmatrix} 8 & 4 & 2 & 1 \\ 4 & 2 & 0 & 0 \\ 2 & 0 & 0 & 0 \\ 1 & 0 & 0 & 0 \end{bmatrix}$$

Figure 4. The Sum of Co-occurrence Matrix with Transpose Matrix

2.2.4. Make a Normalization Matrix

The results of the sum of the co-occurrence matrix with the transpose matrix are then continued into the normalization process. Normalized matrix elements contain the results for each matrix element with the number of pixels.

$$\begin{bmatrix} 8/22 & 4/22 & 2/22 & 1/22 \\ 4/22 & 2/22 & 0 & 0 \\ 2/22 & 0 & 0 & 0 \\ 1/22 & 0 & 0 & 0 \end{bmatrix} = \begin{bmatrix} 0.363 & 0.181 & 0.090 & 0.045 \\ 0.181 & 0.090 & 0 & 0 \\ 0.090 & 0 & 0 & 0 \\ 0.045 & 0 & 0 & 0 \end{bmatrix}$$

Figure 5. Make a Normalization Matrix

2.2.5. Calculate The Feature Extraction

After the normalized matrix is formed, the next step is to calculate the value of each feature. Below is a feature of the normalized matrix.

$$\overset{\longleftarrow\; j \;\longrightarrow}{\begin{bmatrix} 0.363 & 0.181 & 0.090 & 0.045 \\ 0.181 & 0.090 & 0 & 0 \\ 0.090 & 0 & 0 & 0 \\ 0.045 & 0 & 0 & 0 \end{bmatrix}} \; i \updownarrow$$

Figure 6. Normalization Matrix

No.	Fitur	Rumus	
1.	*Contras*	$\sum_i^n \sum_j^m (i-j)^2 C(i,j)$	(2.2.)
2.	*Energy*	$\sum_i^n \sum_j^m C^2(i,j)$	(2.3.)
3.	*Entropy*	$\sum_i^n \sum_j^m C(i,j)\log C(i,j))$	(2.4)
4.	*Homogenity*	$\sum_i^n \sum_j^m \frac{C(i,j)}{1+\lvert i-j\rvert^2}$	(2.5.)
5.	*Correlation*	$\sum_i^n \sum_j^m \frac{(i,j).C(i,j)-\mu_i\mu_j}{\sqrt{\sigma_i^2\sigma_j^2}}$	(2.6.)
		$\mu_i = \sum_i^n \sum_j^n (i)(C(i,j))$	(2.7.)
		$\mu_j = \sum_i^n \sum_j^n (j)(C(i,j))$	(2.8.)
		$\sigma_i^2 = \sum_i^n \sum_j^n ((C(i,j))(i-\mu_i)^2)$	(2.9.)
		$\sigma_j^2 = \sum_i^n \sum_j^n ((C(i,j))(j-\mu_j)^2)$	(2.10.)

Figure 7. The Formula of Feature Extraction

3. Discussion

3.1. *Camera*

In this study, the camera functioned as a device that can capture the image, we used the Logitech C270 Webcam as shown in Figure 8.

Figure 8. Webcam Logitech C270

3.2. Pothole's Dataset

The kind of pothole that will be detected are:

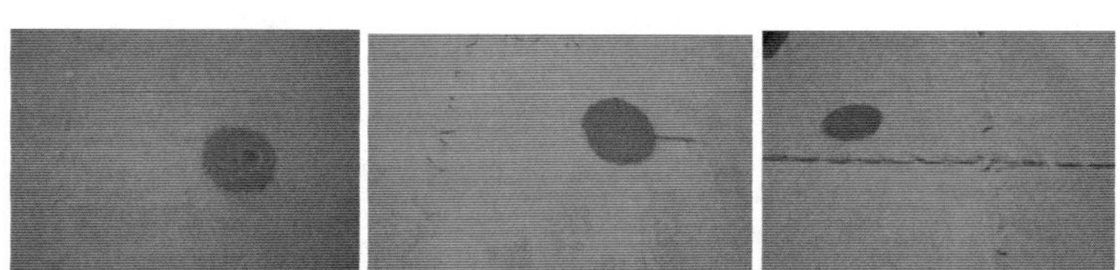

Figure 9. Pothole's Dataset

4. Result

To get good results, it needs testing. in Table 2., this table shows the results of feature extraction of holes in the road. These results are obtained from the calculation of features from the image normalized matrix.

Table 2. Result of Pothole's Features Extraction

No	Real Picture	Feature Extraction
1.	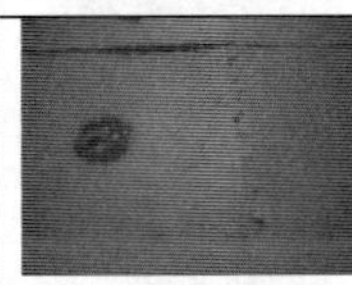	contrast = 0.0008559 dissi = 0.000810243 entro = 0.00853277 homom = 0.99971 energ = 0.998376
2.	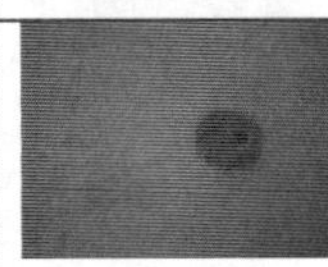	contrast = 0.000026096 dissi = 0.000026096 entro = 0.000319595 homom = 0.999992 energ = 0.999961
3.	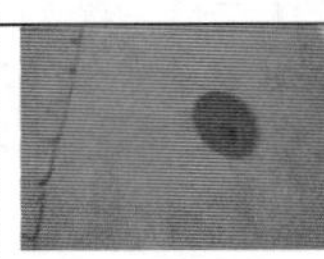	contrast = 0.000104547 dissi = 0.000104547 entro = 0.0016706 homom = 0.999958 energ = 0.999713
4.	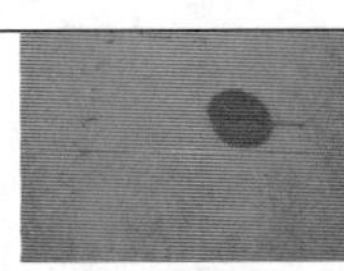	contrast = 0.000143753 dissi = 0.000189492 entro = 0.00241501 homom = 0.999928 energ = 0.999583
5.	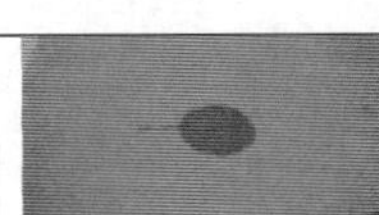	contrast = 0.0000718764 dissi = 0.0000718764 entro = 0.000807436 homom = 0.999984 energ = 0.999896
6.	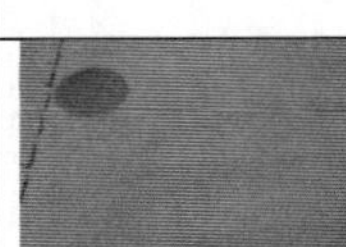	contrast = 0.000287506 dissi = 0.000287506 entro = 0.00306323 homom = 0.999896 energ = 0.999465

No	Real Picture	Feature Extraction
7.	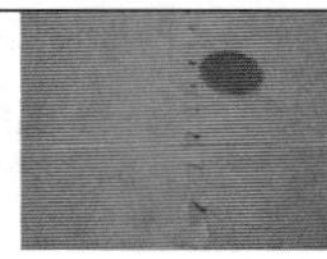	contrast = 0.000853631 dissi = 0.000849281 entro = 0.0091833 homom = 0.999667 energ = 0.998126
8.	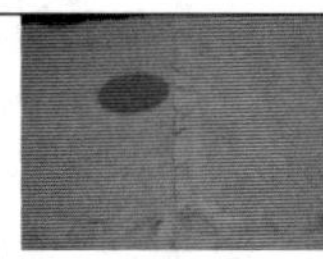	contrast = 0.00914201 dissi = 0.0138669 entro = 0.115225 homom = 0.996471 energ = 0.978065
9.	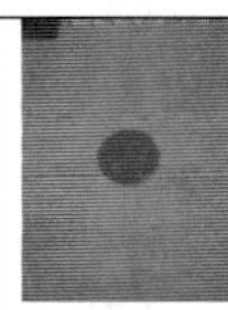	contrast = 0.000931926 dissi = 0.0000718764 entro = 0.0106773 homom = 0.999743 energ = 0.997986
10.	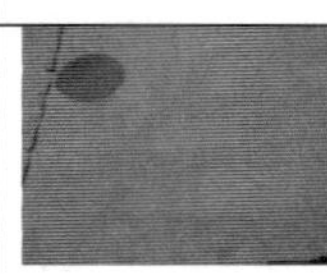	contrast = 0.0000718764 dissi = 0.0000718764 entro = 0.000807436 homom = 0.999984 energ = 0.999896

On Table 3 the results of testing the hole classification on the road using the Neural Network. From the test results obtained a percentage of success rate of 90% from 10 data that have been tested. This shows that the Neural Network method can be used for pothole classification.

Table 3. *Neural Network Testing*

No.	Object	Vektor Input (Biner)	NN Prediction (Biner)	Status
1.	Pothole	[1]	[1]	Correct
2.	Pothole	[1]	[1]	Correct
3.	Pothole	[1]	[1]	Correct
4.	Pothole	[1]	[1]	Correct
5.	Pothole	[1]	[0]	Incorrect
6.	Pothole	[1]	[1]	Correct
7.	Pothole	[1]	[1]	Correct
8.	Pothole	[1]	[1]	Correct
9.	Pothole	[1]	[1]	Correct
10.	Pothole	[1]	[1]	Correct
		Accuration		90%

Table 4. is the result of the Pothole detection test using the Gray Level Co-occurrence Matrix and Neural Network method in real time. From the test results obtained a percentage of success rate of 86.6% from 15 data that have been tested as listed in Table 5.

Table 4. Result of Real-time Testing

No.	Result	Condition	Explanation
1.	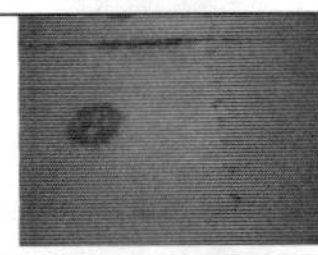	Pothole	Correct
2.	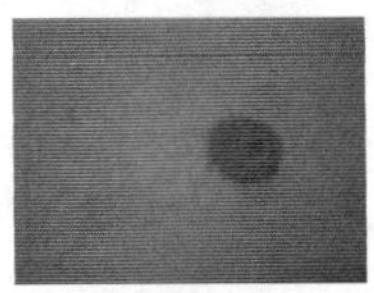	Pothole	Correct
3.	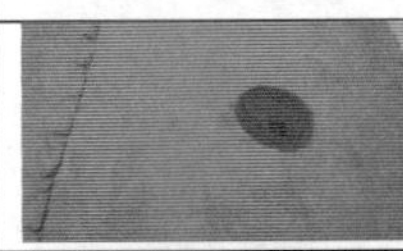	Pothole	Correct
4.	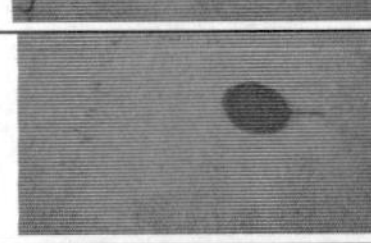	Not Pothole	Incorrect
5.	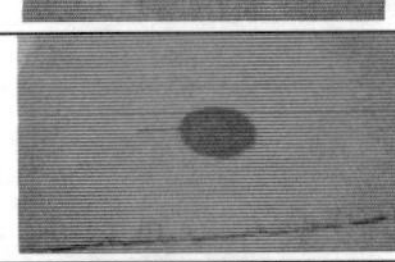	Pothole	Correct
6.		Not Pothole	Correct
7.	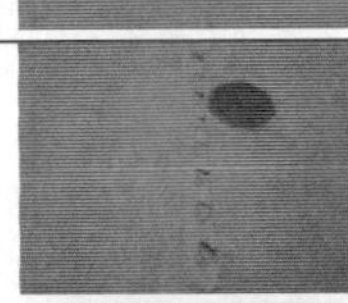	Pothole	Correct
8.	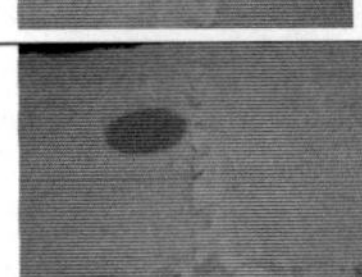	Pothole	Correct
9.		Not Pothole	Correct
10.		Pothole	Correct

No.	Result	Condition	Explanation
	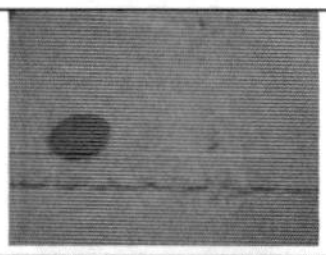		
11.	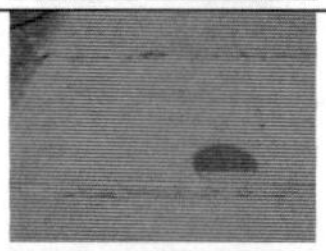	Not Pothole	Incorrect
12.	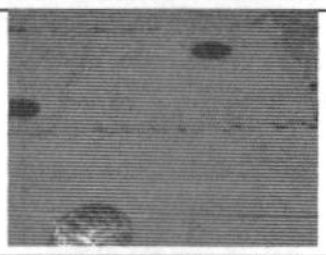	Pothole	Correct
13.	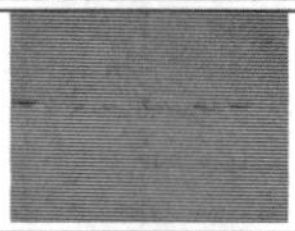	Not Pothole	Correct
14.	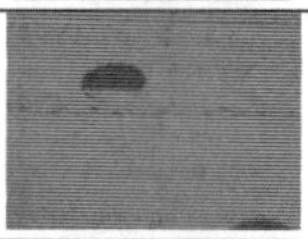	Pothole	Correct
15.	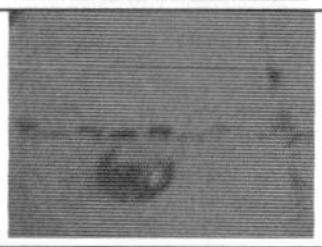	Pothole	Correct

Table 5. Percentage of Success

No.	Category	Total	Result	
			Incorrect	Correct
1.	Pothole	12	2	10
2.	Normal	3	0	3
Total Data		15	2	13
Percentage of Success			86,6%	

5. Conclusion

The conclusion that can be drawn from this experiment is that all component systems can work according to their functions and also this system can detect potholes using GLCM and NN with a system accuracy rate of 86.6%.

References

[1] Clauditta, dkk. 2016. Menghitung Orang dengan Ekstraksi Fitur *Gray Level Co-occurrence Matrix* (GLCM). Palembang (ID): STMIK GI MDP.

[2] Maulina, Erce. 2018. *Sortir Asphalt Drum Pada PT. Pertamina – BPG dengan Ekstraki Fitur Menggunakan Gray Level Co-occurrence Matrix (GLCM)*. Politeknik Perkapalan Negeri Surabaya.

[3] Patikawa, Sammy. 2012. *Greyscale (Konversi RGB menjadi Greyscale)* di https://sammypatikawa.wordpress.com/2012/05/11/greyscale/

[4] Suhartono, Dewin, 2012. *Dasar Pemahaman Neural Network* di http://socs.binus.ac.id/2012/07/26/konsep-neural-network/.

[5] Sutrisno, I., Jami'in, M.A., Hu, J., Modified fuzzy adaptive controller applied to nonlinear systems modeled under quasi-ARX neural network, Artificial Life and Robotics, 2014

[6] Jami'In, M.A., Sutrisno, I., Hu, J., Mariun, N.B., Marhaban, M.H., An adaptive predictive control based on a quasi-ARX neural network model , 2014 13th International Conference on Control Automation Robotics and Vision, ICARCV 2014

[7] Jami'in MA, Sutrisno I, Hu J, Nonlinear Adaptive Control for Wind Energy Conversion Systems Based on Quasi-ARX Neural Network Model

[8] Sutrisno I, Jami'in MA, Hu J, Marhaban MH, Self-organizing quasi-linear ARX RBFN modeling for identification and control of nonlinear systems, 2015 54th annual conference of the society of instrument and control engineers of Japan (SICE). IEEE

[9] Sutrisno I, Jami'in MA, Hu J, Marhaban MH, Mariun N, Nonlinear Model-Predictive Control Based on Quasi-ARX Radial-Basis Function-Neural-Network, 8th Inter. Conference on Mathematical Modelling and Computer Simulation

[10] Sutrisno I, Che C, Hu J, An improved adaptive switching control based on quasi-ARX neural network for nonlinear systems, Artificial Life and Robotics 19 (4), 347–353

[11] Jami'in MA, Sutrisno I, Hu J, The State-Dynamic-Error-Based Switching Control under Quasi-ARX Neural Network Model, in Proc. of the20 th International Symposium on Artificial Life and Robotics (AROB 20 th 2015)

[12] Widodo RB, Quita RM, Amrizal S, Gunawan RS, Wada C, Ardiansyah A, Sutrisno I, Rahmat MB, Arfianto AZ, Handoko CR, Santoso AD, Santosa AWB, Ardhana VYP, Setiawan E, Grasping and Attached Mode in Human-Computer Interaction in the Study of Mouse Substitution, Journal of Physics: Conference Series 1196 (1), 012034

[13] Sutrisno I, Jami'in MA, Hu J, An Improved Fuzzy Switching Adaptive Controller for Nonlinear Systems Based on Quasi-ARX Neural Network, International Seminar on Electrical Informatics and Its Education (SEIE 13)

[14] Sutrisno I, Jami'in MA, Hu J, Implementation of Lyapunov Learning Algorithm for Fuzzy Switching Adaptive Controller Modeled Under Quasi-ARX Neural Network, 2nd International Conference on Measurement, Information and Control (ICMIC)

[15] Sumijan, A. P. Windarto, A. Muhammad, and Budiharjo, "Implementation of Neural Networks in Predicting the Understanding Level of Students Subject," *Int. J. Softw. Eng. Its Appl.*, vol. 10, no. 10, pp. 189–204, 2016.

[16] B. Supriyadi, A. P. Windarto, T. Soemartono, and Mungad, "Classification of natural disaster prone areas in Indonesia using K-means," *Int. J. Grid Distrib. Comput.*, vol. 11, no. 8, pp. 87–98, 2018.

Analysis of Groundwater Contamination Level in Residential Areas Around Cipayung Landfill Using Ground Penetrating Radar

Maryadi Maryadi[1], Supriyanto[1], Fira Mariah Sausan Champai[1], I Nyoman Triananda[1], and Desi Nur Fitriana[2]

[1]Program Studi Geofisika, Fakultas Matematika dan Ilmu Pengetahuan Alam, Universitas Indonesia, Depok City, West Java Province, 16424
[2]Program Studi Geologi, Fakultas Matematika dan Ilmu Pengetahuan Alam, Universitas Indonesia, Depok City, West Java Province, 16424

*maryadi@sci.ui.ac.id

Abstract. Cipayung Landfill is the last disposal location of waste processing in Depok City, West Java. In a day, approximately 800 tons of garbage is dumped, which is then processed by being burned and left to decompose for inorganic and organic waste, respectively. However, poor waste management has resulted in an accumulation that continues to this day. This creates a variety of problems, especially leachate contamination that infiltrates to the groundwater, flows along and threaten the health of residents who live in the vicinity. This liquid contains toxins and hazardous materials. Therefore, it is essential for residents to know the level of contamination occurred and its dispersion, along with actions that must be taken to reduce the impact of pollution. In this study, we map the distribution of contaminated groundwater using the geophysical ground penetrating radar method, together with the chemical analysis of water samples. Ground penetrating radar has been useful to detect the change of permittivity and conductivity of layers below the surface. It is possible to mark off the contaminated groundwater to the fresh one indicated from an extensive amount of electromagnetic signal attenuation. This study suggests that there is contamination occurred in radius up to hundreds of meters from the landfill site. The result of this community-based research gives an advisable policy regarding waste management and groundwater usage in this area.

1. Introduction

Landfill is a land that is built for the disposal of solid waste originating from cities both households and industries [1] which has reached the final stage in waste management starting from the first time the waste is produced, collected, transported, managed and disposed of. According to Government Regulation of the Republic of Indonesia, a landfill must have a location that is well isolated so as not to cause a negative influence on the environment around the landfill, especially residential areas (more than 1 km from the settlement).

Cipayung Landfill is the final disposal site in the waste management stage in Depok City, West Java. The total waste sent to this location reaches approximately 900 - 1000 tons each day [2]. The rubbish is then burned (for inorganic waste) or left to rot (for organic waste). As a result of poor waste management system, the accumulation of waste continues to occur until now reaching 23 m, even though the height limit of the garbage pile is 20 m [3]. This raises a variety of problems, especially those related to cleanliness and health of residents who live around the landfill site. Leachate from the garbage pile is mixed with Pesanggrahan river water which is located on the east side of the landfill, which then pollutes the river water [4]. Groundwater used by residents around the landfill might be contaminated by harmful substances from the leachate. The situation could get worse both in terms of the level of contamination and in terms of distribution. This problem leads to a decline in the health of residents due to the low level of water cleanliness in their area.

One of the causes of pollution around landfills is leachate. Leachate, or wastewater, is a liquid that results from mixing surface water with substances produced from decomposed waste [5]. Leachate can seep into the surface of the soil, then contaminate the ground water underneath. Underground water flow allows the spread of contaminants to other areas around the landfill. Therefore, it is feared that people consciously or unconsciously have used contaminated water for their daily needs. Previously in 2012 a geophysical investigation was carried out using DC resistivitry and induced polarization methods in this area, to map the depths of aquifers that were indicated to have been polluted by leachate [6]. In the study it was found that leachate seepage reached 80 to 120 m from the outer side of the Cipayung landfill with a depth of about 30 m.

Therefore, it is important for residents to know the level of pollution of water resources in their area, know the level of safety or danger, along with the actions that must be taken to reduce the negative impact of the situation. In this study, we are attempting to map the distribution of contaminated groundwater through direct surveys through groundwater content testing using a water test kit and indirect estimation based on geophysical surveys using the ground-penetrating radar method. Ground-penetrating radar method was chosen because it has been proven and can be used to detect contamination of groundwater thanks to its sensitivity to any conductive material below the surface in which related to the contaminants [7-11].

2. Study Area

Cipayung District, located in the southern part of Depok City, West Java Province, has an area of 11.63 km which stretches in low-lying and slightly hilly areas with an elevation of 70 to 100 m above sea level (Badan Pusat Statistik (BPS), 2017). Cipayung Landfill itself is located on the western edge of Cipayung Subdistrict as shown in Figure 1. The subdistrict itself has an area of almost 269 hectares, of which 183.3 hectares are settlements for 9482 households divided into 11 neighborhoods (BPS, 2017).

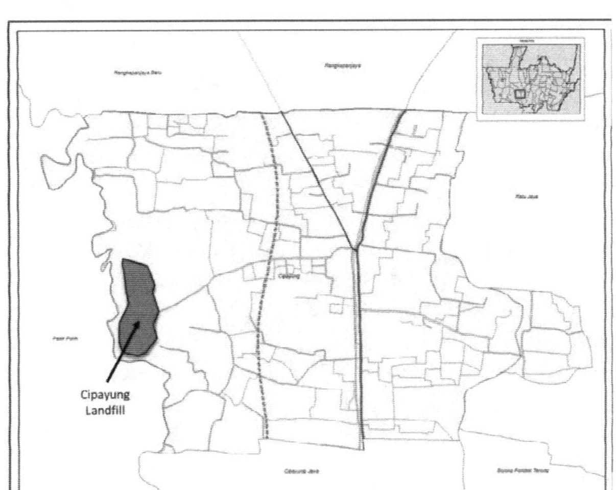

Figure 1. Location of Cipayung Landfill on Cipayung Subdistrict map (modified from BPS, 2017)

The area of the Cipayung landfill is widening from around 4 hectares in the initial year of operation to 20 hectares in 2018 [13] with a height reaching 20 meters. Various problems began to arise due to the prolonged accumulation of waste and management systems that have not been maximized. In reports in various media, residents complained of the increasingly bad odors that flow with the air, an increasingly dirty and polluted environment, lack of clean water for daily necessities, to health problems such as skin diseases and respiratory infections [14]. In order to understand the water contamination level and its distribution, both direct and indirect assessment were performed. The former was done using an instrument for groundwater quality tests while the latter was carried out using ground penetration radar.

3. Methodology

A. Ground Penetrating Radar Method

Ground Penetrating Radar (GPR) is a tool that can be used to detect objects that are in the soil at a certain depth. The term GPR refers to a non-destructive geophysical method that uses electromagnetic techniques to record subsurface characteristics. GPR uses electromagnetic wave vectors in the range of radio frequencies that are transmitted into the ground. The level of depth that can be achieved depends on the frequency used. For surveys at far depths, lower frequency GPR is preferred as used in this research. This frequency has the possibility to detect the presence of underground water sources, as well as study the soil layers or geological basement.

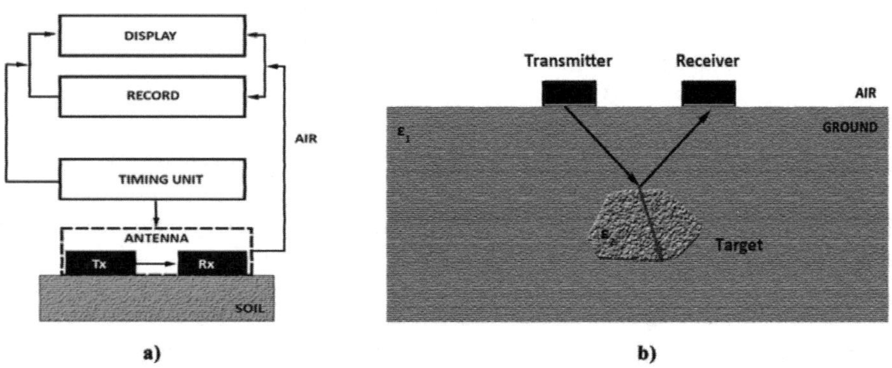

Figure 2. (a) The working scheme of the components of the GPR system and (b) The effects of reflection and refraction experienced by electromagnetic waves due to differences in the electrical properties between the rock and the ground. [15]

In the process, the GPR signal will be fired by the transmitter which is then reflected by the surface layer of soil or rock in the ground before the signal is finally received by the receiver embedded in GPR system. The signals are transmitted into the ground in a very short duration (in the order of 10 seconds) and are propagating until it meets the boundary between two different layers. Some of the waves are reflected, while the remaining energy is transmitted into the next layer. The amplitude of the reflected waveform recorded in the receiver is associated with the wave energy itself. The percentage of the reflected wave (R) depends on the relative dielectric permittivity difference between the host (RDP_1) and the reflector material (RDP_2) as follows,

$$R = \frac{\sqrt{RDP_1} - \sqrt{RDP_2}}{\sqrt{RDP_1} + \sqrt{RDP_2}} \quad (1)$$

so that when the wave progressed from soil (RDP = 9) to water table or water-saturated soil (RDP > 23), strong reflected signal could be recorded.

The transmitted wave, to some extent, will not be able to penetrate any deeper as it runs out of energy due to signal attenuation [15]. Electromagnetic signal attenuation (α) is highly affected by electrical conductivity of the invaded material as formulated as follows,

$$\alpha = \sqrt{\frac{\mu_0}{\varepsilon_0}\frac{\sigma}{2\sqrt{RDP}}} \quad (2)$$

where σ represents electrical conductivity of material, while μ_0 and ε_0 are magnetic permeability and permittivity of vacuum, respectively. Ground water that has been polluted by waste is generally characterized by an increase in conductivity caused by the presence of Fe and Pb contaminants in the water, this can be a useful parameter for detecting areas that experience water pollution using the GPR method. It is expected that surveys will be able illustrates the contrast between polluted water and water that is still in good condition by means of their reflection amplitudes and signel attenuation level.

In this GPR survey process we used electromagnetic waves with a frequency of 50 Mhz with a maximum penetration of 40 m, because the location of the water table is estimated to be located at a depth of 10-30 m. Then the data collection was carried out in 9 lines measuring 100 - 250 m spread over three zone based on its relative distance to the landfill, to find out the extent of the distribution of water contamination. The GPR profiles in the study area are shown in Figure 3. After GPR data is obtained, the data is then processed in 2-D to get a picture related to the distribution of ground water pollution that occurs.

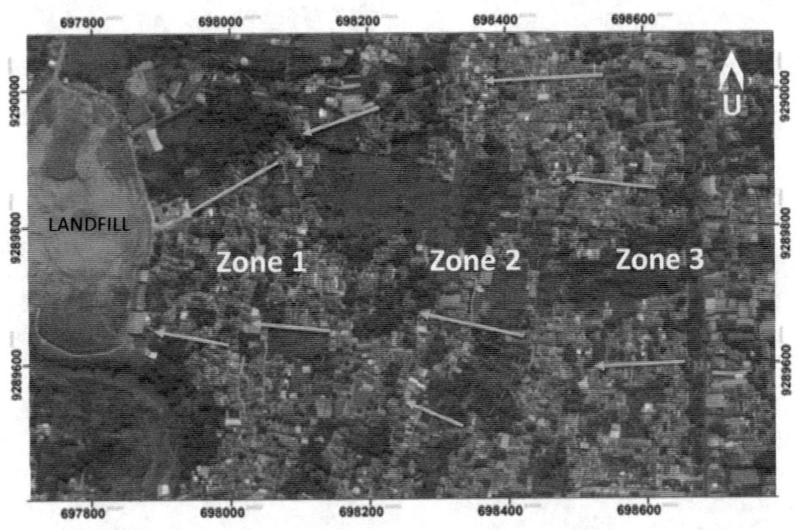

Figure 3. The GPR acquisition profiles (solid lines with arrow) are divided into 3 zones (separated by dashed lines) in respect to each relative distance to the landfill. Arrow represents measurement direction.

B. Water Quality Analysis

Groundwater quality test was carried out against 25 water samples collected from the residents' homes that were selected based on their zones. The collected samples were tested using an AMTAST water test kit to measure the water quality parameter test and obtain the value of TDS (Total Dissolve Solid) to determine the chemical mass of solids dissolved in the water. The higher value of TDS in water means it requires more processing for the water to be drinkable. The next process was determining pH of the water. pH is a measure of how acidic or basic water is. The range goes from 0 to 14, with 7 being neutral, less than 7 indicate acidity and greater than 7 is basic.

The most important parameter in this water quality assessment is conductivity, which can be directly compared to the result of GPR survey. The higher value of conductivity might indicate that the water had been contaminated. Another parameter that was measured is ORP (Oxidation-Reduction Potential) measurement which was done to indicate the level of hygiene in water and to isolate materials that could potentially contaminate the water. These parameters were then taken to determine the quality of groundwater that had been contaminated by waste. By combining the results of all the parameters and methods, an attempt to map the groundwater contamination was accomplished.

4. Results And Discussions

Several physical and chemical parameters from water samples were presented in the distribution maps. Kriging method was used for lateral interpolation of each of the value, since the sample locations are sparsely distributed. The map of conductivity, pH, ORP and TDS are shown in Figure 4.

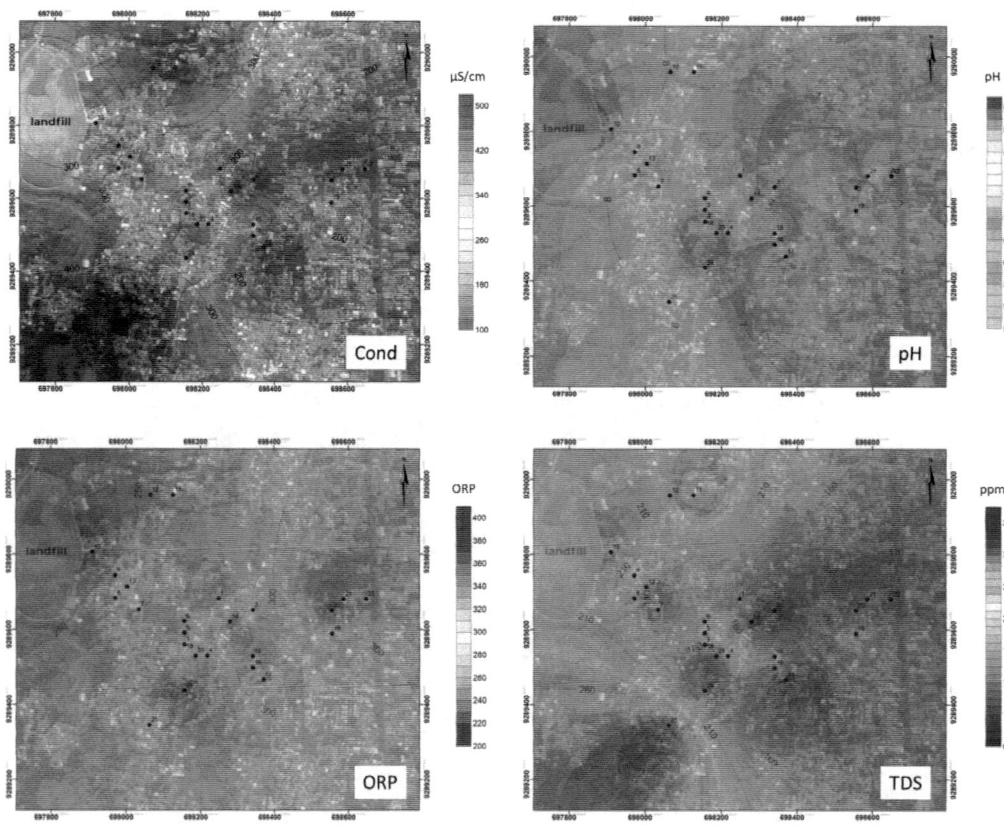

Figure 4. Distribution map of groundwater conductivity (cond), pH, ORP, and TDS around Cipayung Landfill, overlying the aerial image of the study area. The dots are the sampling locations.

Upper left image in Figure 4 shows that there is an area with high conductivity levels in the south and northeast direction from the landfill, where the conductivity levels can reach up to 300-500 μS/cm. Water that is safe to drink has a level of conductivity around 150 to 500 μS/cm [16]. Despite the value is being under 500 μS/cm, does not necessarily mean that the water is safe to drink. Further studies and supervision are still needed, by measuring other parameters to finally signify whether the groundwater of the area is safe for everyday use or not. The other areas are, however, not having the same condition, where the conductivity values are considerably low.

The pH level of drinkable water should be neutral between 6.5 to 8.5, as water with acidic properties would dissolve metallic materials in the water and therefore contaminating it. Upper left image in Figure 4 shows that the distribution of groundwater pH near the Cipayung landfill is varying from 8 to 9, which indicates that the water is considered basic and, therefore, unsafe to drink. Whereas in the western areas, the groundwater is considered neutral as the pH is differing around 7. There are also several anomalies in the middle of the area where high pH zones are encountered. This is probably occurred because this area was presumably a swamp that could isolate the contaminated water.

In the lower left area of Figure 4, the ORP distribution map is presented. The ORP of drinking water should be around 250 mV. The lower the value of ORP in an area indicates that the water is contaminated and therefore unsafe to use [17]. Fig. 4 clearly shows that some areas near the Cipayung landfill have low ORP values, no higher than 250 mV. This result suggests that that the groundwater in these areas need to be supervised as they are most likely contaminated.

Total Dissolve Solids (TDS) is the term used to describe the inorganic salts and small amounts of organic matter present in solution in water. The palatability of drinkingwater has been rated by panels of tasters in relation to its TDS level as follows: excellent, less than 300 mg/litre; good, between 300 and 600 mg/litre; fair, between 600 and 900 mg/litre; poor, between 900 and 1200 mg/litre; and unacceptable, greater than 1200 mg/litre [18]. The TDS values in Fig. 4 are various, in between 60 to 320 mg/L (60 to 320 ppm). The result indicates that there was no indication of water contamination due to hazardous solids in most of the area, which, however, contradicts the other results.

On the other hand, each GPR raw data was processed through filtering, noise removal, and gain control to get a clearer image of the subsurface condition. The results are presented in two-dimensional vertical cross-section up to 40 m depth with varying horizontal distance. Both water quality and GPR data were combined to explain about water contamination level and its distribution. To ease the understanding, the interpretation of the results is divided for each zone.

A. Zone 1

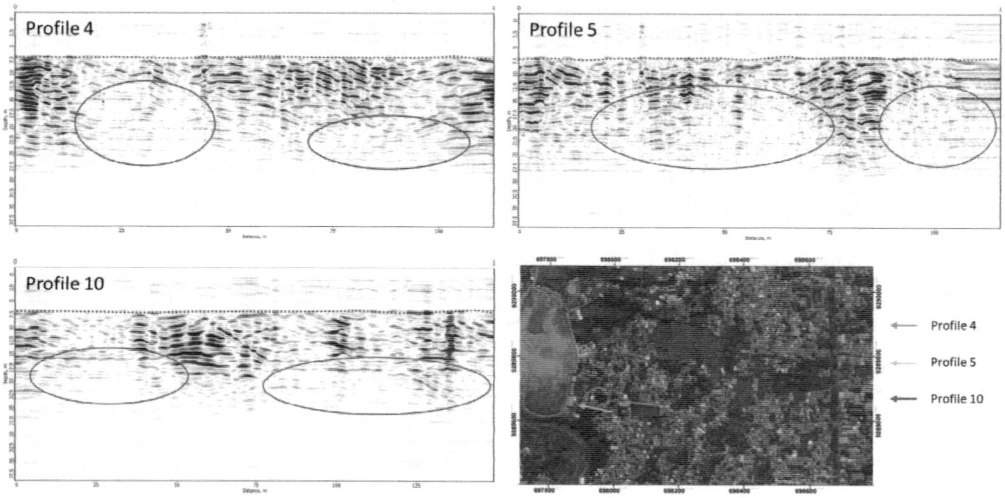

FIGURE 5. GPR cross-sectional results from Profile 4, 5, and 10 whose data were measured in Zone 1. Red circles show the electromagnetic signal attenuation due to the presence of conductive bodies in the subsurface which is deduced to be correlated with the contaminated water.

There are 3 GPR profiles in this zone: Profile 4, 5, and 10. Vertical cross-sectional view of GPR results from these profiles and their respective locations are shown in Figure 5. Profile 4 is located in about 320 m from the Cipayung Landfill. Based on the GPR image, a reflection contrast was found at a depth of 7.5 m which is most probably occurred due to the difference in electrical permittivity between the soil and water table. Another important finding is that there are occurrences of signal attenuation zones at a depth of 12.5 to 25 m which are suspected to be caused by the presence of conductive substances in the area. Similar situation is also found in profile 5 which is situated in about 215 m to the southeast from the garbage dump. Line 10, the closest profile to the landfill, is in about 150 m from the main entrance of the city waste dumping site. In this profile, the sign of signal attenuation is very distinct and massive. The conductive anomaly each profile is detected in the deeper part of water table since the density of contaminated water is higher than the unaffected ones.

It is obvious that these results is, to some extent, in agreement with the water quality assessment which conclude that in Zone 1, the conductivity is relatively high. This finding is also consistent with other chemical parameters mapped in Fig. 4. It can thus be suggested that there is some level of groundwater contamination occurred in this zone and further action must be taken.

B. Zone 2

In Zone 2, three GPR profiles are exist (Profile 1, 2, and 9) and the results of GPR survey in this zone are shown in Figure 6. The results of GPR survey in these profiles indicate that there are chances of water impurity in this area. However, this outcome is contrary to the water quality studies (Fig. 4) which have suggested that there is no evidence of water contamination in the same spots. One unanticipated finding is that the signal attenuation is uneven and detected in the relatively shallow depth. A possible explanation for this might be that these conductive sections have no relation with the leachate from the landfill but, instead, the anomaly is found after the existence of an archaic swamp in this area. This interpretation is, however, need to be evaluated further.

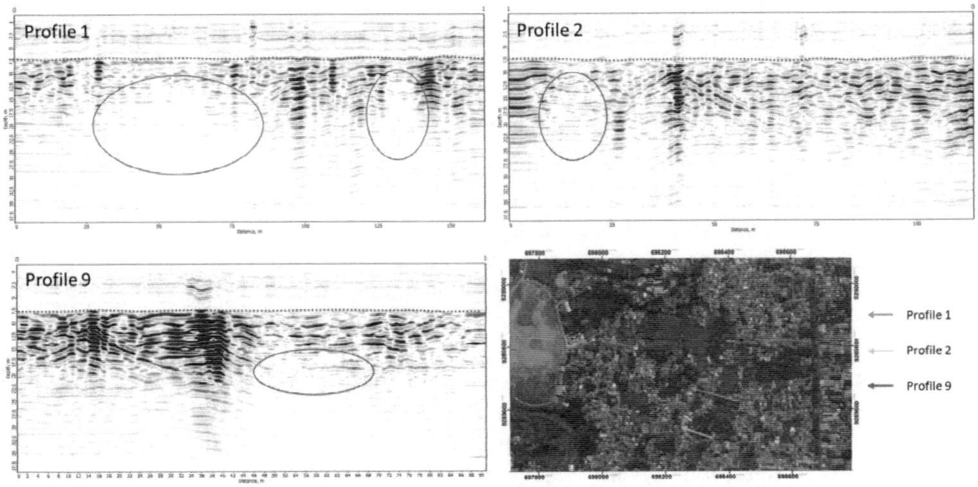

FIGURE 6. GPR cross-sectional results from Profile 1, 2, and 9 located in Zone 2. Red circles show the electromagnetic signal attenuation due to the conductive bodies in the subsurface which is probably related to the presence of water contaminants.

C. Zone 3

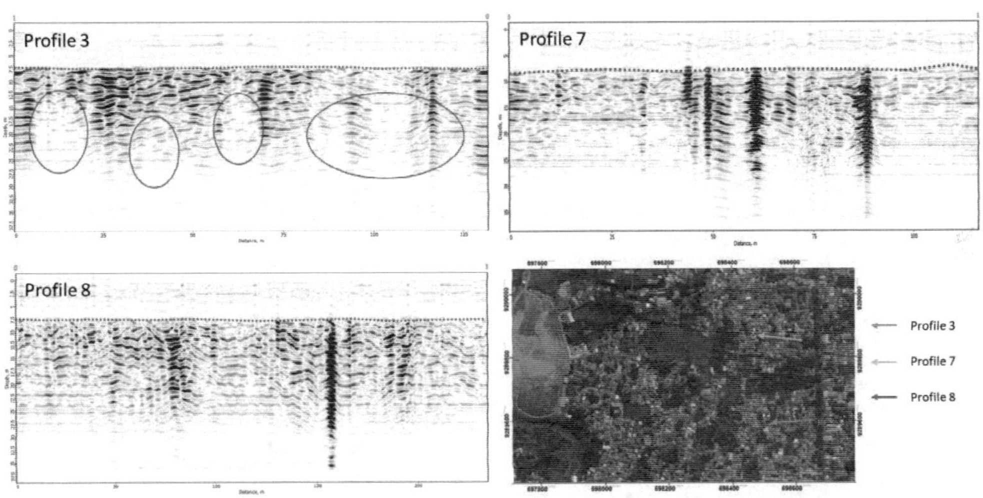

FIGURE 7. GPR cross-sectional results from Profile 1, 2, and 9 located in Zone 2. Red circles show the electromagnetic signal attenuation due to the conductive bodies in the subsurface which is probably related to the presence of water contaminants.

There are four GPR Profiles measured in Zone 3 (Profile 3, 6, 7, and 8). However, Profile 6 is neglected due to high level of noise which could distort the result and interpretation. The results of GPR survey in this zone are shown in Figure 7. The GPR investigation in this area generally did not detect any evidence for water contamination as this area is quite far away from the garbage dump. These results support the findings from water quality assessment in the same zone. One inconsistency is found

IOP Conf. Series: Materials Science and Engineering **874** (2020) 012013 doi:10.1088/1757-899X/874/1/012013

in the result of Profile 3, where the EM signal attenuations are detected in some spot of the cross-section. This rather contradictory result may be due to the presence of past drainage swamps which probably holds conductive substance.

All the results may help us to understand that there are groundwater contaminations occurred in radius up to hundreds 400 of meters from the landfill site as the GPR survey and water analysis conducted in Zone 1 suggest the same conclusion. However, further work is required to confirm the findings in Zone 2 and 3. To develop a clearer picture of water contamination level, more water sample analysis from denser sample location is needed. It would be better if the data is measured in the rainy season. It is also highly suggested that the water sample location should be adjacent to the GPR profiles so that the data comparison and interpretation would be more explicitly explained.

5. Conclusions

This study is set to understand the groundwater contamination level around landfill using GPR and direct water quality assessment. Both methods have found that the occurrence of contamination is happened within a radius of 400 m from the dump, which could affect the civilians living in that area regarding their water consumption. Overall, this study strengthens the idea that ground penetrating radar could be useful for detecting the groundwater table and contamination in terms of its conductivity. Although small number of GPR profiles and water samples did not allow a comprehensive interpretation, this study gives the idea about groundwater assessment using geophysical method, especially GPR. A limitation of this study is that the data was taken in dry season and therefore leachate contamination is minimal. Comparison study carried out rainy season is highly suggested.

Acknowledgments

We acknowledge the Universitas Indonesia as this study is funded through Universitas Indonesia Community Engagement Grant Program number NKB-1372/UN2.R3.1/HKP.05.00/2019. We also thank the government of Cipayung District and Subdistrict for the cooperation during data acquisition processes.

References

1. A. C. Tse and C. I. Adamu, "*Assessment of Anthropogenic Influence on Quality Of Groundwater In Hand-Dug Wells In Parts Of Makurdi Metropolis, North Central Nigeria*", Ife Journal of Science **14**, pp. 123 - 135 (2012).
2. Windarto, "*Kondisi TPA Cipayung Di Depok Kritis*" (Suara Karya, June 11, 2019), Accessed on August 23, 2019
3. R. Kisar, "*Depok Akui Pengolahan Sampah di TPA Cipayung Buruk*" (Media Indonesia, March 4, 2019), Accessed on August 23, 2019.
4. J. Armanto, *"Kali Pesanggrahan Diduga Tercemar TPA Cipayung, Pemkot Depok Terancam Dilaporkan"* (Indopos, June 11, 2018), Accessed on August 23, 2019
5. J. Soemirat, *Kesehatan Lingkungan (Cetakan Keempat)* (Gadjah Mada University Press, Yogyakarta, 2000).
6. Y. S. Arbi, *Pemetaan Sebaran Air Lindi di Daerah TPA Depok dengan Menggunakan Metode Resistivity dan IP*, Thesis, Universitas Indonesia, 2012.
7. N. S. Wijewardana, L. W. Galagedara and M. I. M. Mowjood, "*Assessment of groundwater contamination by landfill leachate with ground penetrating radar*" (14th International Conference on Ground Penetrating Radar (GPR), Shanghai, 2012), pp. 728-732.
8. J. J. Daniels, R. Roberts and M. Vendl, "*Ground penetrating radar for the detection of liquid contaminants*" (1994) (Journal of Applied Geophysics 33 1995), pp. 195-207.

9. L. Capozzoli, V. Giampaolo, E. Rizzo, M. Votta, P. L. Cucci, M. D. Biase and S. Straface, "*Ground Penetrating Radar as A Powerful Tool for The Study and The Monitoring Of Lnapl-Contamination In The Subsoil"* (GNGTS, 2012), pp. 60-67.
10. J. L. Porsani, W. M. Filho, V. R. Elis, F. Shimeles, J. C. Dourado and H. P Moura, "*The use of GPR and VES in delineating a contamination plume in a landfill site: a case study in SE Brazil"*, Journal of Applied Geophysics **55**, pp. 199-209 (2004).
11. K. K. Singh, "*Application of Ground Penetrating Radar for Hydo-Geological Study",* Journal of Scientific & Industrial Research **65**, pp. 160-164 (2006).
12. Badan Pusat Statistik, "*Kecamatan Cipayung Dalam Angka, Badan Pusat Statistik (BPS) Kota Depok"* (2017) ISBN: 978-602-0925-56-1.
13. B. Arifianto, "*Pengelolaan TPA Cipayung Buruk, Warga Gugat Pemkot Depok*". (Kompas, 2018), Accessed on February 20, 2019.
14. I. Sarwindaningrum, "*Sampah Tak Terkelola Baik, Warga TPA Cipayung Mengeluh Kian Sengsara",* (Kompas, 2018), Accessed on February 21, 2019.
15. J. J. Daniels, "*Ground Penetrating Radar Fundamentals",* Appendix to U.S.EPA, Region V, Ohio, 2000.
16. S. Behar, "*Testing the Waters: Chemical and Physical Vital Signs of a River"*, VT: River Watch Network, Montpelier, 1997, ISBN 0787234923.
17. T. Bastian and J. Brondum, "*Do Traditional Measures of Water Quality in Swimming Pools and Spas Correspond with Beneficial Oxidation Reduction Potential?",* Public Health Rep., **124**, pp. 255–261 (2009).
18. W. H. Bruvold and H. J. Ongerth, "*Taste quality of mineralized wate*r", Journal of the American Water Works Association, pp. 61:170 (1969).

IOP Conf. Series: Materials Science and Engineering **874** (2020) 012014 doi:10.1088/1757-899X/874/1/012014

Design of Information System to Detect Antenatal Depression

Alfiyana Yuliasari[1] and Artha Prabawa[2]

[1]Master Study Program of Health Informatics, Faculty of Public Health, Universitas Indonesia
[2]Department of Biostatistics and Population Studies, Faculty of Public Health, Universitas Indonesia
Building A 2nd Floor Faculty of Public Health Kampus UI Depok 16424, Indonesia

alfiyana.yulia@gmail.com

Abstract. Depression is a major cause of disease burden worldwide in women of childbearing age and 5-10% of all pregnant women experience antenatal depression. depression in pregnancy is associated with the incidence of stillbirth, premature, low birth weight, asphyxia, and congenital abnormalities in newborns. Although antenatal depression has a very detrimental effect, anxiety and depression during pregnancy are still less detected so that the treatment has not been maximized. The aim of this study is to develop model of antenatal depression detection system. This research will design an information system on detection of antenatal depression based on Android mobile using prototyping method. The needs analysis of this information systems is carried out by interviews, observations, and literature studies aimed at producing an effective and efficient system. Antenatal depression detection test uses the Edinburgh Postnatal Depression Scale (EPDS) questionnaire consisting of 10 questions. The results of this study is the form of antenatal depression detection information system design in android and web based.

Keywords : information system, antenatal depression, pregnancy

1. Introduction

Not only being an exciting and joyous period, pregnancy can also be a hard and stressful period due to the physical and psychosocial changes experienced by pregnant women. Therefore, pregnancy increases the risk of mental health disorders in pregnant women [1]. Unstable emotional state can increase stress in some pregnant women. The response to stress experienced by pregnant woman also varies, it can be clearly seen or not [2].

Depression is a major cause of disease burden worldwide in women of childbearing age and 5-10% of all pregnant women experience antenatal depression [3]. Depression in general is more common in women (5.1%) than in men (3.6%) [4]. In a study conducted by Mak et al., (2019), it was found that the prevalence of antenatal depression was 17.5% of 1440 pregnant women in China. Another study found that during pregnancy, the incidence of antenatal depression mostly occurs at 12-16 weeks' gestation with a prevalence of 5.2%, at 22-26 weeks' gestation at 2.6% and at 32-36 weeks' gestation at 3.5% [5]. The symptoms of antenatal depression are associated with low bio-psycho-social-cognitive in newborns until the neonatal period. Depression in pregnancy affects the fetus through hyperactivation of the maternal hypothalamus-pituitary-adrenal axis and increases inflammatory signs that can affect cortisol levels in the fetus and maternal placental genes [6]. Furthermore, depression and anxiety that occur in pregnant women are associated with the incidence of intra uterine fetal death (IUFD), premature, low birth weight, asphyxia, and congenital abnormalities in newborns. Antenatal depression is also known to be a major trigger for postpartum depression and postpartum depression is a strong predictor of parental stress and lack of maternal and child bonding [1].

Although antenatal depression has a very detrimental effect, anxiety and depression during pregnancy are still less detected so that the treatment has not been maximized [7]. There are several reasons why antenatal depression is less commonly detected than postpartum depression. They are including hormonal changes during pregnancy, community stigma, the tendency of health workers and pregnant women to only focus on physical health care during pregnancy, and early symptoms of depression are often ignored because they are similar to normal pregnancy symptoms such as nausea, weakness, changes in appetite and sleep disorder [1].

In Indonesia, the data from Basic Health Research (Riskesdas) in 2010 shows that prevalence of postpartum depression is 2.32% (440 / 18,937) [8]. Although the incidence of antenatal depression is unknown nationally, antenatal depression that is not treated properly has a great chance to becoming postpartum depression. Research conducted by Annerangi (2013) shows that the prevalence of depression in pregnant women in the Pasar Minggu District Health Center is 14.8% with categories mild 12.2%, moderate 1.7%, and severe rate 0.9% of 107 pregnant women [9]. Another study by Rachmawati (2013) conducted at Sukmajaya and Cimanggis District Health Center showed that 3% of 98 third trimester pregnant women experienced depression before delivery [10]. Another research also conducted at Sukmajaya and Cimanggis District Health Center by Rahmawati (2016) showed that 0.9% of 110 third trimester pregnant women experienced mild depression while 97.3% of third trimester pregnant women experienced mild to moderate anxiety [11].

Several studies have suggested that the main risk factors for antenatal depression are maternal anxiety disorders, stress, a history of previous depression, lack of social support, family violence, and unwanted pregnancy. These risk factors can help health workers to identify mothers with a high risk of depression but do not show any symptoms of depression so that health workers can provide preventive intervention [1]. Antenatal depression is a condition that needs to be identified and treated as early as possible [3]. Detection of antenatal depression that is carried out routinely at least in the first trimester and third trimester will increase the chances of successful interventions in mothers suffering with mental health problems [12].

Smartphone is the most often used device by public to doing communication and almost 27% of smartphone users carry out online activities. With so many smartphone users, the use of an Android-based detection tool can be used in broaden reach [13]. The android-based screening service model can quickly detect mental health disorders so that person who suffering from depression can get proper treatment by professionals [14].

The aim of this study is to design an antenatal depression detection information system, hereinafter referred to as "Care Bumil" (Cegah Depresi Ibu Hamil). Through this information system, it is expected that depressive symptoms in pregnant women can be detected as early as possible so that maternal and neonatal complications can be reduced. the impact of this study is to decrease maternal and infant mortality rates.

2. Method

The method in designing this system uses the prototyping method but at this time only developed in the modeling stage. This system focuses only on the stages of requirements analysis, design and implementation. Data collection for system identification is done by interviewing, observing, and studying literature through searching several books, articles, and journals related to this research.

3. Results and Discussion

The results of this study will describe how the process of implementing antenatal depression detection is performed. This condition is obtained from qualitative study results by interview, observation, and literature study methods. It also explained the system development stage which refers to the prototyping method.

3.1. System Requirements Analysis

System requirements analysis is described in terms of input, process and output to categorize the problems found in accordance with the data obtained.

Table 1. System Requirements Analysis

	Current Conditions	Needs
Input	There is no standard form of antenatal depression detection, depression detection examination has not become a routine examination on antenatal care, and Stigma in the society that depressed mothers are mothers who fail to rise their children.	Designing a standard antenatal depression detection tool that can be accessed personally in pregnant women to reduce stigma.
Process	The service to the patients is not optimal because a lot of task in clinical care and administrative matter (recording and reporting) that should be done by midwives	Developing antenatal depression detection applications that can be done independently in pregnant women and the results can be recorded in the midwife database.
Output	There is no recording and reporting of antenatal depression detection in health care professionals.	Designing a reporting form in an attractive and easy-to-understand output view or interface.

3.2. System Planning

The system planning process will be based on the results of literature studies on antenatal depression management procedures at the National Institute for Health and Care Excellence (NICE) guidelines on clinical management and service guidance of antenatal and postnatal mental health [15]. The process starts from pregnant women filling in the self-assessment data of depression risk factors and filling out the Edinburgh Postnatal Depression Scale (EPDS) questionnaire, then the test results are analyzed by midwives. if antenatal depression is found, the midwives conduct information and counseling, provide support to pregnant women and families, doing coordinate care, monitoring and improve communication with pregnant women. After that, the pregnant woman is examined whether the depression is reduced or lost. If depression decreases, the course of antenatal depression is complete. If depression persists, pregnant women are referred to a psychiatrist / psychologist for further intervention and pharmacology therapy.

3.3. Design System

System design involves identifying and describing the underlying software system and its relationships. The design is divided into several sections, namely:

3.3.1. Context Diagram.

Context diagram is a diagram that shows system boundaries, the interaction of all external entities with the system, and information that generally flows between entities and systems. The context diagram of the application of antenatal depression screening can be seen in the figure 1.

3.3.2. Entity Relational Diagram (ERD).

Entity Relationship Diagram illustrates the relationship between entities in the system. The relationship between entities in detection depression in pregnancy can be seen in the figure 2.

3.3.3. Table Relational Diagram (TRD).
In the development of antenatal depression detection system there are several entities that connected to each other. Relationships between these entities can be described in figure 3.

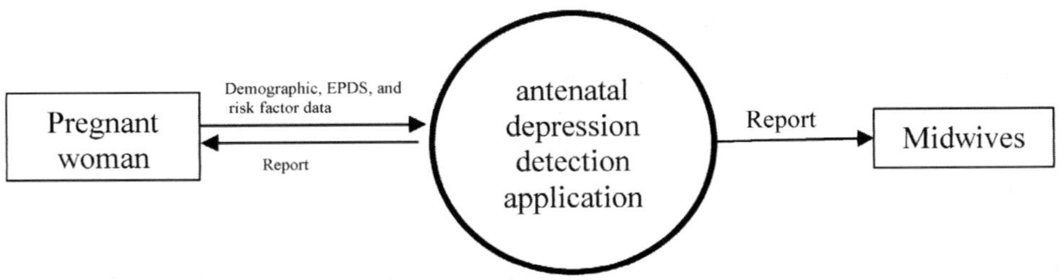

Figure 1. Context Diagram

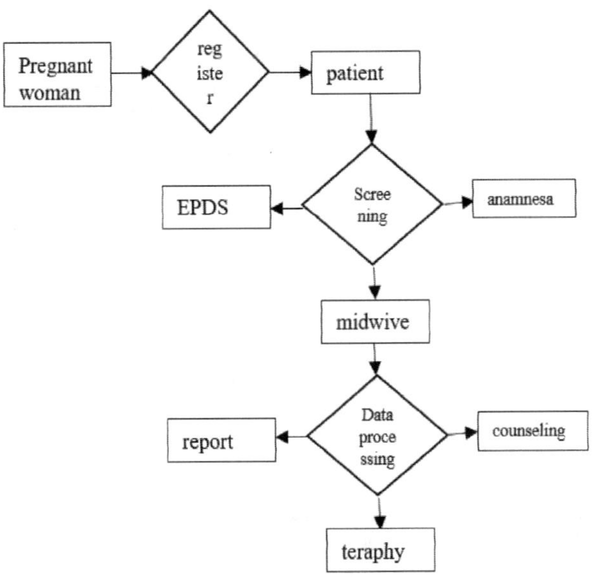

Figure 2. Entity Relational Diagram

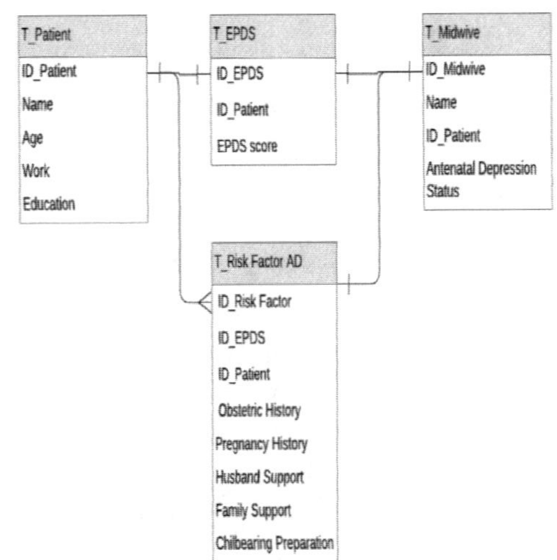

Figure 3. Table Relational Diagram

3.4. Model Development for Android-Based Application
Depression and anxiety often occur in pregnant women, especially in third trimester. therefore, detection of depression in pregnancy is needed to reduce the risk of postnatal depression. The key to managing depression is immediate identification through routine detection of midwifery care services. Antenatal depression detection information system is an android-based application designed to record, process, detect, and report the mental health of pregnant women. An overview of Care Bumil application can be seen in the interface system that has been made. Care Bumil application is divided into 2 application models namely the android application model and the web application model. While the android application model contains a menu Tentang Saya, antenatal depression tests using the Edinburgh Postnatal Depression Scale test, a history of EPDS test results, and info, the web application model contains data from the results of detection of depression in pregnant women and their risk factors. the web application model is designed for health care professional especially midwives and the android application model is designed for pregnant woman. The interface form of each menu is as follows:

a) Register and Login Menu
Register and login forms can be accessed after the user downloads the Care Bumil application. Pregnant women register with their full name, e-mail and password. After completing the registration form, pregnant women will get an email containing a patient ID that will be used to log in to the application.

b) Tentang Saya Menu
Menu of tentang saya contains depression risk factor assessment consisting of demographic data on pregnant women, current pregnancy history, past obstetric history, husband support, family support, and delivery preparation. The interface of this menu can be seen in figure 4.

c) Test Menu
This menu contains 10 questions from Edinburgh Postnatal Depression Scale (EPDS) about perinatal depression. The questionnaire in EPDS have been commonly used to detect antenatal and postnatal depression. This tool is reliable, valid, has been translated into several languages and has been used in various populations. The interface of this menu is showed in figure 5.

d) Test Result Menu
The test results menu contains the score of the EPDS test results showing the level of depression in pregnant women. An EPDS score higher than 13 is used as a reference for estimating antenatal and postnatal depression. this user interface can be seen in figure 6.

e) History Menu
History menu contains a history of antenatal depression checks performed by pregnant women. In this menu, pregnant women can see a graph of the state of their mental health during pregnancy. The picture of this menu can be seen in figure 7.

f) Info Menu
The Info menu contains information about the health of pregnancy, childbirth, postnatal, and newborns to increase the knowledge of pregnant women about their health.

g) Web Interface
The Web Interface used by midwives contains data on antenatal depression detection in pregnant women and their risk factors for further use as a basis in giving patient intervention.

Figure 4. History Menu

Figure 5. Test Result Menu

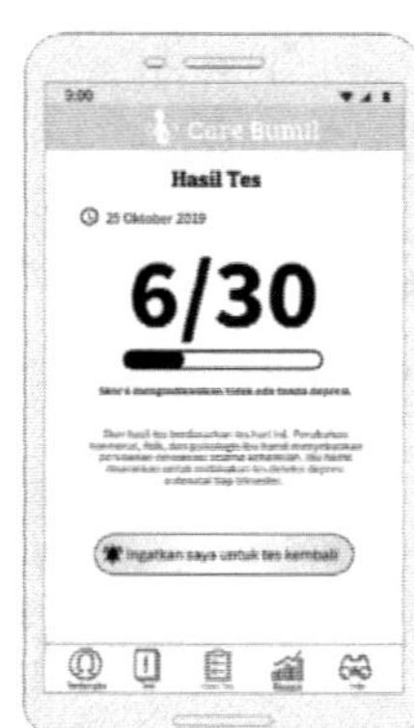

Figure 6. Test Menu

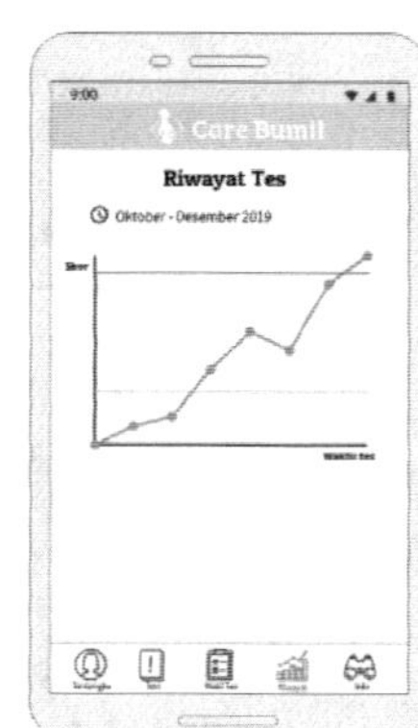

Figure 7. Tentang Saya Menu

4. Conclusion
The design of antenatal depression detection information system with this prototyping method supports the improvement of the mental health of pregnant women. The Android-based application approach is based on user needs, namely the need for an effective, efficient, personal depression detection tool that will not compound the workload of midwives. The results of this study identified four of the most commonly used and influential framework methods in system development stage of prototyping method. They are system requirements analysis, system planning, design system, and model development for android-based application. This information system can be developed by organizations involved in improving maternal and child health so that it can be utilized by the general public. This antenatal depression detection information system needs further development so that the system can be applied for future needs.

IOP Conf. Series: Materials Science and Engineering **874** (2020) 012014 doi:10.1088/1757-899X/874/1/012014

References

[1] Biaggi A, Conroy S, Pawlby S, Pariante CM. Identifying the women at risk of antenatal anxiety and depression: A systematic review. J Affect Disord [Internet]. 2016;191:62–77. Available from: http://dx.doi.org/10.1016/j.jad.2015.11.014

[2] Kurniawan ES, Ratep N, Bagian WW, Smf /, Fakultas P, Universitas K, et al. Factors Lead To Depresion During Antenatal Care Every Trimester of Pregnant Mother. E-Jurnal Med Udayana [Internet]. 2013;(perinatal Depressive):1–13. Available from: http://ojs.unud.ac.id/index.php/eum/article/viewFile/4936/3726#page=3&zoom=auto,-12,508

[3] Forsell E, Bendix M, Holländare F, Szymanska von Schultz B, Nasiell J, Blomdahl-Wetterholm M, et al. Internet delivered cognitive behavior therapy for antenatal depression: A randomised controlled trial. J Affect Disord. 2017;221(June):56–64.

[4] WHO. Depression and Other Common Mental Disorders Global Health Estimates [Internet]. Geneva, Swiss: World Health Organization; 2017. 8 p. Available from: https://apps.who.int/iris/bitstream/handle/10665/254610/WHO-MSD-MER-2017.2-eng.pdf

[5] Pereira AT, Bos SC, Marques M, Maia BR, Soares MJ, Valente J, et al. The postpartum depression screening scale: Is it valid to screen for antenatal depression? Arch Womens Ment Health. 2011;14(3):227–38.

[6] Lim HA, Chua T-E, Malhotra R, Allen JC, Teo I, Chern BSM, et al. Identifying trajectories of antenatal depression in women and their associations with gestational age and neonatal anthropometry: A prospective cohort study. Gen Hosp Psychiatry [Internet]. 2019; Available from: https://doi.org/10.1016/j.genhosppsych.2019.09.001

[7] Loughnan SA, Sie A, Hobbs MJ, Joubert AE, Smith J, Haskelberg H, et al. A randomized controlled trial of 'MUMentum Pregnancy': Internet-delivered cognitive behavioral therapy program for antenatal anxiety and depression. J Affect Disord [Internet]. 2019;243(May 2018):381–90. Available from: https://doi.org/10.1016/j.jad.2018.09.057

[8] Idaiani S, Basuki B. Postpartum depression in Indonesia women: a national study. Heal Sci J Indones. 2012;3(1 Jun):3–8.

[9] Annerangi W. Prevalensi dan Determinan Ansietas dan Depresi Antenatal di Puskesmas Kecamatan Pasar Minggu Tahun 2013. Universitas Indonesia; 2013.

[10] Rachmawati U. Gambaran tingkat kecemasan dan depresi menjelang persalinan pada ibu hamil trimester tiga. Universitas Indonesia; 2013.

[11] Rahmawati D. Hubungan Kecemasan dan Depresi dengan Kualitas Tidur Ibu Hamil Trimester Tiga. Universitas Indonesia; 2016.

[12] Stark EL, Shim J, Ross CM, Miller ES. The Association between Positive Antenatal Depression Screening and Breastfeeding Initiation and Continuation. Am J Perinatol. 2019;1(212).

[13] Linda Yani A. Layanan Berbasis Android Untuk Melakukan Deteksi Dini Depresi Pada Remaja. 2018; Available from: http://eprints.unipdu.ac.id/1085/1/jurnal android depresi.pdf

[14] Andriyani A, Widigdo J. Model Layanan Screening Dalam Konseling Kesehatan Mental Berbasis Android (E-Couns) Untuk Mendukung Gerakan Revolusi Mental. Pros Semin Bimbing dan Konseling [Internet]. 2017;1(1):96–103. Available from: http://pasca.um.ac.id/conferences/index.php/snbk/article/view/211

[15] Excellent NI for HC. Antenatal and postnatal mental health : clinical management and service guidance. In 2014. Available from: https://www.nice.org.uk/guidance/cg192

IOP Conf. Series: Materials Science and Engineering **874** (2020) 012015 doi:10.1088/1757-899X/874/1/012015

Design and Development of Parking Motor Parking Information System at Muhammadiyah University, Sidoarjo

Arief Wisaksono[1], Catur Aswar Ragil[2]
Electrical Engineering Study Program, Muhammadiyah University, Sidoarjo

[1]Ariefwisaksono@umsida.ac.id, [2]Catur.aswar@gmail.com

Abstract. The problem that often occurs in high-rise motorbike parking lots is the lack of information on the availability of motorbike parking lots. Currently at Muhammadiyah University in Sidoarjo there is no information system in the motorbike parking lot so users are often forced to surround parking lots to find empty parking lots, in the face of problems this requires a good information system so that motorbike parking users can find out information on the availability of parking spaces. Through this research entitled the design of a motorbike parking information system in a multi-storey building, it is expected to make it easier for users to park at Muhammadiyah University in finding a parking space. This research is done by making a design simulation by utilizing Arduino module components, Ethernet shield modules & infrared obstacle sensors, the working principle of the system is the infrared obstacle sensor installed at the entrance and exit on each floor whose function is to provide data to the Arduino then Arduino processes the sensor by sending data to the database, the database is used to store data and sensor values in the database that will be displayed in the Web browser which will be displayed on the monitor installed on the ground floor to notify the driver of the empty parking slot so as to produce efficiency BBM and time efficiency.

1. Introduction

The parking lot is one of the vehicle stop facilities that must be available in a workplace, lecture center, mall, and others that have many vehicle users. There are also various types of parking lots, one of which is the parking lot in a multi-storey building, the parking area, especially motorcycle parking in this building, often makes it difficult for motorists to find an empty parking space so that it forces motorists to surround the parking lot to find a parking space which is still empty and this often makes the driver disappointed.[1][2]. Problems that occur when searching for parking slots can occur anywhere, one of them is at Muhammadiyah Sidoarjo University, motorcycle parking area in the form of a five-storey building is very difficult for motorcyclists when searching for parking slots because there is no definitive information system, using only a crossbar notification with one security to steer manually, this is very ineffective for users of motorcycle vehicles.[3] The globalization, technology is developing very fast along with the enormous human needs and the internet is a solution of current technological developments, all work can be overcome with internet technology and its supporting devices, internet of things is technology that uses the internet as a means to do something, the internet of things works by transferring data with networks without requiring interaction from human to human or from human to computer.[4] In its application IOT aims to facilitate human activities such as booking tickets online, Live Streaming, GPS Tracking, Remote sensors, Health Monitoring and so on by utilizing the internet. The communication data from IoT and API modules will be stored in the cloud server, which will be visualized by the application server to be used by the user who is the owner of each machine. So that users can monitor the activities with web applications.[5]

IOP Conf. Series: Materials Science and Engineering **874** (2020) 012015 doi:10.1088/1757-899X/874/1/012015

The above background, the writer has an idea by combining information systems on motorcycle parking availability with the internet, with the title "Design of building information systems for parking on multi-storey buildings in Muhammadiyah Sidoarjo University based on internet of things", with this title it is expected that all citizens of Muhammadiyah Sidoarjo University get information motorcycle parking space, before entering the building and it is hoped that motorists will no longer be confused in finding empty parking slots so as to produce fuel efficiency and time efficiency.[6][7]

2. Materials And Methods

2.1 Research Tools and Materials

The research tools and materials used to design the building of a motorbike parking information system based on internet of things .

The tools used in making this research are as follows:

1. Solder
2. Avo meter
3. Screwdriver +.
4. Peeled Pliers.

The materials used in making this thesis are as follows:

1. Arduino Mega 2560
2. Web Server.
3. Infrared sensors
4. Ethernet Shield Module
5. 0.75 cable.
6. Acrylic thick 2mm: 30 x 30 x 100 cm

2.2 System Analysis

2.2.1 Diagram Blok

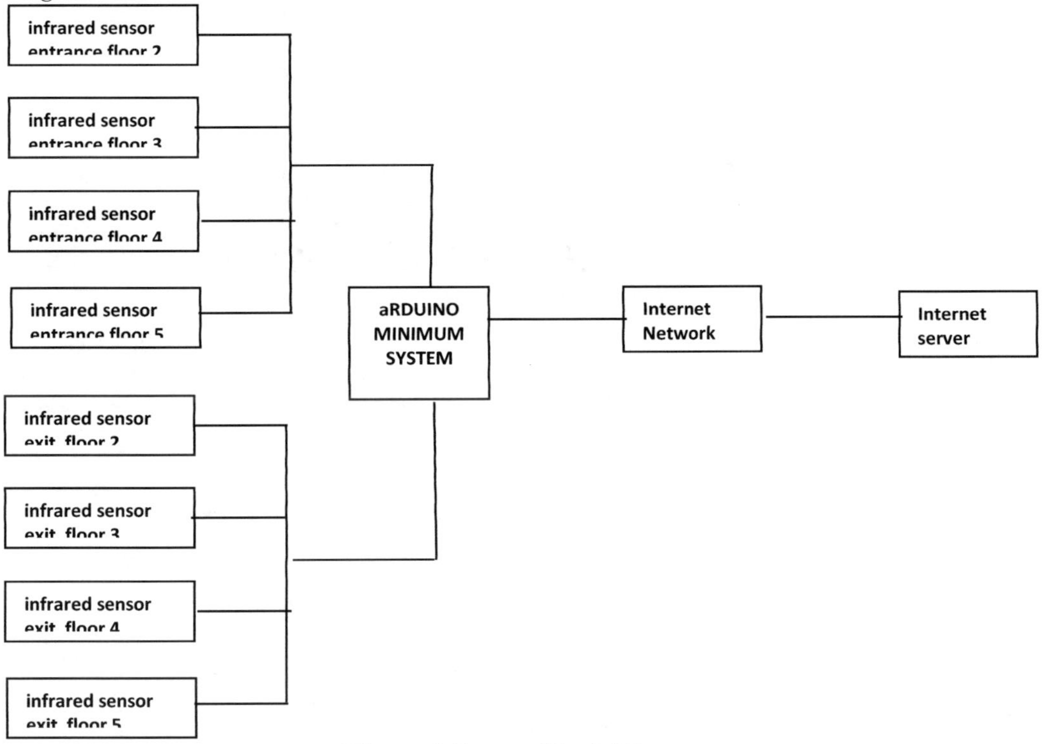

Figure 1 System Block Diagram

IOP Conf. Series: Materials Science and Engineering **874** (2020) 012015 doi:10.1088/1757-899X/874/1/012015

Based on the block diagram above, it is explained that the infrared sensor on each floor detects incoming and outgoing motorbikes. If an infrared sensor that is in the entry and exit stairs is detected, the sensor will send analog data that will be processed by the arduino minimum system with an up-down system counter, the results of data processing by the arduino minimum system will output the remaining parking slots on the web that will be displayed with an LCD monitor with arduino communication with the Ethernet shield. The design and manufacture of Internet Based Building Information Parking System of Things Based on the University of Muhammadiyah Sidoarjo includes software and hardware, including:

1. Arduino As a control or control center of all infrared sensors and outputs on analog data output.
2. Infrared sensor As a sensor to detect motorcycle vehicles.
3. Web As a process to display parking slot information from a data base.
4. LCD monitor as a display for parking slot information.
5. Data Base As data storage.

3. Results And Discussion

The results of this Final Project were applied in a miniature form using acrylic material with a width of 30cm x10cm x2 mm with a pole made of MS long drat 8 mm.

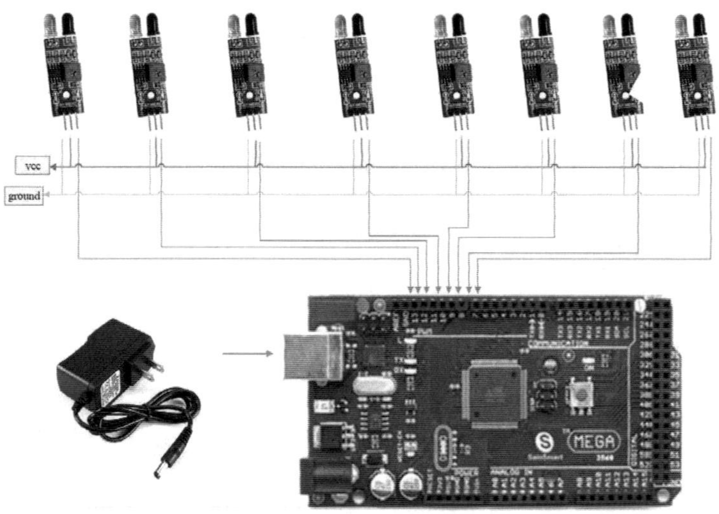

Figure 2. circuit diagram

And to find out the system is running well then testing is done in every part of the system, Testing every part of the system is carried out in order to know the principles and how the tool works that are in accordance with the plan or not. Retrieval of test data is carried out from the system every part. The tests are as follows:

1. Infrared Sensor Testing In

Table Infrared Sensor testing

trial-	Infrared Sensor In floor 2		Infrared Sensor In floor 3		Infrared Sensor In floor 4		Infrared Sensor In floor 5		Voltage Average		Standart Deviasi voltage		Deviasi	% accuracy
	measurement (V)	V (In)	measurement (V)	V (In)	measurement (V)	V (In)	measurement (V)	V (In)	measurement (V)	V (In)	measurement (V)	V (In)		
1	4,04	4,11	4,04	4,11	3,85	3,53	3,84	3,94	3,924	3,98	0,0118	0,0258	0,056	99,49
2	4,03	4,11	4,04	4,11	3,85	3,53	3,84	3,94	3,921	3,968	0,015	0,015	0,046	100
3	4,03	4,11	4,04	4,11	3,85	3,53	3,84	3,94	3,921	3,968	0,015	0,015	0,046	100
4	4,03	4,11	4,04	4,11	3,85	3,53	3,84	3,94	3,921	3,968	0,015	0,015	0,046	100
5	4,03	4,11	4,04	4,11	3,85	3,53	3,84	3,94	3,921	3,968	0,015	0,015	0,046	100

average =	3,922	3,97	0,0137	0,01736	0,04825	99

2. Infrared Sensor Testing out

Table Infrared Sensor testing

tri al-	Infrared Sensor Out floor 2		Infrared Sensor Out floor 3		Infrared Sensor Out floor 4		Infrared Sensor Out floor 5		Voltage Average		Standart Deviasi voltage		Devi asi	% accur acy
	measure ment (V)	V (In)	measu rement (V)	V (In)	measu rement (V)	V (In)	meas urem ent (V)	V (In)	measur ement (V)	V (In)	meas urem ent (V)	V (In)		
1	4,08	4,11	3,89	3,9	3,84	3,84	3,81	3,9	3,924	3,98	0,0118	0,0258	0,056	99,49
2	4,08	4,11	3,89	3,9	3,84	3,84	3,81	3,8	3,921	3,968	0,015	0,015	0,046	100
3	4,08	4,11	3,89	3,9	3,84	3,84	3,81	3,8	3,921	3,968	0,015	0,015	0,046	100
4	4,08	4,11	3,89	3,9	3,84	3,84	3,81	3,8	3,921	3,968	0,015	0,015	0,046	100
5	4,08	4,11	3,89	3,9	3,84	3,84	3,81	3,8	3,921	3,968	0,015	0,015	0,046	100
average =									3,922	3,97	0,0137	0,01736	0,04825	99

Based on the Infrared Sensor testing table, using calculations as much as five times the data collection. And also the sensor input voltage to compare the sensor reading results whether the sensor output voltage is the same as the sensor input voltage or not. Where is explained in the table written below the Infrared sensor reading that uses the Infrared Obstacle sensor with the difference made measuring the input voltage has a pretty good reading accuracy that is equal to 99.94% to 99.26% accuracy of the reading. With a deviation of 0.04 v this explains that the reading level of the reading is not too different from the input voltage. Where in the table above is the calculation formula as below.

Average calculation formula: $\mu = \frac{x_1+x_2+x_3...+x_n}{n}$

Standard deviation calculation formula: $\sigma = \sqrt{\frac{\sum_{i=1}^{n} (x_i - \mu)}{n-1}}$

Deviasi formula:

$$Deviasi = |\ \mu^{Sensor\ Suhu} - \mu^{Alat\ Standar}\ |$$

Proportion Percentage Formula: $\%\ Ketepatan = |1 - \left|\frac{Yn-Xn}{Xn}\right||\times 100\ \%$

Information :	x_1 = Data 00-1
μ = Average Value = Standard Deviation	x_n = Data ke-n
n = Amount of data	x_i = Data ke-i
Yn = Measurement results with a Volt meter	
Xn = Value read on the sensor output	

Calculation of Average Infrared Sensor Measurement

$$\mu = \frac{x_1+x_2+x_3+x_4+x_5}{n}$$

$$= \frac{4{,}04+4{,}04+3{,}85+3{,}84+4{,}08+3{,}89+3{,}84+3{,}81}{5}$$

$$= 3{,}924$$

Calculation of Average V (in)

$$\mu = \frac{x_1+x_2+x_3+x_4+x_5}{n}$$

$$= \frac{4{,}11+4{,}11+3{,}93+3{,}93+4{,}11+3{,}9+3{,}84+3{,}8}{5}$$

$$= 3{,}968$$

IOP Conf. Series: Materials Science and Engineering **874** (2020) 012015 doi:10.1088/1757-899X/874/1/012015

Calculation of Standard Deviation of Infrared Sensor measurements

$$\sigma = \sqrt{\frac{\sum_{i=1}^{n} (x_i - \mu)^2}{n-1}}$$

$= 0{,}011$

Calculation of Standard Deviation on V (in)

$$\sigma = \sqrt{\frac{\sum_{i=1}^{n} (4{,}11-3{,}968)^2+(4{,}11-3{,}968)^2+(3{,}93-3{,}968)^2+(3{,}93-3{,}968)^2+(4{,}11-3{,}968)^2+(3{,}9-3{,}968)^2 +(3{,}84-3{,}968)^2+(3{,}81-3{,}968)^2}{5-1}}$$

$$\sigma = \sqrt{\frac{0{,}02+0{,}02+0+0+0+0{,}02+0+0}{4}}$$

$$\sigma = \sqrt{\frac{0{,}06}{4}}$$

$\sigma = 0{,}015$

a) Calculation of Deviation

$$Deviasi = |\ \mu^{V\ input} - \mu^{pengukuran}\ |$$
$$Deviasi = |0{,}015 - 0{,}010|$$
$$Deviasi = 0{,}05$$

b. Calculation of Percentage of Accuracy

$$\%\ \text{accuracy} = \left|1 - \left|\frac{Yn-Xn}{Xn}\right|\right| \times 100\%$$
$$\%\ \text{accuracy} = |1 - \left|\frac{3{,}968 - 3{,}964}{3{,}964}\right|| \times 100\ \%$$
$$\%\ \text{accuracy} = |1 - 0{,}00100908| \times 100\ \% = 99{,}89$$

1. Infrared Sensor Linearity Testing of distances

Tabel Infrared Sensor Linearity Testing of distances

No.	Testing	distances (cm)	With Object	Without Object	Vin
1	Infrared Sensor In floor 2	1 cm	0	4,04	4,11
		3 cm	0	4,03	4,11
		4 cm	0	4,04	4,11
		5 cm	0	4,03	4,11
		6 cm	4,04	4,04	4.11
2	Infrared Sensor In floor 3	1 cm	0	4,04	4,11

No.	Testing	distances (cm)	With Object	Without Object	Vin
		3 cm	0	4,04	4,11
		4 cm	0	4,05	4,11
		5 cm	0	4,05	4,11
		6 cm	4,05	4,05	4.11
3	Infrared Sensor In floor 4	1 cm	0	3,85	3,93
		3 cm	0	3,85	3,93
		4 cm	0	3,86	3,93
		5 cm	0	3.85	3,93
		6 cm	3,85	3,85	3,93
4	Infrared Sensor In floor 5	1 cm	0	3,84	3,93
		3 cm	0	3,84	3,93
		4 cm	0	3,84	3,93
		5 cm	0	3.84	3,93
		6 cm	3,84	3,84	3,93
5	Infrared Sensor Out floor 2	1 cm	0	4,03	4,08
		3 cm	0	4,03	4,08
		4 cm	0	4,03	4,08
		5 cm	0	4,03	4,08
		6 cm	4,03	4,04	4.08
6	Infrared Sensor Out floor 3	1 cm	0	3,89	3,90
		3 cm	0	3,89	3,90
		4 cm	0	3,89	3,90
		5 cm	0	3.89	3,90
		6 cm	3,89	3,89	3,90
7	Infrared Sensor Out floor 4	1 cm	0	3,83	3,84
		3 cm	0	3,83	3,84
		4 cm	0	3,83	3,84
		5 cm	0	3.83	3,84
		6 cm	3,83	3,83	3,84
8	Infrared Sensor Out floor 5	1 cm	0	3,81	3,80
		3 cm	0	3,81	3,80
		4 cm	0	3,81	3,80
		5 cm	0	3.81	3,80
		6 cm	3,81	3,81	3,80

Note: From the results of the linear sensor infrared test against distance can be concluded that the sensor is linear between a distance of 5 cm - 6 cm, if there are objects in the range of 1 cm - 4 cm, the output voltage becomes 0 v and in the range of 5 cm - 6 cm the output voltage is linear

2. Ethernet Shield Module Connectivity Testing

	Connectivity Testing Modul Ethernet Shield dengan Device (Local Host)			
	Device (Lenovo B490)	Device (Asus X453 M)	Device (Compact CQ43)	Device (Lenovo B520)
Status Connectiing	Connectiing	Connectiing	Connectiing	Connectiing
Ip Address	192.168.137.41	192.168.137.67	192.168.137.130	192.168.137.99

IOP Conf. Series: Materials Science and Engineering **874** (2020) 012015 doi:10.1088/1757-899X/874/1/012015

Note: From the test results it can be concluded that the ethernet module can be used to connect arduino to the internet through different devices but the IP address must be reset because the ethernet shield module that is connected to arduino itself does not have a fixed IP address.

Testing Website Display

Tabel Testing Serial Monitor at Website

No.	Serial Monitor	Website Display	Information
1.	Connected Parking Lot parking available 2 = 0 Parking Lot parking available 3= 6 Parking Lot parking available 2 = 0 Parking Lot parking available 2 = 6 (Real Time)	Parking Lot parking available 2 = 0 Parking Lot parking available 3= 6 Parking Lot parking available 2 = 0 Parking Lot parking available 2 = 6 (Delay 3-5 detik)	There is a time difference between the monitoring results on the website and the serial monitor
2.	Connected Parking Lot parking available 2 = 2 Parking Lot parking available 3= 4 Parking Lot parking available 2 = 1 Parking Lot parking available 2 = 6 (Real Time)	Parking Lot parking available 2 = 2 Parking Lot parking available 3= 4 Parking Lot parking available 2 = 1 Parking Lot parking available 2 = 6 (Delay 3-5 detik)	There is a time difference between the monitoring results on the website and the serial monitor
3.	Connected Parking Lot parking available 2 = 0 Parking Lot parking available 3= 6 Parking Lot parking available 2 = 0 Parking Lot parking available 2 = 6 (Real Time)	Parking Lot parking available 2 = 0 Parking Lot parking available 3= 6 Parking Lot parking available 2 = 0 Parking Lot parking available 2 = 6 (Delay 3-5 detik)	There is a time difference between the monitoring results on the website and the serial monitor
4.	Connected Parking Lot parking available 2 = 7 Parking Lot parking available 3= 7 Parking Lot parking available 2 = 7 Parking Lot parking available 2 = 7 (Real Time)	Parking Lot parking available 2 = 7 Parking Lot parking available 3= 7 Parking Lot parking available 2 = 7 Parking Lot parking available Parking Lot parking available 2 = 7 (Delay 3-5 detik)	There is a time difference between the monitoring results on the website and the serial monitor

After testing and retrieval of data for several times, it can be concluded that the Design of Building Information Systems for Multi-storey Building Motorcycle Parking at IOT-based Muhammadiyah University can be concluded as follows:

1. Infrared sensor reading to detect motorcycle, with a comparison between input and measurement voltages with a volt meter at the output voltage. The accuracy of the sensor readings in two data retrieval successfully reads a range of 99.94%. So the sensor works well.

2. The Arduino Mega 2560 module with Ethernet Shield is used as the sender of data to a database that will be accessed by a web browser, where this module will be coded using the Arduino IDE software so that it can be used. In this experiment a smooth internet network is needed so that the process of transferring data to a database is expected to be faster.

References

[1] Z. A. A. Maulana, "Simulasi Sistem Informasi Tempat Parkir Berbasis WEB," *Tek. Elektron.*, pp. 1–132, 2018.
[2] R. A. . Putra, "Sistem Informasi Ketersediaan Slot Parkir menggunakan Arduino Uno," p. 19, 2017.
[3] A. Ahfas, D. Hadidjaja, S. Syahrorini, and A. Wicaksono, "Sound indicators as safety of motorcycle," *J. Phys. Conf. Ser.*, vol. 1402, no. 4, 2019.
[4] A. Wisaksono, A. Farihah, and N. Ariyanti, "Aplikasi Tematis Al-Qur'an Berbasis Web," *AKSIOLOGIYA J. Pengabdi. Kpd. Masy.*, vol. 2, no. 2, p. 139, 2018.
[5] R. Dijaya, E. A. Suprayitno, and A. Wicaksono, "Integrated Point of Sales and Snack Vending Machine based on Internet of Things for Self Service Scale Micro Enterprises," *J. Phys. Conf. Ser.*, vol. 1179, no. 1, 2019.
[6] A. Imron and J. Jamaaluddin, "Rancang Bangun Sistem Informasi Parkir Mobil Otomatis Pada Gedung Bertingkat Berbasis Arduino Mega 2560," *J. Tek. Elektro dan Komput. TRIAC*, vol. 4, no. 2, 2017.
[7] A. B. Warsito, M. Yusup, and M. Aspuri, "Penerapan Sistem Monitoring Parkir Kendaraan Berbasis Android Pada Perguruan Tinggi Raharja," *Technomedia J.*, vol. 2, no. 1, pp. 82–94, 2017.

IOP Conf. Series: Materials Science and Engineering **874** (2020) 012016 doi:10.1088/1757-899X/874/1/012016

Classification Using C4.5 Algorithm in Election Participation Prediction

Arif Senja Fitrani[1], Mochamad Alfan Rosid[2], Cindy Taurusta[3], Indah Fauzia[4]
Universitas Muhammadiyah Sidoarjo, Sidoarjo, Indonesia

asfjim@umsida.ac.id[1], alfanrosid@umsida.ac.id[2], cindytaurusta@umsida.ac.id[3], Indahfauzia09@umsida.ac.idi[4]

Abstract. The General Election of the President, DPD, DPR, Provincial DPRD and Regency / City DPRD in 2019, in Indonesia, is carried out by an independent institution namely the General Election Commission (KPU), where there is a measure of the success of the holding by holding the principle of direct, general, free of secrecy. Another component of the election will be the implementation of contestants and voters. In the electoral factor, this is also a measure for success in the overall implementation process, which is a success if the participation of the community is high in the holding of elections. And conversely, if the community participation is low, one of them is the level of public trust in the organizer (government) decreases. Analysis of classification mining data with C4.5 algorithm places the prediction class "Present" and "Not Present" on the permanent voter list (DPT) taken from form C7 (voter attendance list form). C7 data was obtained from polling stations (TPS) located in the village of Wonokasian, Wonoayu District, Sidoarjo Regency, East Java. The number of datasets is 4249 instances, and the number of attributes is 10. With split datasets, for training data are 2965 instances, and testing data are 1284 instances. From testing, the testing data obtained the results of true or appropriate accuracy of 1151 instances (90%) and obtained incorrect or incompatible data as many as 134 instances (10%). This shows a good value for the results of 90% of the corresponding data in the prediction class in the Presidential Election, DPD, DPR, Provincial DPRD, and Regency / City DPRD in 2019.

Keywords: Data Mining, Classification, Algorithm Classification, Election

1. Introduction

General Election, abbreviated as PEMILU, is a people's party in determining the representation of both legislative, senator (DPD) and executive (president, governor, and district/city). The implementation of the ELECTION of the President, DPD, DPR, Provincial DPRD, and Regency / City DPRD is carried out in 2019. ELECTION is an important political event to determine leaders in a democratic country. The Permanent Voters List (DPT), which is then determined to be a DP4, is the right of citizens to be protected by law to choose their representation in the executive, senator and excesses. Residents who have been recorded to channel their aspirations to the General Election are conducted at the polling station (TPS). Requirements for citizens to have the right to vote are regulated by law. The voter segment group is divided into three namely beginner voters, millennial voters and post menial voters. Where in the previous research the focus was on the scope of results, namely, predictions of legislative elections (C4.5 algorithm to predict the results of legislative selection of DKI

Jakarta DPRD, Techno Nusa Mandiri Vol. IX No.1, March 2013). Here, we try to take the other side, which is related to voter participation in the democratic party process that takes place.

The involvement of the community for participation in the democratic process is determined by the presence of the community in the booth (TPS), where residents come to be able to vote. The absence of the public in the ELECTION process at polling stations was termed golput (white group) or indeed was not present at the time of the democratic party that was held. Many factors determine the absence of the public at the democratic party, among others, based on the age level that is identical to the millennial voters (ages 17 to 32), beginner voters (aged 17 to 20) who have no experience in a democratic party being held, apathy will representations that have been legally registered and other factors that generally allow the age level to not be able to participate in the democratic party this time.

The study took a sample of data in Wonokasian Village, Wonoayu District, Sidoarjo Regency, East Java Province at the 2019 ELECTION simultaneously in Indonesia. Classification method (Naïve Bayes algorithm, C4.5, and kNN), attribute analysis on attendance data and voter data is the focus of this research so that it can be known whether the determination of the effect of the prediction class is produced. Because there may be many attributes, but they have no influence on the prediction class (results) at all. Many attributes are irrelevant and will place unnecessary computational overhead on the data mining algorithm. At worst, it can cause the algorithm to be a poor outcome [23].

Classification methods in several algorithms are used in the implementation of research for quality attribute analysis to be tested on several different attributes. The hope is, can it provide a significant percentage change with the same number of instances.

2. Classification

Classification method that is studying a set of data that can be produced by a new classification of data. The classification process in data mining techniques is a set of data that can produce a classification model (target function). So, it requires a dataset on the set for the classification process. The dataset used is attributes and features using training data and testing data. Classification techniques can be grouped into two categories. Namely, the classification technique globally calculates all training data and classification locally taking into account some training data [6].

3. C4.5 algorithm

At the learning stage of the data, the C4.5 algorithm constructs decision trees from training data, in the form of cases or records (tuples) in the database. The three working principles of the C4.5 algorithm at the learning stage of the data are:

1. Making a decision tree.
2. The decision tree pruning and evaluation (optional).
3. Making rules from decision trees (optional).

The formula for the C4.5 algorithm is:

$$Gain(S,A) = Entropy(S)\sum_{i=1}^{n}\frac{Si}{S} * Entropy(Si) \quad (1)$$

Where:
S: set of cases
A: attribute
n: number of partition attributes A
Si: number of cases on the i-partition
S: number of cases in S

$$Entropy(S)\sum_{i=1}^{n}\frac{Si}{S} * pi * log2\, pi \quad (2)$$

Where :
S: set of cases
A: feature
n: number of S partitions
pi: proportion of Si to S

4. Methods

Diagram for determining how to select characteristics by determining attendance class (present and absent) in the election process. And determine the number of attributes for the best value that results in a comparison of the three conditions for the number of attributes (figure 1).:

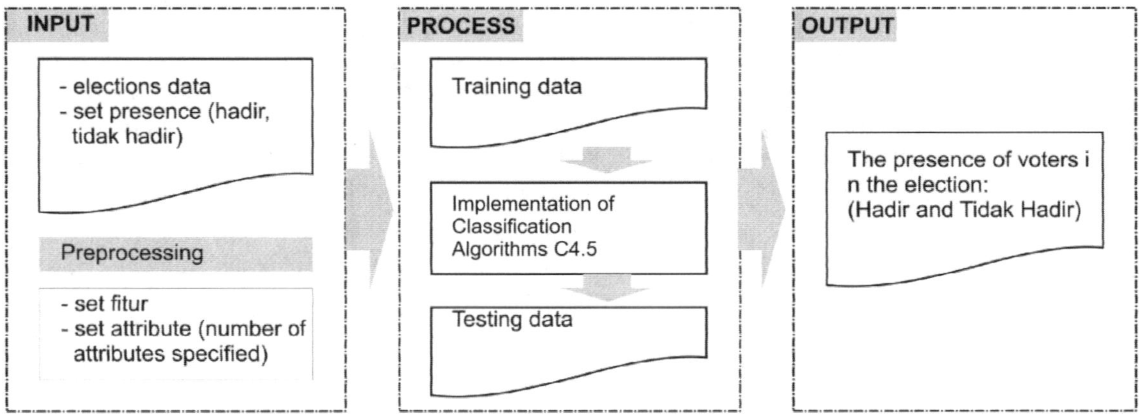

Figure 1. classification diagram.

For the 'train and test' method, the available data is split into two parts called a training set and a test set (Figure 2). First, the training set is used to construct a classifier (decision tree, neural net, etc.). The classifier is then used to predict the classification for the instances in the test set. If the test set contains N instances of which C are correctly classified the predictive accuracy of the classifier for the test set is p = C/N. This can be used as an estimate of its performance on any unseen dataset. In cases where the dataset is only a single file, we need to divide it into a training set and a test set before using Method 1. This may be done in many ways, but a random division into two parts in proportions such as 1:1, 2:1, 70:30 or 60:40 would be customary [23].

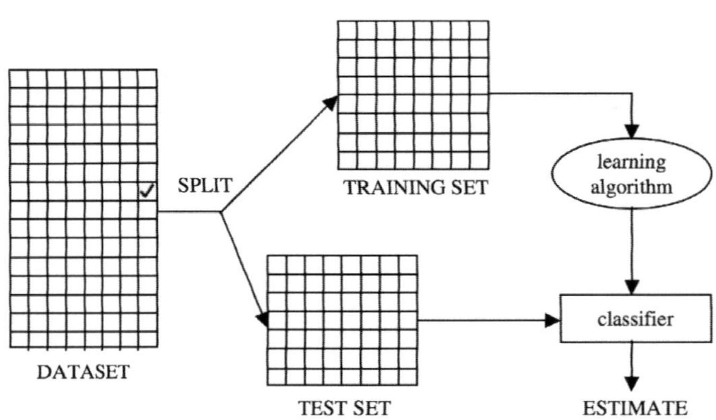

Figure 2. split train and test.

5. Data Processing

Data warehouse originating from 2 data sources, namely C7 data (attendance list) and DP4 data (DPT data). From 2 references, a merger process is conducted (C7 data and DP4 data) based on the voter ID, to determine the presence of voters at the polling station. And changes in some attributes to produce new information, to produce better predictive results. Attributes from the results of the changes are the class, origin, category, and status of the KK. For the addition of a new attribute the location_TPS is raised based on the address (rt) of the voter and the name of the tips, so that it can know whether the location of the polling station is in the address "rt" or outside, this determines the distance from the voter to the polling station location. The data warehouse has been set as many as 11 attributes, and the dataset is 4249 instances, then divided into two parts for further processing. Data train in the capacity of 70% (2973 instance) and test data of 30 (1276 instances). Selecting data that is dismissed information to produce the best value in the testing process.

5.1 Data Collection

The dataset is obtained from the final voter list (DPT) in Wonokasian village, Wonoayu sub-district, Sidoarjo, East Java. DPT is used in the General Elections of the President, DPD, DPR, Provincial DPRD and Regency / City DPRD in 2019. The set datasets are 4249 instances, and 10 attributes are taken from form C7 (voter attendance list). From these datasets scattered in 15 polling stations (TPS), where polling stations are the polling implementation unit table 1.

Table 1. Dataset Description.

Dataset	No. of attribute	No. of instance
Form C7 at 15 polling stations, in the election of the President, DPD, DPR, Provincial DPRD and Regency / City DPRD in Wonokasian village, Wonoayu District, Sidoarjo Regency	10	4249

Form C7 data is sourced from the List of Electoral Potential Population (DP4), where the data summarizes the three conditions, namely probability data, geographic data, and attendance data. Personal and geographical data are found in DP4 data and attendance data in form C7 data. The description of processing 2 data sources (form C7 and DP4) is table 1:

Table 2. Attribute and Future dataset.

No	Attribute	Variable	Information
1	jmlKK	{dua,>dua,singgle}	Number of family members
2	asal	{SIDOARJO,LUAR,'LUAR PROV'}	Origin of voters
3	kategori	{Milenial,PascaMilenial,Pemula}	Type of society
4	kawin	{S,B,P}	Marriage Status (S = Already, B = Not yet and P = Ever)
5	jk	{P,L}	Gender (P = Female and L = Male)
6	alamat	{WONOKASIAN,LENGKONG,NGGODEK,NGEMPLAK,KRAMAT,NDOKOH,KRESAN,KLITIH}	Address Village and or hamlet in one village
7	rt	{rt01,rt02,rt03,rt04,rt05,rt06,rt07,rt08,rt09,rt10,rt11,rt12,rt13,rt14,rt16,rt15,rt17,rt19,rt18,rt20,rt21}	Address RT (Neighborhood) in one village
8	tps	{TPS-1,TPS-2,TPS-3,TPS-4,TPS-5,TPS-6,TPS-7,TPS-8,TPS-9,TPS-10,TPS-11,TPS-12,TPS-13,TPS-14,TPS-15,TPS-16}	Name of TPS (polling station) in one village
9	lokasi_TPS	{dalam,luar}	Distance between polling station locations and voter address
10	statusKK	{TSH,HS}	Attendance at polling stations in one family
11	hadir	{TH,H}	Individual attendance at TPS (class prediction)

5.2 Dataset

At the preprocessing stage, a dataset is determined, which states the data is in normal condition. To prepare for the next process, split two data conditions, namely, training data and testing data with prediction classes present and not present, Table 5.

Table 3. dataset, 10 attribute

No	Data election
1	dua,SIDOARJO,Milenial,S,P,WONOKASIAN,rt01,TPS-1,dalam,TSH,TH
2	dua,SIDOARJO,Milenial,S,L,WONOKASIAN,rt01,TPS-1,dalam,TSH,H
3	>dua,LUAR,Milenial,B,L,WONOKASIAN,rt01,TPS-1,dalam,TSH,TH
4	singgle,SIDOARJO,Milenial,S,L,WONOKASIAN,rt01,TPS-1,dalam,HS,H
…	…

5.3 Ratting Attribute

To see the best weight of 10 attributes, it can be done by calculating the gain ratio, so that the best and worst attribute sequence can be known, namely:

Table 4. Ranked attributes:

Rangking	Gain Ratio	No	Attribute
1	0.390750	10	statusKK
2	0.054964	8	tps
3	0.036399	7	rt
4	0.018774	6	alamat
5	0.013883	9	lokasiTPS
6	0.004621	1	jmlKK
7	0.004580	4	kawin
8	0.001704	2	asal
9	0.001436	3	kategori
10	0.000126	5	jk

5.4 Training Data

The dynamic condition characteristics of the dataset for the prediction class are present and not present, specifying the instance conditions for each attribute and variable. In figure 3, it also shows visually the number of each variable, where each instance in each variable shows a dynamic condition with 70% training data (2973 instances) out of a total of 4249 data sets. The condition in the prediction class for features is present in 614 instances (20.65%), and 2359 instances (79.35%) are absent.

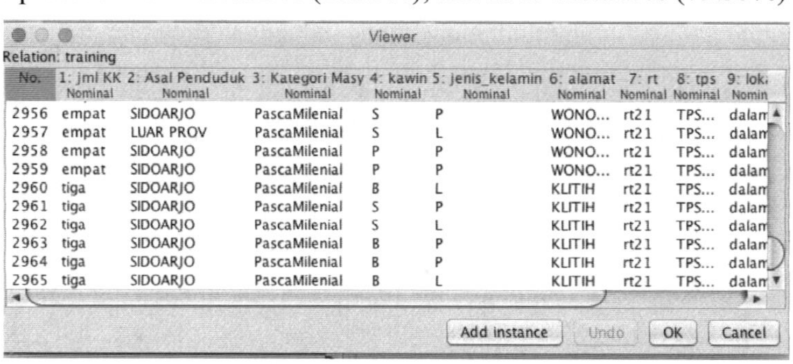

Figure 3. Training data.

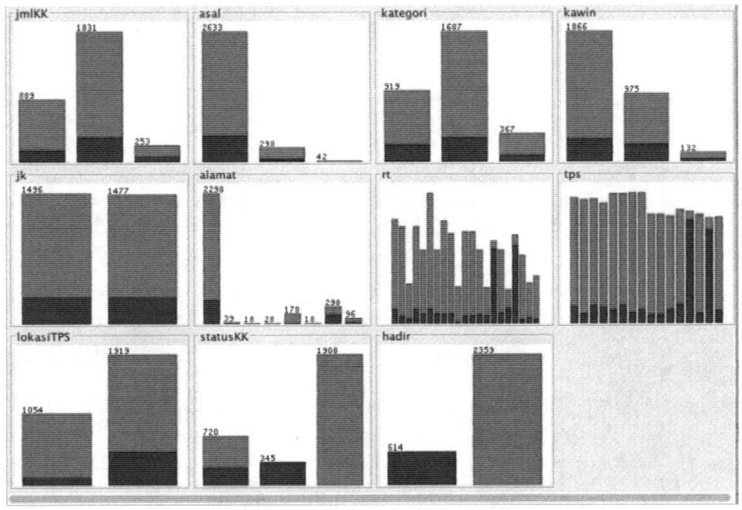

Figure 4. Visual all attribute.

Table 5. Classifier model (full training set)

No	Rule	No	Rule
1	hadir kel = THS: TH (329.0)	17	| | | | jenis_kelamin = P: TH (2.0)
2	hadir kel = HS: H (1906.0)	18	| | | tps = TPS-6: H (3.0/1.0)
3	hadir kel = TSH	19	| | | tps = TPS-7: H (3.0/1.0)
4	| kawin = S: H (425.0/118.0)	20	| | | tps = TPS-8: TH (0.0)
5	| kawin = B	21	| | | tps = TPS-9: H (3.0/1.0)
6	| | jml KK = tiga: TH (80.0/28.0)	22	| | | tps = TPS-10: TH (0.0)
7	| | jml KK = dua: TH (26.0/11.0)	23	| | | tps = TPS-11: TH (6.0/2.0)
8	| | jml KK = empat	24	| | | tps = TPS-12: H (9.0/2.0)
9	| | | Kategori Masy = PascaMilenial: TH	25	| | | tps = TPS-13: TH (6.0)
10	(10.0/5.0)	26	| | | tps = TPS-14
11	| | | Kategori Masy = Milenial: TH (66.0/30.0)	27	| | | | jenis_kelamin = L: H
12	| | | Kategori Masy = Pemula: H (15.0/5.0)	28	(4.0/1.0)
13	| | jml KK = lima	29	| | | | jenis_kelamin = P: TH (2.0)
14	| | | tps = TPS-1: TH (6.0/2.0)	30	| | | tps = TPS-15: TH (0.0)
15	| | | tps = TPS-2: TH (2.0/1.0)	31	| | | tps = TPS-16: TH (0.0)
16	| | | tps = TPS-3: H (1.0)	32	| | jml KK = satu: TH (2.0)
	| | | tps = TPS-4: H (3.0/1.0)		| | jml KK = enam: H (23.0/7.0)
	| | | tps = TPS-5		| | jml KK = tujuh: TH (0.0)
	| | | | jenis_kelamin = L: H (6.0/2.0)		| kawin = P: H (27.0/12.0)

From the process for calculating training data, the rule base is obtained with Leaves 30 and Size of the tree 37, and Table 6 sets the rule base, which will lead to all the testing processes carried out for test data in addition to the results obtained for the rule base. Figure 4, visual classification tree visual, encourages to make it easier to visually understand the rule base on the results of the Algorithm C4.5.

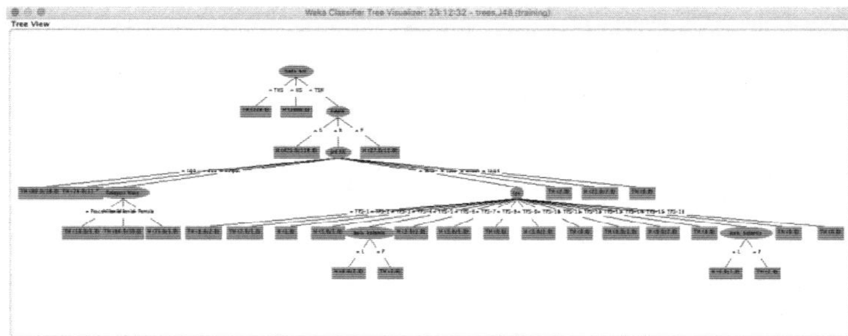

Figure 5. Tree view.

6. Testing Data Results

The results of the classification algorithm C4.5 to determine the prediction class "Hadir and Tidak hadir". The results of table 8, explain the comparison of class predictions in both the number of attributes and algorithm groups. The highest value is in 11 attributes, namely, in the Naïve Bayes algorithm with a percentage value of 89.3417% or 1140 instances. In this analysis, try to see the results in the reduction of attributes. Subtraction is done for attributes that have the first ranking "statusKK" (10 attributes), second "TPS" and third rank "rt" (9 attributes). For the highest ranking of Gini ratio table 3. For prediction class 9 attributes, where the results of the prediction class show the highest value of 3 algorithms namely C4.5 with a percentage of 99.6364% or 1131 instances, then the highest percentage prediction class for ten attributes is 10.7555 % or 1107 instances with algorithms that have the same nine attributes, namely C4.5. The best ranking is based on two groups of attributes (9 attributes and ten attributes), namely on nine attributes.

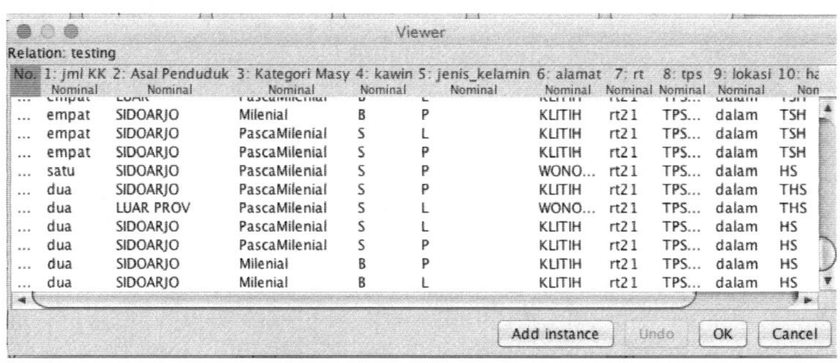

Figure 6. Testing data.

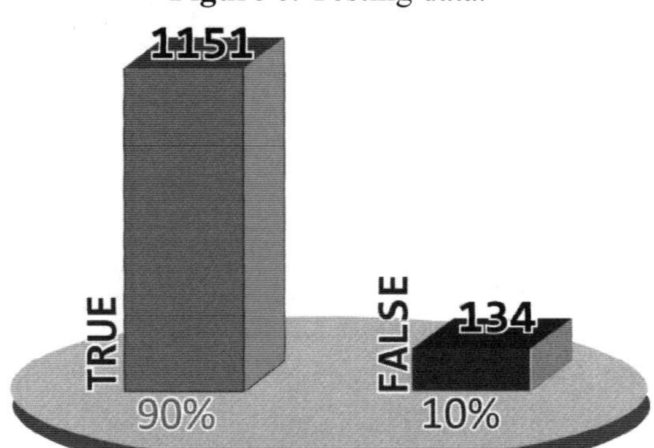

Figure 7. Graph of prediction class results

Table 6. Result Testing Data

Attribute	Status	C45	
11 attribute	True	1151	90 %
	False	134	10 %

5. Conclusion

Analysis of the three C4.5 algorithms in the test contained ten attributes and with the Present and Not Present classes of predictions. The total dataset uses 4249 instances, which are divided into 2965 instances of training data and 1284 instances of testing data. The dataset used is C7 form in the presidential election, DPD, DPR, Provincial DPRD, and Regency / City DPRD held by KPU in Wonokasian village, Wonoayu sub-district, Sidoarjo Regency, East Java. Testing the classification method on the C4.5 algorithm obtained prediction results that are as many as 1151 instances or 90% of the conditions declared correct and 134 instances or 10% of the terms said inappropriate or incorrect. Then from the results of the true percentage value, which is significant, the testing process in this dataset can prove the level of voter attendance in the election implementation and the presentation model of attributes that support or have considerable weight as well. It is hoped that the testing process can be carried out in the implementation of the upcoming elections.

Acknowledgments

We hereby thank you to Universitas Muhammadiyah Sidoarjo for supporting the publication of this research.

References

[1] M. (School of C. of P. Bramer, *Data for Data Mining. In: Principles of Data Mining*. London: Springer, 2016.
[2] J. Zhang, S. O. Williams, and H. Wang, “Intelligent computing system based on pattern recognition and data mining algorithms,” *Sustain. Comput. Informatics Syst.*, vol. 20, pp. 192–202, 2018.
[3] J. Xie and F. Burstein, “Using machine learning to support resource quality assessment: An adaptive attribute-based approach for health information portals,” *Lect. Notes Comput. Sci. (including Subser. Lect. Notes Artif. Intell. Lect. Notes Bioinformatics)*, vol. 6637 LNCS, pp. 526–537, 2011.
[4] S. S. Nikam, “A Comparative Study of Classification Techniques in Data Mining Algorithms,” *Int. J. Mod. Trends Eng. Res.*, vol. 4, no. 7, pp. 58–63, 2017.
[5] G. S. Xia *et al.*, “AID: A benchmark data set for performance evaluation of aerial scene classification,” *IEEE Trans. Geosci. Remote Sens.*, vol. 55, no. 7, pp. 3965–3981, 2017.
[6] J. Novaković, P. Strbac, and D. Bulatović, “Toward optimal feature selection using ranking methods and classification algorithms,” *Yugosl. J. Oper. Res.*, vol. 21, no. 1, pp. 119–135, 2011.
[7] M. A. Rosid, G. Gunawan, and E. Pramana, “Centroid Based Classifier With TF – IDF – ICF for Classfication of Student’s Complaint at Appliation E-Complaint in Muhammadiyah University of Sidoarjo,” vol. 1, no. 1, 2015.
[8] I. Abdallah *et al.*, “Fault diagnosis of wind turbine structures using decision tree learning algorithms with big data,” *Saf. Reliab. - Safe Soc. a Chang. World - Proc. 28th Int. Eur. Saf. Reliab. Conf. ESREL 2018*, no. December, pp. 3053–3062, 2018.
[9] Y. Yang and W. Chen, “Taiga: Performance optimization of the C4.5 decision tree construction algorithm,” *Tsinghua Sci. Technol.*, vol. 21, no. 4, pp. 415–425, 2016.
[10] G. D. Rashmi, A. Lekha, and N. Bawane, “Analysis of efficiency of classification and prediction algorithms (Naïve Bayes) for Breast Cancer dataset,” *2015 Int. Conf. Emerg. Res. Electron. Comput. Sci. Technol. ICERECT 2015*, pp. 108–113, 2016.
[11] S. L. B. Ginting, J. Adler, Y. R. Ginting, and A. H. Kurniadi, “The Development of Bank Application for Debtors Selection by Using Naïve Bayes Classifier Technique,” *IOP Conf. Ser. Mater. Sci. Eng.*, vol. 407, no. 1, 2018.
[12] K. Balasaravanan and M. Prakash, “Detection of dengue disease using artificial neural network based classification techniquetion,” *Int. J. Eng. Technol.*, vol. 7, no. 1, pp. 13–15, 2018.
[13] S. D. Jadhav and H. P. Channe, “Comparative Study of K-NN, Naive Bayes and Decision Tree Classification Techniques,” *Int. J. Sci. Res.*, vol. 5, no. 1, pp. 1842–1845, 2016.
[14] S. Pouriyeh, S. Vahid, G. Sannino, G. De Pietro, H. Arabnia, and J. Gutierrez, “A comprehensive investigation and comparison of Machine Learning Techniques in the domain of heart disease,” *Proc. - IEEE Symp. Comput. Commun.*, no. Iscc, pp. 204–207, 2017.
[15] A. P. Windarto *et al.*, “Analysis of the K-Means Algorithm on Clean Water Customers Based

on the Province," *J. Phys. Conf. Ser.*, vol. 1255, no. 1, 2019, doi: 10.1088/1742-6596/1255/1/012001.

[16] W. Katrina, H. J. Damanik, F. Parhusip, D. Hartama, A. P. Windarto, and A. Wanto, "C.45 Classification Rules Model for Determining Students Level of Understanding of the Subject," *J. Phys. Conf. Ser.*, vol. 1255, no. 012005, pp. 1–7, 2019, doi: 10.1088/1742-6596/1255/1/012005.

[17] Sudirman, A. P. Windarto, and A. Wanto, "Data mining tools | rapidminer: K-means method on clustering of rice crops by province as efforts to stabilize food crops in Indonesia," *IOP Conf. Ser. Mater. Sci. Eng.*, vol. 420, no. 1, 2018, doi: 10.1088/1757-899X/420/1/012089.

IOP Conf. Series: Materials Science and Engineering **874** (2020) 012017 doi:10.1088/1757-899X/874/1/012017

Improving Text Preprocessing For Student Complaint Document Classification Using Sastrawi

Mochamad Alfan Rosid*, **Arif Senja Fitrani, Ika Ratna Indra Astutik, Nasrudin Iqrok Mulloh, Haris Ahmad Gozali**

Program Studi Informatika, Universitas Muhammadiyah Sidoarjo, Jl. Raya Gelam 250 Candi Sidoarjo, Jawa Timur, Indonesia

*alfanrosid@umsida.ac.id

Abstract. In the text mining there are stages that must be passed namely the text preprocessing stage. Text preprocessing is the stage to do the data selection process in each document, including case folding, tokenizing, filtering, and stemming. The results of the preprocessing process can affect the accuracy of document classification. In documents Bahasa Indonesia, there are still often over-stemming and under-stemming, so improvements are needed in the stemming process. In this study, it is proposed to use sastrawi libraries to improve the results of previous studies that are still not optimal in the results of preprocessing, especially in the filtering and stemming process. From the results of the study, the sastrawi library is able to reduce over stemming and under stemming and a faster processing time compared to using a Tala stemmer.

1. Introduction

Text Mining is a method of mining unstructured data in the form of text located at sources such as documents, attachments or sentences that will later find the words of the document so that analysis can be done related to documents[1]. Text mining is usually used to retrieve important words that will be classified which will later be analyzed the relationship between documents. With text mining we can find out the information contained in a document that we did not know before.

The process of text mining is more or less the same as data mining, but there are different, data mining is designed to handle structured data while text mining can handle unstructured or semi-structured datasets such as HTML email files and text documents. Structured data is data that is in a fixed field in a record or file. This data is contained in relational databases and spreadsheets. Unstructured data usually refers to information that is not in a traditional row column database and that is the opposite of structured data. Semi-structured data is data that is not raw data, or data that is typed in a conventional database system[2].

Before carrying out operations on text documents, both for classification purposes, looking for similarities, sentiment analysis etc. then the text document must go through preprocessing[3], This stage aims to structure the text and extract features[4], in general there are four steps such as tokenization, stop-word removal and lower case conversion (case foding) and stemming[5]. The results of this preprocessing affect the level of classification accuracy and computational time required.

In a study conducted by Mochamad Alfan Rosid et al in 2015, produced a machine to classify student complaints using the centroid base classifier method and the tf-idf-icf feature with an accuracy rate of 79% [6], but in this study found weaknesses in the stemming process using Tala porters to process student complaint documents Bahasa Indonesia, there are terms that experience under stemming and over stemming, so this in my opinion greatly affects the level of accuracy. For this reason, another stemming method is needed to solve the problem, in this study we propose the use of sastrawi libraries, in addition to the stemming process we also apply to the stopword remover process.

Sastrawi library is a stemmer library that is used to overcome the problem of changing words with words into basic words[7]. Sastrawi stemmer applies an algorithm based on Nazief and Adriani, then enhanced with the CS (Confix Stripping) algorithm, the ECS Algorithm (Enhanced Confix Stripping), further enhanced with Modified ECS. It is expected that by applying this sastrawi library to the preprocessing process of student complaints documents, especially the stemming and stopword remover stages, can make significant non-important words so as to improve the accuracy of the classification of student complaints.

2. Literature Review

Porter's algorithm was first created by Martin Porter (1980), used for the English stemming process. Then this algorithm was developed by W.B. Frakes in 1992 to be implemented in Indonesian. Then Fadilah Z Tala develops again using five steps by simulating the inflection and derivation of words. At each step, a certain ending is deleted by substitution. The design of Porter Stemmer for Indonesian can be seen in Figure 1[8].

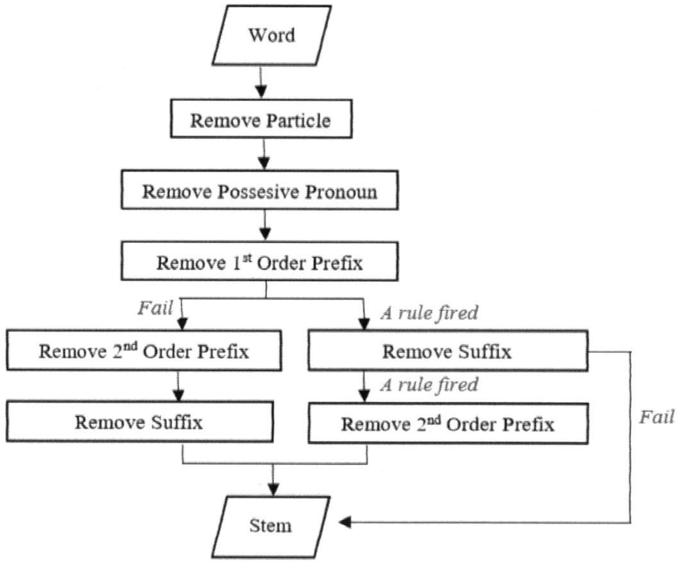

Figure 1: The Algorithm of a Porter stemmer for Bahasa Indonesia.

In this stemmer porter there is a shortage of high levels of over stemming, this is because the stemming process removes the longest first string in each step, potter stemming also produces many ambiguous words[9]. Besides these shortcomings, the porter algorithm also has a problem of dependence on the basic word dictionary in the stemming process.

The Nazief-Adriani algorithm was first developed by Bobby Nazief and Mirna Adriani. This algorithm is based on broad Indonesian morphological rules that are collected into a group encapsulated in the affixes that are allowed and the prefixes that are not allowed. This algorithm uses basic words to support recoding, which is rearranging words that get excess stemming. Compared to porters, this algorithm has a lower level of over stemming[7]. The Nazief and Adriani algorithms were further developed into sastrawi libraries with several improvements. In basic word formation, the sastrawi library does not need a base word database like porters. Sastrawi libraries provides for the programming languages PHP, Python, Java, C, Go and Ruby. The steps in this algorithm are as follows:

1. The first step is to check whether the word is the root word (root) contained in the list of root words (root). We say the word is the root word, then the process is stopped at this first stage.
2. Eliminating Inflection Suffixes ("-lah", "-lah", "-ku", "-mu", or "-nya"). If the word is in the form of particles ("-lah", "-lah", "-tah" or "-pun") then this step is repeated again to remove the Positive Pronouns ("-ku", "-mu", or "-nya").
3. Eliminating the derivational suffix. Eliminate the affixes -i, -kan, -an.
4. Remove derivational prefixes. Eliminate the prefix be-, di-, ke-, me-, pe-, and te-.
5. If from step 4 above you haven't found it yet. Then analyze whether the word is included in the last column's diambiguity table or not.
6. If from step 4 above you haven't found it yet. Then analyze whether the word is included in the last column's diambiguity table or not.
7. If all the above processes fail, then the algorithm returns the original word.

3. Methods

Preprocessing method is a very important stage in the technique and application of text mining[10][11]. This is the first step in the text mining process. In this paper, we discuss the four main preprocessing steps, namely, Case Folding, Tokenizing, Stop-word removal and stemming (Figure 2). In this study the stop-word removal and stemming stages use sastrawi libraries.

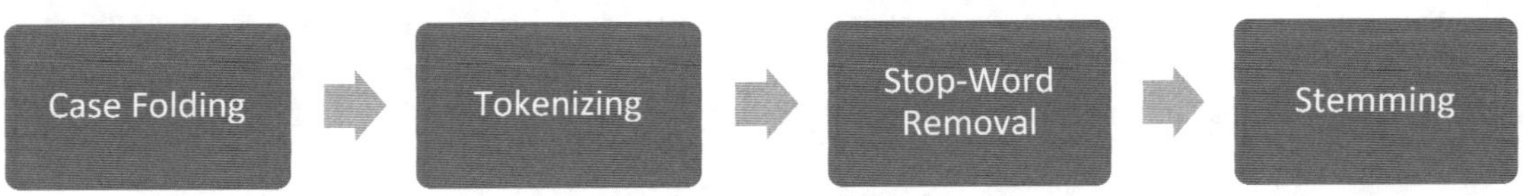

Figure 2: Text Mining Preprocessing Step

3.1. Case Folding

In a document that uses capital letters or the like sometimes it does not have in common, this can be due to writing errors. In text preprocessing the case folding process aims to change all the letters in a text document into lowercase letters, for example the word "Wifi" becomes "wifi".

3.2. Tokenizing

A text document consists of a set of sentences, the tokenization process breaks the document into parts of words called tokens[12].

3.3. Stop-Word Removal

The purpose of erasing words is to eliminate words that are not important, this process can reduce the dimensions of space that looks heavy. Common words in text documents such as clothing, prepositions and nouns etc., which do not give meaning to the document. Examples of the words "which", "and", "you", "until" are deleted from the document because they are not measured as keywords in a text mining application. In this study using sastrawi libraries. Figure 3 is a stop-word source code snippet.

```
require_once __DIR__ . '/vendor/autoload.php';
$stopwordFactory = new \Sastrawi\StopWordRemover\StopWordRemoverFactory();
$stopword  = $stopwordFactory->createStopWordRemover();
$hasilstop  = $stopword->remove($hasilcasefolding);
```

Figure 3: Source Code Stop-Word Removal Snippet

IOP Conf. Series: Materials Science and Engineering **874** (2020) 012017 doi:10.1088/1757-899X/874/1/012017

3.4. Stemming

Stemming is the process of mapping and breaking down the form of a word into its basic word form. Easier to say, the process of changing affixed words into basic words. This stemming process is the most important process in the text mining stage. Good stemming results can affect whether or not the text mining application. Figure 4 is a piece of source code using the sastrawi library. Sastrawi library for the PHP programming language can be installed with the following steps:

1. Open a terminal (command line) and navigate to your project directory
2. Download Composer so that the composer.phar file is in that directory.
3. Add sastrawi to your composer.json file:

php composer.phar require sastrawi/sastrawi:^1

```
require_once __DIR__ . '/vendor/autoload.php';
$stemmerFactory = new \Sastrawi\Stemmer\StemmerFactory();
$stemmer  = $stemmerFactory->createStemmer();
$hasilstem= $stemmer->stem($hasilstop);
```

Figure 4: Source Code Stemming Piece

4. Result and Discussion

The results of this study show a very significant difference between the tala porter algorithm and Nazief and Andriani algorithms that use sastrawi libraries, in terms of time, in our study the porter algorithm takes 241.6 seconds while the sastrawi library only takes 0.6 seconds. This long porter processing time is influenced by the use of a base word data dictionary taken from the base word database, whereas sastrawi does not use a base word database. Figure 5 preprocessing applications using tala stemmer porters, Figure 6 preprocessing applications using sastrawi libraries.

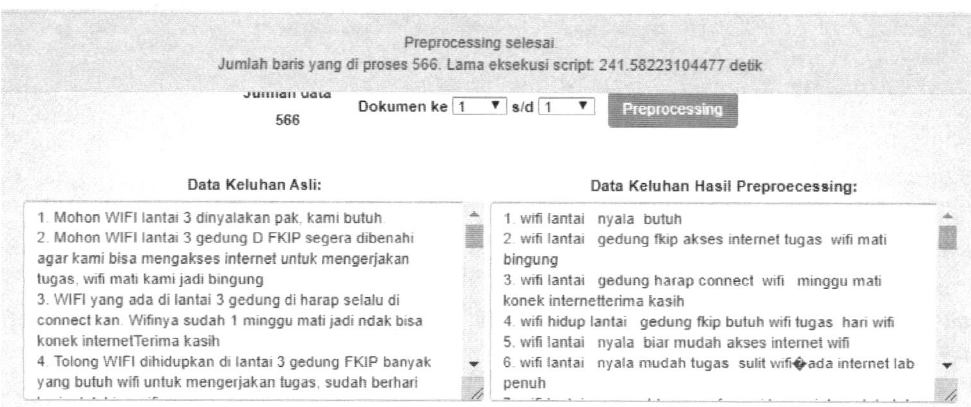

Figure 5: Preprocessing application with Tala Stemmer Porters

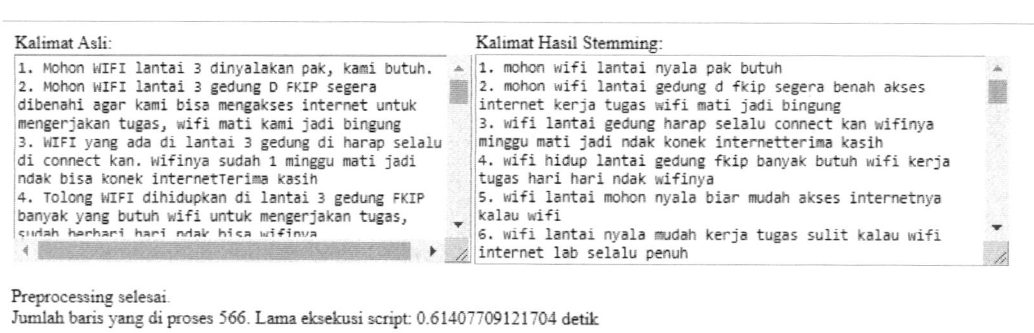

Figure 6: Preprocessing application with Sastrawi Library

From the results shown in Figures 5 and 6 it can also be seen that the results obtained at the porter still appear to be unreadable characters, while the character's sastrawi has been removed. In the research results also found several words that are over stemming in porters such as the word "Allah" into the word "al" while in sastrawi remains "allah" and in porters found under stemming "as soon as" in sastrawi is appropriate to produce the word "fast" etc. Table 1 shows the comparison obtained in 50 affixed words that are processed by the porter algorithm and sastrawi library in this study.

Table 1. Comparison of Porter Stemming Processes and Sastrawi Library

Stemmer Results Categories	Porter		Sastrawi	
	Total	Percentage (%)	Total	Percentage (%)
Exact Match	41	82	46	92
Unchange	2	4	3	6
Over Stemming	5	10	0	0
Under Stemming	2	4	1	2
Total	**50**	**100**	**50**	**100**

5. Conclusion

From the results of trials conducted with 50 samples of Indonesian text word-influenced text documents that have been selected by the author, resulting in the category of Exact Match stemmer results of 82% for Porter stemmer and 92% for Sastrawi, Unchange of 4% for Porter stemmer and 6% for Sastrawi, Under Stemming by 4% for Porter stemmer and 2% for sastrawi, and Overstemming by 10% for Porter stemmer and 0% for Sastrawi. Based on the processing time, sastrawi libraries are faster which is an average of 0.6 seconds while porters require an average time of 241.6 seconds. So it can be concluded that the use of sastrawi libraries in the preprocessing stage of student complaint documents produces better results compared to previous studies using the porter algorithm.

Acknowledgements
We hereby thank you to Universitas Muhammadiyah Sidoarjo for supporting the publication of this research.

References

[1] Jing L P, Huang H K and Shi H B 2002 Improved feature selection approach TFIDF in text mining *Proc. 2002 Int. Conf. Mach. Learn. Cybern.* **2** 944–6
[2] Vijayarani S, Ilamathi M J, Nithya M, Professor A and Research Scholar M P Preprocessing Techniques for Text Mining -An Overview **5** 7–16
[3] Durairaj M and Alagu Karthikeyan A 2018 Modified Porter's Algorithm For Pre -Processing Academic Feedback Data *Int. J. Pure Appl. Math.* **118** 3009–15
[4] Sugumar R and Priya M R 2018 Journal of Global Research in Computer Science Available Online at www.jgrcs.info IMPROVED PERFORMANCE OF STEMMING USING EFFICIENT STEMMER The resultant stemmer to which we will refer as Enhanced Porter Stemmer algorithm includes four **9** 1–5
[5] Rani Spr, Ramesh B and Anusha M 2015 Evaluation of stemming techniques for text classification *J. Comput. ...* **43** 165–71
[6] Rosid M A, Gunawan G and Pramana E 2015 Centroid Based Classifier With TF – IDF – ICF for Classfication of Student's Complaint at Appliation E-Complaint in Muhammadiyah University of Sidoarjo **1**
[7] Purnamasari K K and Suwardi I S 2018 Rule-based Part of Speech Tagger for Indonesian Language *IOP Conf. Ser. Mater. Sci. Eng.* **407**
[8] Tala F Z 2003 A Study of Stemming Effects on Information Retrieval in Bahasa Indonesia *M.Sc. Thesis, Append. D* **pp** 39–46
[9] Magriyanti A A 2018 Analisis Pengembangan Algoritma Porter Stemming Dalam Bahasa Indonesia *Attrib. 4.0 Int.*
[10] Torayev A, Magusin P C M M, Grey C P, Merlet C and Franco A A 2019 Text mining

assisted review of the literature on Li-O 2 batteries *J. Phys. Mater.* **2** 044004
[11] Virmani D and Taneja S 2019 *A text preprocessing approach for efficacious information retrieval* vol 669 (Springer Singapore)
[12] Gupta G 2015 Text Document Tokenization for Word Frequency Count using Rapid Miner (Taking Resume as an Example) *Int. J. Comput. Appl.* 24–6

IOP Conf. Series: Materials Science and Engineering **874** (2020) 012018 doi:10.1088/1757-899X/874/1/012018

Application of a combination between Principal Component Analysis and Logistic Regression Based on Support Vector Machine on Educational Data Mining with Overlapping Data Problem

Siti Mutrofin[1*], Maisarah Maisarah[1], Slamet Widodo[1], Raden Venantius Hari Ginardi[2], Chastine Fatichah[2]

[1]Universitas Pesantren Tinggi Darul Ulum, Jombang, Indonesia
[2]Institut Teknologi Sepuluh Nopember, Surabaya, Indonesia

[*]sitimutrofin@ft.unipdu.ac.id

Abstract. In 2019, the government of the Republic of Indonesia issued a zoning-based policy for New Student Admissions (PPDB) from the level of elementary school (SD) to high school (SMA), especially for public schools. The policy is documented in Permendikbud No.51 / 2018. The government policy aims to ensure the equality of education and make prospective students not focus only on favorite schools. However, this policy raises new problems. One of them is that if the potential student has got a medium UN (National Examination) score or medium distance of the house to the destination school, then his potential to be accepted at the destination school is very small. It is even worse if the potential students do not know the lowest score and the farthest distance the destination school can accept. Thus, potential students will choose schools by only guessing without basing on valid data, so their chances of being accepted will be very small. This current research focused on Educational Data Mining at PPDB Public High School (SMA) in Jombang in the academic year 2019/2020 which aims to accommodate the needs of potential students to predict the destination schools based on their own grades and home distances using classification techniques of data mining. However, another problem emerged in this study. An overlapping data occurred where one data was also owned by more than one class. For example, a potential student of SMA Negeri 2 Jombang (SMAN 2 Jombang) has got a score of 80 in Bahasa Indonesia subject, which is the same as that of a student from SMA Negeri 3 Jombang (SMAN 3 Jombang). Data overlapping does not only occur in one data but almost all of the data. The data used in this study were 600 data, consisting of 308 from PPDB 2019 of SMAN 2 Jombang, and the rest were from SMAN 3 Jombang. The attributes used were the home distance from the destination school, overall UN scores, UN scores of Mathematics, Natural Sciences, Bahasa Indonesia, and English subjects. The algorithm used was a combination of Principal Component Analysis (PCA) with Logistic Regression (LR)-based Support Vector Machine (SVM) with Anova kernel. The validation applied 10-fold cross-validation and the evaluation of algorithm performance used the aspects of accuracy, precision, and recall. The results of this current study showed an accuracy of 94.33%, a precision of 96.28%, and a recall of 92.53%. The results were better than those that did not apply PCA (70.83% accuracy, 69.62% precision, and 76.62% recall). By PCA, data could be seen from another angle that could separate or differentiate one class from the others. Even though there were 100% overlapping data, none of them, from all attributes, was 100% exactly the same.

1. Introduction

In 2019, the Republic of Indonesia government issued a zoning-based policy for New Student Admission (PPDB) process starting from elementary schools (SD) to senior high schools (SMA), especially for public schools. The policy is documented in Permendikbud No.51 / 2018 [1]. The policies are made to make the quality of education in Indonesia equitable and there will be no favorite school anymore which commonly relied heavily on the best scores of the students in the admission process. This policy raises new problems for the potential students whose scores are moderate and home distance from the destination school is not so close, especially in densely populated areas such as urban areas. Their potential to be accepted in the destination school is very small. If potential students' homes are close to public schools, they will tend to underestimate the result of the National Examination (UN). An example of complaints came from one of students' parents, Albert Mercelino, whose son registered in SMA Negeri 1 Kuta Utara, who said, "Keep using general zoning, not a radius. Use the score ranking too. That's fair. The zoning policy is just like we teach our children not to struggle. The fools will relax. Learning or not, the most important thing is close to (school)" [2]. The second complaint came from Fitri Suhermin, one of the parents of a potential student of SMP Negeri 8 Surabaya, as quoted from the Indopos daily, "Disappointment because her child could not be accepted in SMP Negeri 8 which was only 700 meters from his home, which was due to the zoning system." [3].

Table 1. Examples of some data that have overlapping data problems

Distances	National Exam	Mathematics	Science	English	Indonesian	Class
491	248.5	50	42.5	64	92	SMADAJO
1020	243	62.5	62.5	38	80	SMADAJO
511	242	60	60	54	68	SMADAJO
839	228	57,5	52.5	46	72	SMADAJO
491	334.5	80	82,5	94	78	SMAGAJO
524	248.5	45	67,5	60	76	SMAGAJO
451	220.5	50	52.5	38	80	SMAGAJO
507	181	32.5	42.5	38	68	SMAGAJO

This current research was conducted to predict the potential students who were potentially accepted in the Public Senior High Schools in Jombang Regency, East Java Province, Indonesia by implementing Educational Data Mining with classification techniques. The general goal of this current research was to support the government policy and help provide recommendations to potential students in choosing the potential schools. However, in this current research overlapping data problems occurred, in which each datum overlaps one another among classes (note Table 1 printed in red), which were the data accepted in SMA Negeri 2 Jombang (SMADAJO) and those accepted in SMA Negeri 3 Jombang (SMAGAJO). Overlapping data problem is not only in one attribute of each data, but all the attributes of all data. Any algorithm undergoing overlapping data problems will not produce a good classification [4]. The solution is by removing dimensions or attributes. If not removed, the existing data may not be able to distinguish among classes, because there are the exact same data in both classes [4]. According to Gu and Cheng, overlapping data problems become one of the obstacles in data mining and pattern recognition because they affect the accuracy of the classification and the ability to generalize directly [5]. The specific purpose of this current research is to find out an algorithm that is suitable for overcoming the overlapping data problems. The solution to overlapping data problems is like Suyanto's statement, [4] "To well understand a problem, look at it from another perspective". For example, by looking at data from a certain angle, they would look normal, not overlap.

To realize the objectives of this research, several algorithms were chosen according to the characteristics of the research data. One of the algorithms commonly used to overcome an overlapping data problem is the Principal Component Analysis (PCA) algorithm [4] [6] [7]. According to Suyanto [4]], "PCA is a mathematical method that transforms data into a new domain which results in many more important principle components". Whereas for overcoming overlapping data problems, the most suitable one is Logistic Regression [8] based on Support Vector Machine (SVM) which refers to myKLR [9], because it has been proven to have good resistance [10]. The kernel used on both PCA and SVM is the Anova kernel, because according to Stitson et al. [11] ANOVA has proven to have good multi-dimensional performance.

2. Proposed method

The method proposed involved collecting datasets, dimensionality reduction, classification, validation, and evaluating the performance of classification algorithms. Illustrations related to the proposed method are presented in Figure 1.

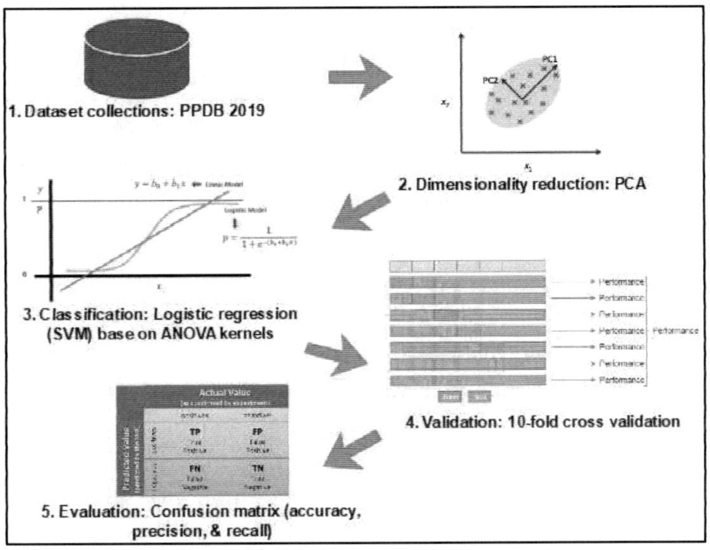

Figure 1. Proposed of classification models

2.1. Data

1.1. Data The data in this study were obtained directly from PPDB of the Government of the East Java Province of Indonesia in the academic year 2019/2020. The data sources were two public high schools in Jombang: SMADAJO and SMAGAJO. Both of them were used as classes in this study. The number of data was 600 data, consisting of 308 students of SMADAJO, and the rest were those accepted in SMAGAJO. Six attributes were applied: The home from the chosen school, the total scores of the National Exam (UN), the UN scores of Mathematics, Natural Sciences, English, and Bahasa Indonesia. All attributes for all data underwent overlapping data problems, but not all had exactly the same data between those in the SMADAJO and SMAGAJO classes. Data could only be accessed when the PPDB registration was opened. The East Java Provincial PPDB system could be accessed on https://ppdbjatim.net. All data were numeric both integer and real. The illustration of the dataset in this study can be seen in Table 1.

2.2. Dimensionality reduction

Overlapping data problem is one of the common problems in data mining and pattern recognition [5]. Adopting Suyanto [4], the illustration of data undergoing overlapping data problems were presented in Table 2 in the form of five objects with three attributes or dimensions (length, width, and height in meters) grouped into two classes (Tables and Chairs); Table 3 was in the form of five objects with two attributes or dimensions (length and width in meters) grouped into two classes (Tables and Chairs). The height dimension was removed because they could not distinguish between classes. In Table 2 and Table 3 there were overlapping data problems marked with red-printed scores. In figure 2 visualizing the data of five objects in a two-dimensional space (length and width), the position of a circle was close to two triangles; In figure 3 visualizing the data of five objects in one dimension of width (the dimension of length was removed), two overlapping objects occurred because both had the same width of 1.5; In figure 4 visualizing the data of five objects in one dimension of length (the dimension of width was removed), two overlapping objects occurred because they had the same length of 2.1; In figure 5 visualizing the data from another perspective. Tilting head to the right or left to see whether the data were randomly scattered or rather normal? The visualization of data could be handled by PCA, where the first principal component (PC1) and the second principal component (PC2) were the new dimensions in the PCA realm. The five objects could be separated into two different classes (triangles and circles) with one dividing line (dashed).

Table 2. Five objects with three attributes or dimensions (length, width, and height in meters) [4]

Object	Length	Width	Height	Class
Object 1	2.1	1.5	0.8	Table
Object 1	2.3	1.7	0.8	Table
Object 1	2.1	1.3	0.8	Chair
Object 1	1.6	1.5	0.8	Chair
Object 1	2.5	1.9	0.8	Table

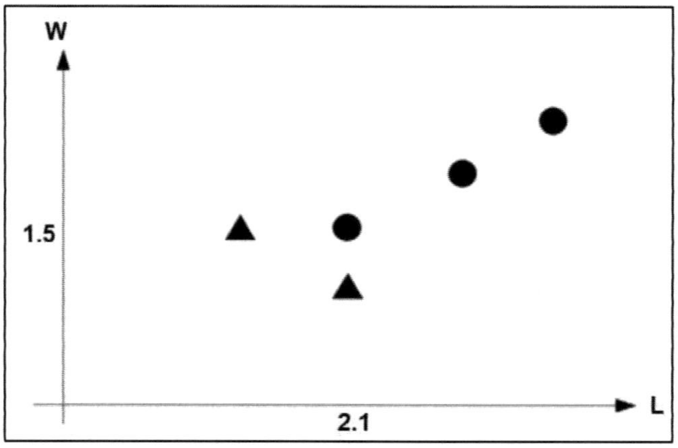

Figure 2. Visualization of data of five objects in a two-dimensional space (length and width) [4]

Table 3. Five objects with two attributes or dimensions (length and width in meters) [4]

Object	Length (L)	Width (W)	Class	Symbol
Object 1	2.1	1.5	Table	●
Object 1	2.3	1.7	Table	●
Object 1	2.1	1.3	Chair	▲
Object 1	1.6	1.5	Chair	▲
Object 1	2.5	1.9	Table	●

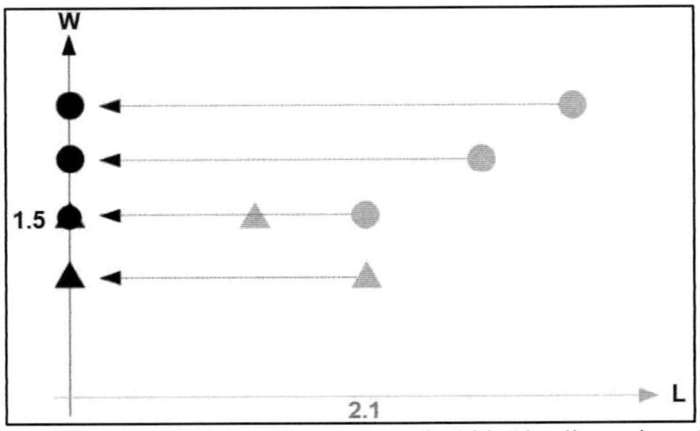

Figure 3. Data visualization of five objects in one dimension of width (the dimension of length was removed) [4]

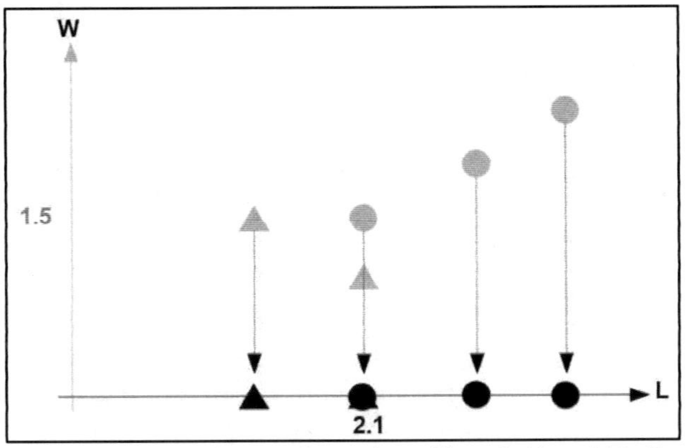

Figure 4. Data visualization of five objects in one dimension of length (the width dimension was removed) [4]

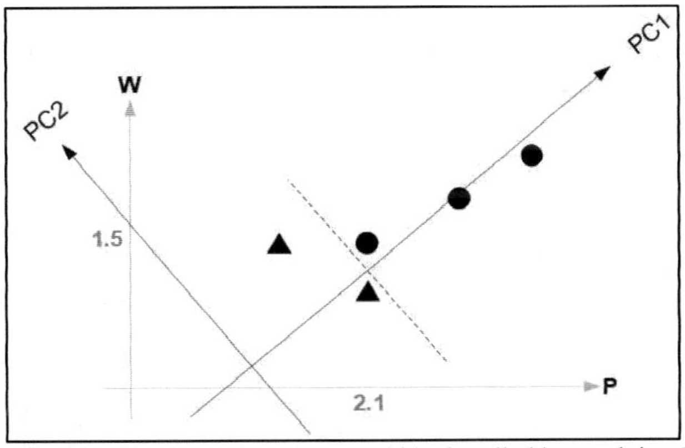

Figure 5. The visualization of the data could be handled by applying PCA [4]

In this study, all selected algorithms were run by using RapidMiner with ANOVA kernel and the gamma score of 1 and a degree of 3. The scores were the default scores of RapidMiner. This ANOVA kernel was also used in the SVM kernel.

2.3. Classification

The classification algorithm involved three algorithms: Logistic Regression, Logistic Regression based on Support Vector Machine (SVM) (in RapidMiner known as Logistic Regression (SVM) referring to myKLR by Rueping [12]), and PCA with Logistic Regression (SVM). Three algorithms were chosen to confirm the expectation of this research that the combination of PCA, Logistic

Regression (SVM) and ANOVA kernel would result in better results compared to other algorithms. All algorithms in this study, from beginning to end of the process, applied RapidMiner.

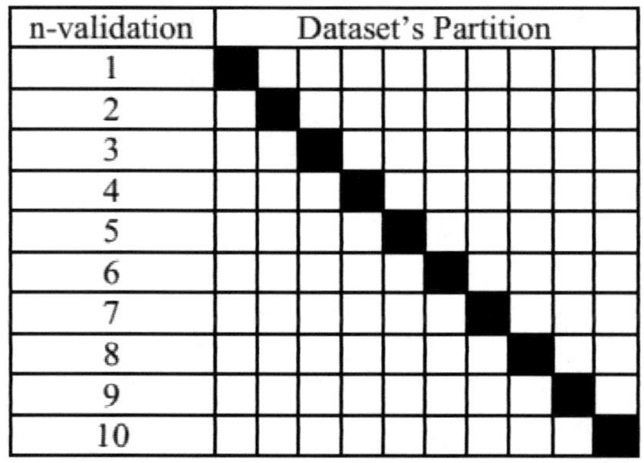

Figure 6. Illustration of the stratified 10-fold cross validation [13]

2.4. *Validation*
Validation in this study used the default from RapidMiner, namely 10-fold cross-validation with Stratified random sampling. The stratified random sampling was chosen because the distribution between training sets and testing sets can be spread evenly compared to other random samplings. The illustration of stratified 10-fold cross-validation is presented in Table 4 refers to Wahono [13].

2.5. *Evaluation*
The evaluation of algorithms applied confusion matrix which may produce accuracy, precision, and recall. These three evaluations were chosen because accuracy is not enough; another evaluation was needed to ensure that the performance of the algorithm in this study was really good. The illustration of the difference between accuracy, precision, and recall is presented in Figure 7 which refers to Data [14]. The illustration of the confusion matrix is presented in Figure 8 [15]. In True Positive (TP), a datum is predicted to be wrong because the actual datum is wrong. In False Positive (FP) a datum is predicted to be wrong but the actual datum is not wrong. In True Negative (TN), a datum is predicted not to be wrong because the actual datum is not wrong. In False Negative (FN), a datum is predicted not to be wrong but the actual datum is wrong. The formulas for accuracy, precision, and recall are shown in Table 5.

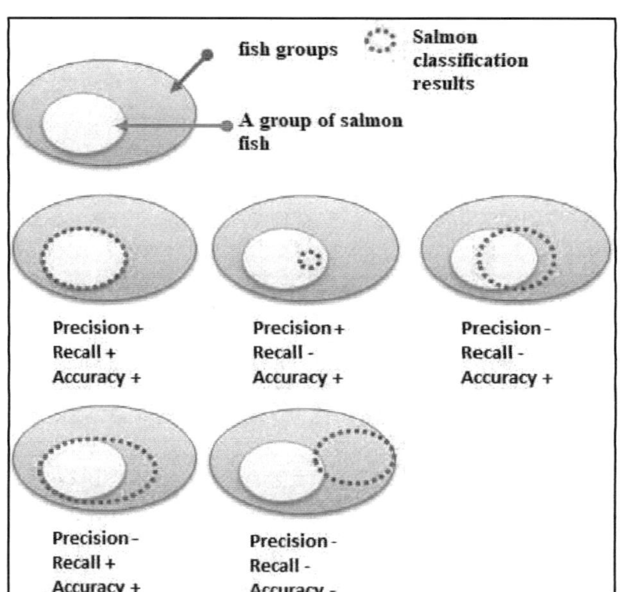

Figure 7. Illustration of the difference between accuracy, precision, and recall according to Data [14]

Table 4. Confusion matrix [15]

Object	True SMADAJO	True SMAGAJO
Prediction SMADAJO	TP	FP
Prediction SMAGAJO	FN	TN

Table 5. Compound performance measures [15]

Measures	Defined as
Accuracy / Correct Classification Rate (CCR)	$\frac{TP+TN}{TP+TN+FP+FN}$
Precision	$\frac{TP}{TP+FP}$
Recall / True Positive Rate / Sensitivity / Probability of Detection (pd)	$\frac{TP}{TP+FN}$

3. Experimental Results

The experiments in this study were carried out three times, using Logistic Regression, Logistic Regression (SVM), and PCA with Logistic Regression (SVM). The results of this study are presented in the form of a confusion matrix, as in Figure 8 which presents the results of Logistic Regression. Figure 9 presents the results of the Logistic Regression (SVM). Figure 10 presents the results of a combination of PCA with Logistic Regression (SVM). A more complete result is presented in Table 6.

accuracy: 70.83% +/- 4.10% (micro average: 70.83%)

	true SMADAJO	true SMAGAJO	class precision
pred. SMADAJO	236	103	69.62%
pred. SMAGAJO	72	189	72.41%
class recall	76.62%	64.73%	

Figure 8. Illustration of the confusion matrix as a result of Logistic Regression

accuracy: 72.33% +/- 5.10% (micro average: 72.33%)

	true SMADAJO	true SMAGAJO	class precision
pred. SMADAJO	229	87	72.47%
pred. SMAGAJO	79	205	72.18%
class recall	74.35%	70.21%	

Figure 9. Illustration of the confusion matrix as a result of Logistic Regression (SVM) with ANOVA kernel

accuracy: 94.33% +/- 1.96% (micro average: 94.33%)

	true SMADAJO	true SMAGAJO	class precision
pred. SMADAJO	285	11	96.28%
pred. SMAGAJO	23	281	92.43%
class recall	92.53%	96.23%	

Figure 10. Illustration of the confusion matrix as a result of a combination of PCA and Logistic Regression (SVM) with the ANOVA kernel ANOVA

Table 6. Comparison of the three algorithms [15]

Aalgorithms	Accuracy	Precision	Recall
Logistic Regression	70.83%	69.62%	76.62%
Logistic Regression (SVM)	72.33%	72.47%	74.35%
PCA dan *Logistic Regression* (SVM)	94.33%	96.28%	92.53%

Based on the experimental results shown in Figure 8, Figure 9, Figure 10, and Table 6 it was proven that the combination of PCA and Logistic Regression (SVM) with ANOVA kernel had better performance, as printed in blue; whereas the Logistic Regression had very poor performance, as printed in red. Therefore, it can be concluded that overlapping data problems can be overcome by PCA with Logistic Regression (SVM) with ANOVA kernel for predicting potential students to be potentially accepted in selected high schools. For further research, especially in the case of multi classes, it is better to know whether they can be solved by applying the same method, or whether the method proposed will only be suitable for binary classification cases. This is because there are more than two Public high schools in Jombang, and the potential students can choose not only SMADAJO and SMAGAJO.

4. Conclusion

The problem of this research is related to the overlapping data problems that occurred in the PPDB dataset of Public High Schools in Jombang which could be overcome by applying a combination of PCA algorithm and Logistic Regression (SVM) algorithm with ANOVA kernel. Based on this study, the proposed method successfully outperformed compared to not using PCA, with an accuracy of 94.33%, precision 96.28%, and 92.53% recall. For further research, it is better to prove whether the proposed algorithm can also be applied to multiclass cases, not just binary classes as in this study.

5. Acknowledgment

The researcher is grateful to the Directorate of Research and Community Service, Director General of Research and Development Reinforcement, Ministry of Research, Technology and Higher Education, for financing this research on Inter-University colaboation in 2018 with titled “Educational Data Mining *Berbasis Klasifikasi untuk Menganalisis Mahasiswa Berpotensi Berhenti Kuliah pada Kasus* Imbalanced Class Distribution”.

References

[1] L. P. Wardhana and K. Supriyoko, "Manajemen Penerimaan Peserta Didik Baru Secara Online Berbasis Zonasi, Prestasi dan Perpindahan," *MMP (Media Manajemen Pendidikan),* vol. 2, no. 2, pp. 228-235, 2019.

[2] P. Suyatra, "PPDB, Belajar Nggak Belajar, yang Penting Rumah Dekat Sekolah," Bali Express, 2 July 2019. [Online]. Available: https://baliexpress.jawapos.com/read/2019/07/02/144253/ppdb-belajar-nggak-belajar-yang-penting-rumah-dekat-sekolah. [Accessed 2 November 2019].

[3] J. Armanto, "Nah, Jokowi Akui Pelaksanaan PPDB Sistem Zonasi Banyak Masalah," Indopos, 21 June 2019. [Online]. Available: https://indopos.co.id/read/2019/06/21/178778/nah-jokowi-akui-pelaksanaan-ppdb-sistem-zonasi-banyak-masalah/. [Accessed 2 November 2019].

[4] S. Suyanto, Data Mining untuk Klasifikasi dan Klasterisasi Data, Bandung: Informatika, 2019.

[5] Y. Gu and L. Cheng, "Classification Of Class Overlapping Datasets By Kernel-MTS Method," *International Journal of Innovative Computing, Information and Control,* vol. 13, no. 5, pp. 1759-1767, 2017.

[6] N. Zhang, K. Leatham, J. Xiong and J. Zhong, "PCA-K-means Based Clustering Algorithm for High Dimensional and Overlapping Spectra Signals," in *Ninth International Conference on Intelligent Control and Information Processing (ICICIP)*, Wanzhou, China, 2018.

[7] A. Füzy, R. Kovács, I. Cseresnyés, I. Parádi, T. Szili-Kovács, B. Kelemen, K. Rajkai and T. Takács, "Selection of plant physiological parameters to detect stress effects in pot experiments using principal component analysis," *Acta Physiologiae Plantarum,* vol. 41, no. 5, pp. 55-56, 2019.

[8] A. Nöu, Logistic Regression versus Support Vector Machines, Stockholm: Stockholms universitet, 2018.

[9] S. Rüping, "myKLR - kernel logistic regression," Technische Universität Dortmund, 2003.

[Online]. Available: https://www-ai.cs.uni-dortmund.de/SOFTWARE/MYKLR/index.html#Keerthi/etal/2003a. [Accessed 2 November 2019].

[10] A. Christmann, I. Steinwart and M. Hubert, "Robust learning from bites for data mining," *Computational Statistics & Data Analysis,* vol. 52, no. 1, pp. 347-361, 2017.

[11] T. Hofmann, B. Schölkopf and A. J. Smola, "Kernel Methods in Machine Learning," *The Annals of Statistics,* vol. 36, no. 3, pp. 1171-1220, 2008.

[12] V. V. Raghavan, V. N. Gudivada, V. Govindaraju and C. R. Rao, Cognitive computing: Theory and applications, Elsevier, 2016.

[13] R. S. Wahono, N. S. Herman and S. Ahmad, "A comparison framework of classification models for software defect prediction," *Advanced Science Letters,* vol. 20, no. 10-12, pp. 1945-1950, 2014.

[14] S. Mutrofin, M. M. Machfud, D. H. Satyareni, R. V. H. Ginardi and C. Fatichah, "Komparasi Kinerja Algoritma C4.5, Gradient Boosting Trees, Random Forests, dan Deep Learning pada Kasus Educational Data Mining," *Jurnal Teknologi Informasi dan Ilmu Komputer (JTIIK),* vol. 7, no. 3, 2020.

[15] D. Bowes, T. Hall and D. Gray, "Comparing the performance of fault prediction models which report multiple performance measures: recomputing the confusion matrix," in *Proceedings of the 8th International Conference on Predictive Models in Software Engineering (PROMISE '12),* New York, USA, 2012.

[16] B. Supriyadi, A. P. Windarto, T. Soemartono, and Mungad, "Classification of natural disaster prone areas in Indonesia using K-means," *Int. J. Grid Distrib. Comput.*, vol. 11, no. 8, pp. 87–98, 2018.

[17] A. P. Windarto *et al.*, "Analysis of the K-Means Algorithm on Clean Water Customers Based on the Province," *J. Phys. Conf. Ser.*, vol. 1255, no. 1, 2019, doi: 10.1088/1742-6596/1255/1/012001.

[18] D. Hartama, A. Perdana Windarto, and A. Wanto, "The Application of Data Mining in Determining Patterns of Interest of High School Graduates," *J. Phys. Conf. Ser.*, vol. 1339, no. 1, 2019, doi: 10.1088/1742-6596/1339/1/012042.

[19] Sudirman, A. P. Windarto, and A. Wanto, "Data mining tools | rapidminer: K-means method on clustering of rice crops by province as efforts to stabilize food crops in Indonesia," *IOP Conf. Ser. Mater. Sci. Eng.*, vol. 420, no. 1, 2018, doi: 10.1088/1757-899X/420/1/012089.

IOP Conf. Series: Materials Science and Engineering **874** (2020) 012019 doi:10.1088/1757-899X/874/1/012019

Augmented reality using Natural Feature Tracking (NFT) method for learning media of *makharijul huruf*

Syita Ulfah, Diena Rauda Ramdania, Uwes Fatoni*, Khoiruddin Mukhtar, Hajir Tajiri, Ahmad Sarbini

UIN Sunan Gunung Djati Bandung

*kanguwes@uinsgd.ac.id

Abstract. *Makharijul huruf* is the place of the letter *hijaiyah* (Arabic letters) from *alif* to *ya*. If someone mistakes in reciting a letter on al-Qur'an then the meaning of the word or sentence may be different from the meaning of the original word. Few learning media that provides information about *makharijul huruf* such as school academic books, books of *tajwid al-Qur'an*, etc. However, to make terms of visual and audio clearer we can use augmented reality technology as a learning media. Augmented reality for *makharijul huruf* requires a marker of *hijaiyah* in printing form. Tracking marker in this research using Natural Feature Tracking (NFT) method. In the application of learning this *makharijul huruf* will produce a 3D object visualization of the oral organs and audio sounds *hijaiyah* letters and explanations about *makharijul huruf*. The results of black box testing and beta testing states that this application can be accepted by users by obtaining a percentage of 91%. It can be concluded that applications built functionally have produced results as expected.

1. Introduction

Makharijul huruf is the place where letters of *hijaiyah* come out when someone recited letters in al-Qur'an. Learning about *makhorijul huruf* is very important, as much as possible must learn from an early age. The introduction of *hijaiyah* letters is usually done at the kindergarten level.

Children are generally exploratory and adventurous in nature. Encouraged by a strong curiosity, children are usually happy to explore, try, and learn new things. They like to dismantle pairs of toy tools that they bought. Children sometimes are involved intensively in the activities of watching, playing with, and doing things with the things he has. The nature of this child is actually the foundation and basic capital to be creative and innovative human beings. Therefore, teachers must creatively stimulate child development with a variety of activities as well as media and learning resources so that children are encouraged to develop their abilities and think creatively, constructively and innovatively. Teachers must avoid making monotonous activities and make children bored. The choice of learning methods that enable many children to be one of the solutions so that children are facilitated by their explorative nature, for example through experiments, projects and field trips [1].

Augmented Reality or AR is a technology that allows people to visualize the virtual world as part of the real world that is around effectively so that it makes the real world seem to be able to connect with the virtual world and an interaction can occur [2].

AR has several methods that can be used as a marker detection algorithm or an object. One method that can be used as a marker tracking is Natural Feature Tracking (NFT) which is an algorithm in the Vuforia library to detect image patterns.

At present there are many fields that implement AR technology, one of which is in the fields of education and health. In the world of education, AR is used for media presentations. This will certainly attract the attention of children when learning about *makhorijul* letters and *hijaiyah* letters by using AR technology, where at this time they may also get used to the presence of AR because it is not uncommon for children's toys to be accessed via mobile phones and using AR technology. For example, gift cards with AR technology when buying chocolate sticks can display cartoon characters that kids enjoy doing.

2. Theoritical Frameworks

Makharij has a root word from the verb *kharaja* which means out. The origin of the word is then made into a form of eating *isim* (which indicates a place), so that it becomes *makhraj* which means a place out. While *makharij* is the plural form of *makhraj*. What is meant by *makhorijul* letters are places where letters come out of the reader's mouth. All letters have a place of origin issued by the reader, thus forming a certain sound. If the letter is not removed from its original place, then it makes it obscure for the reader himself and the listener, and cannot be distinguished from one letter to another [3]. The letter of hijaiyah is the plural form of *al-harfu* which means the smallest part of the pronunciation which cannot form its own meaning unless it must be strung together with other letters. Whereas *Hijaiyah* is spelling. The purpose of spelling here is Arabic spelling as the native language of al-Qur'an. Therefore, what is meant by *hijaiyah* letters are Arabic spelling letters as the native language of the Qur'an [3].

Augmented reality is one of the breakthroughs used lately in the field of interaction. The use of this technology will be very helpful in conveying information to the user. Augmented Reality is an interaction technology that combines the real world and the virtual world. The purpose of using Augmented Reality technology is to add understanding and information to the real world where the Augmented Reality system takes the real world as a basis and combines several technologies by adding contextual data so that one's understanding becomes clear [4].

Unity 3D is a game engine or software that was created to make a game application, which has been equipped with an IDE (Integrated Development Environment) or in other words unity does not require software development such as Delphi or Ms. Visual C ++ in terms of application development, because Unity already has its own code editor and compiler [5]

The Rational Unified Process (RUP) method is a software engineering method developed by collecting various best practices contained in the software development industry. The main characteristic of this method is to use use-case driven and iterative approaches for the software development cycle. RUP uses the concept of object oriented, with activities that focus on developing models using the Unified Model Language (UML). The picture below shows that the RUP has, namely: The first dimension is drawn horizontally. This dimension represents the dynamic aspects of software development. This aspect is elaborated in the development stage or phase. Each phase will have a major milestone that marks the end of the beginning of the next phase. Each phase can stand up from one iteration. This dimension consists of Inception, Elaboration, Construction, and Transition. And he second dimension is described vertically. This dimension represents the static aspects of the software development process which are grouped into several disciplines. The software development process described in several disciplines consists of four important elements, namely who is doing, what, how and when. These dimensions consist of: Business Modeling, Requirement, Analysis and Design, Implementation, Test, Deployment, Configuration and Change Management, Project Management, Environment [6].

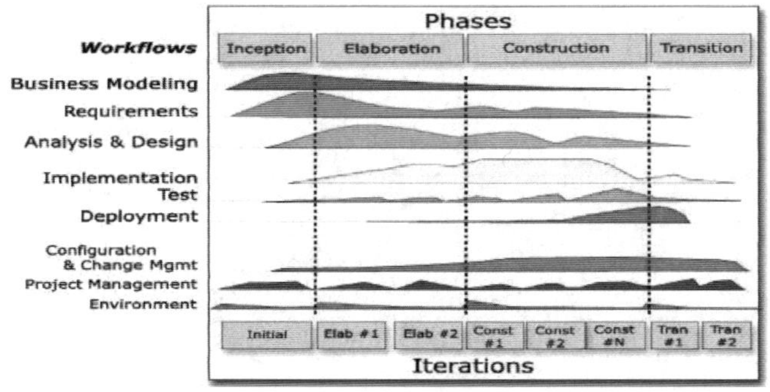

Figure 1. Rational Unified Process (RUP) [6].

Vuforia is an Augmented Reality Software Development Kit (SDK) for mobile devices that enables the creation of Augmented Reality applications. Vuforia uses Computer Vision technology to recognize and track markers or target images and simple 3D objects, such as boxes, in real-time [4].

Vuforia is the world's most used platform for AR development, with support for leading mobile phones, tablets and eyewear. Developers can easily add sophisticated computer vision functions to Android, iOS and UWP applications, to create an AR experience that realistically interacts with objects and environments. More than 50,000 Vuforia-powered applications have been published on the App Store and Google Play. Vuforia also supports an active developer ecosystem with more than 375,000 registered developers and more than 45,000 projects under development [7].

UML orf Unified Modeling Language means a standard modeling language. UML has syntax and semantic. When we make a model using the UML concept there are rules that must be followed. How the elements in the models relate to one another must follow the existing standards. UML is not just a diagram, but also tells the context.

3. Analysis And Design

3.1. System Architecture

System architecture in the application *makharijul* augmented reality letters can be seen in the following figure 3.

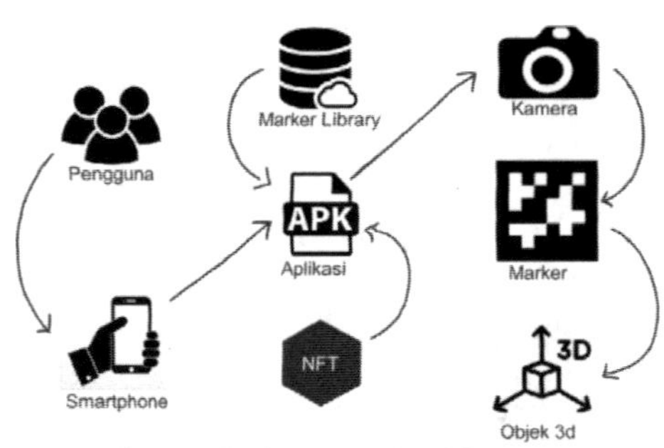

Figure 2. System Architecture

Figure 3 explains how the process of making this application is built by using Augmented Reality technology. The following are the components forming the system, including: (a) 1. Users are users of the Augmented Reality application *Makharijul* Letters and *Hijaiyah* Letters. (b) 2. Mobile or smartphone is the media used as a liaison between the user and the application of Augmented Reality *Makharijul* Letters and *Hijaiyah* Letters. (c) Marker library is a marker storage database that has been provided by vuforia. (d) NFT is a marker recognition method used in the augmented reality method in the vuforia library. (e) The camera is used to scan marker images that have been provided previously, so that tracking markers can be done. (f) Marker is a picture of *hijaiyah* letters that has been provided by the author. (g) 3D objects are visualizations displayed on the Augmented Reality application *Makharijul* Letters and *Hijaiyah* Letters.

3.2. Use Case Diagram

Use Case Diagram is modeling to illustrate the interaction between actors with the system to be created. Use Case Diagrams on this menu combination application can be seen in Figure 4.

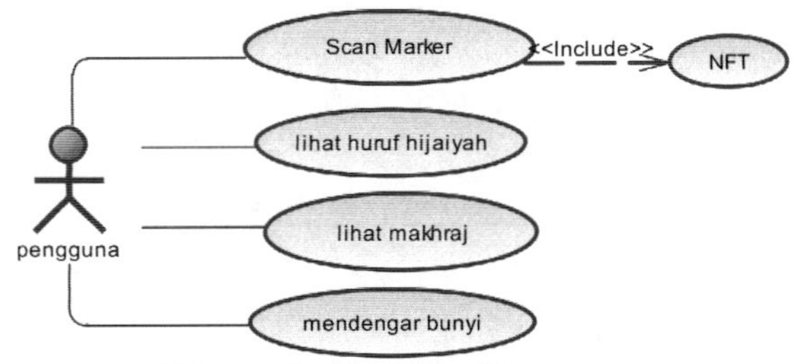

Figure 3. Use Case Diagrams

3.3. Activity Diagram Application Augmented Reality Makharijul Letters And Letters Hijaiyah

Activity diagrams illustrate the various activities flow in the system that is being designed, how each flow starts, decisions that might occur, and how they end. Activity diagrams can also illustrate parallel processes that might occur in several executions.

3.4. Sequence Diagram Application Augmented Reality Makharijul Letters

Sequence Diagram is one of the diagrams in UML, this sequence diagram is a diagram that illustrates dynamic collaboration between a number of objects. Its purpose is to indicate the sequence of messages sent between objects as well as interactions between objects. Something that happens at a certain point in the system execution. In Figure 6 the following is a sequence diagram in this application.

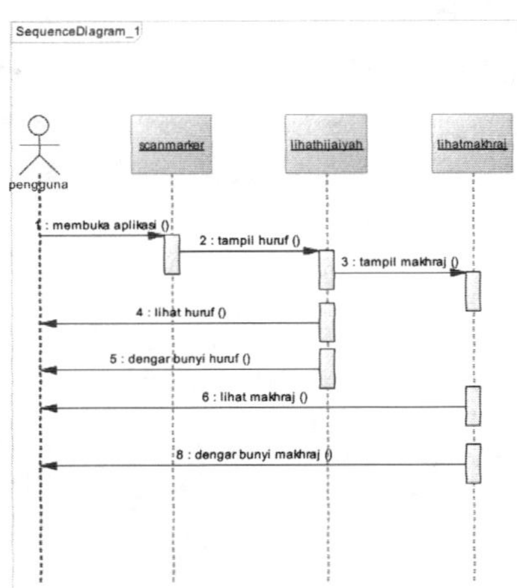

Gambar 4. Sequence Diagram

3.5. Natural Feature Tracking

Natural Feature Tracking (NFT) contained in the vuforia library functions to determine whether an image can be recognized or not based on pattern recognition by detecting the angular points of the pattern in the image. The goal is to make detecting feature patterns easier. Basically the Vuforia library uses NFT (Natural Feature Tracking) in which there is a natural feature tracking approach namely SIFT (Scale Invariant Feature Transform). SIFT is used to detect "feature points" and determine the scale of the marker (target image) by mapping the coordinates value [17]. Figure 7 below shows the steps to detect markers in Natural Feature Tracking (NFT).

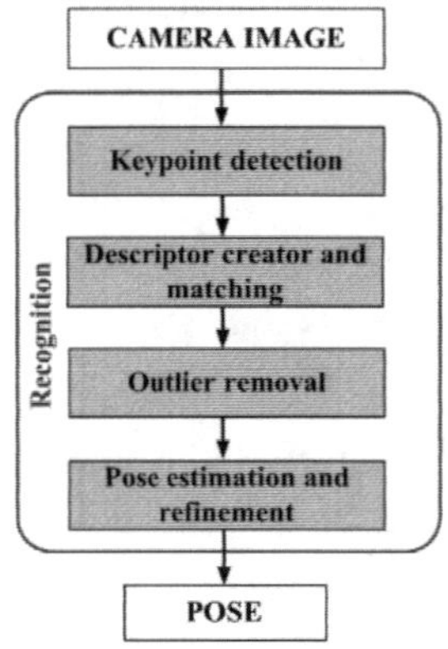

Figure 5. steps for detecting markers [18]

3.6. Interface Design

Interface design is one of the most important parts in software development, because the first impression of a user is how the interface or appearance of a software is presented. Here is the interface of the software that will be built : Main Interface Display, *Hijaiyah* Letters Scan Page Interface, *Makharijul* Hurricane Information Interface Page and Interface page See *Makhraj*.

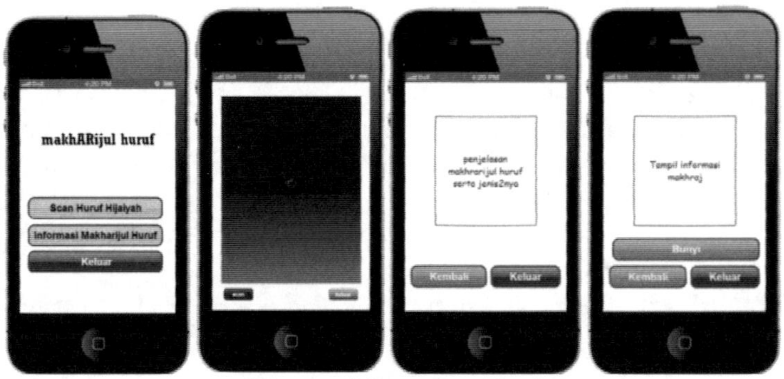

Figure 6. Interface Design

4. Implementation and Testing

Implementation of the interface aims to make the interface and function functions in accordance with the previous design. Next in this section the results of the interface that have been made and the explanation are explained.

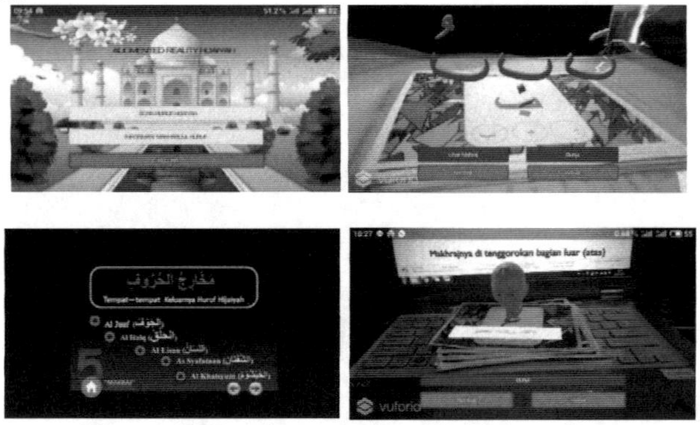

Figure 7. Implementation of the interface

4.1.1. Main interface implementation
Figure 12 shows the main menu interface in the augmented reality application *makharijul* letters and *hijaiyah* letters. On this page there are also several menus available including, scan button *hijaiyah* letters, *makharijul* letter information button and exit button.

4.1.2. Implementation of hijaiyah letter scan display
Figure 13 shows a page that displays a visualization of 3D *hijaiyah* letters, where *hijaiyah* letters on a previously scanned marker will appear in 3d.

4.1.3. Implementation of the display of makharijul letters
Figure 14 is the *Makharijul* letters information interface page on the augmented reality application of *Makharijul* letters and *hijaiyah* letters which serves to show brief information about the types of letters.

4.1.4. Implementation of the display Makharijul *of makharijul letters*
Figure 15 shows an interface page to see *makhraj* in the application of augmented reality makharijul letters and *hijaiyah* letters that serve to show information about the location where the *makhraj* letters have been scanned before. On this page there is also a sound button that serves to display the audio location of the letter *makhraj*.

5. Conclusion
After going through the stages in accordance with the Rational Unified Process (RUP) software development method and the Natural Feature Tracking algorithm in the development of this augmented reality application, it can be concluded (a) The application has been designed and built using augmented reality technology with the Natural Feature Tracking (NFT) method for learning makharijul letters and hijaiyah letters for android. Learning makharijul letters and hijaiyah letters using augmented reality technology on an android device can be an alternative media introduction that is more interesting because it can display the hijaiyah and makhraj letters in three dimensions, and can play audio audio sound pronunciation of hijaiyah letters clearly so that it can be easily understood. (b) Based on the results of the test on each letter sound and makhraj as well as the object letters and makhraj are all appropriate and can be displayed properly. So it can be concluded that the application that was built functionally has issued results that are as expected.

References
[1] Badru Zaman, *Esensi Sumber Belajar dalam Pembelajaran Anak Usia Dini*. Jakarta: Pustaka UT
[2] Paul E Jacobs, in *www.qualcom.co.id/*products/augmented-reality. 2012,
[3] Tahali Ahmad, "PENGEMBANGAN APLIKASI IAR (IQRA' AUGMENTED REALITY) BERBASIS ANDROID SEBAGAI MEDIA BELAJAR MAKHORIJUL HURUF HIJAIYAH PADA MATA PELAJARAN PAI DI SMK NEGERI 1 MAGELANG," in *Perpustakaan FT UNY*, Yogyakarta, 2017.
[4] K. I. Satoto, K. T. Martono I. S. Nugraha, "Pemanfaatan Augmented Reality untuk Pembelajaran Pengenalan Alat Musik Piano," in *Jurnal Teknologi dan Sistem Komputer*, Vol. 2, No. 1, pp. 62-70, 2014.
[5] Rickman Roedavan, "Unity Tutorial Game Engine Edisi Revisi," 2017.
[6] S. M. Rosa, *Rekayasa Perangkat Lunak Terstruktur dan Berorientasi Objek*. Bandung, 2013.
[7] Vuforia Developer Library | Portal Pengembang. (2018) Vuforia. [Online]. https://library.vuforia.com/
[8] S. Taryana, "Metode RUP. Bandung: STMIK LIKMI BANDUNG," 2007.
[9] Badru Zaman, *Esensi Sumber Belajar dalam Pembelajaran Anak Usia Dini*. Jakarta: Pustaka UT.

Smart campus attempt using green transportation mobile-based tracking app

M Rachmaniah[1*], AI Suroso[2*], M Syukur[3], I Hermadi[4], RW Putra[4], and MRG Tirani[4]

[1] Computer Science Department, IPB University, Bogor, West Java, Indonesia
[2] School of Business, IPB University, Bogor, West Java, Indonesia
[3] Agronomy and Horticulture Department, Faculty of Agriculture, IPB University, Bogor, West Java, Indonesia
[4] Computer Science Department, Faculty of Mathematics and Natural Sciences, IPB University, Bogor, West Java, Indonesia

meuthiara@apps.ipb.ac.id; arifimamsuroso@apps.ipb.ac.id

Abstract. Green campus is the creation of a green ecologically friendly environment that leads to smart campus establishment. In this study, green campus status is revealed by applying cross-campus community collaboration under green transportation bus. The purpose of this study was to develop a real-time Android-based tracking system, using the V model approach, aimed at green transportation movements and accessibility. The global positioning system of green transportation is interconnected via a back-end server that utilizing web-socket to incorporate client and server communications. The results are passenger app that enabling the passenger to track green transportation movement, acquire green transportation meta-information, perceive campus routes, and send awaiting notification. Further results are driver app for green transportation drivers to choose which vehicle to drive, change driving status, recognize other active coaches, locate nearer passengers stops, and identify campus routes. The created mobile app has successfully experimented on Android smartphones such as Samsung Galaxy Mega J5, Lenovo A6000, Xiaomi Redmi 3, Asus 2 Zenfone Laser, and Samsung J7 Prime.

1. Introduction

Self and information services and environment platforms are the three main characteristics of a smart campus. The smart campus is an outcome of the rapid development of GIS and computer network technology, and the internet of things (IoT) [1]. A smart campus is a campus that can adapt to every new situation that occurs in daily operations [2] and of three campus pillars that are administration, students, and lecturers [3]. The smart campus comprises of six components: iLearning, iGovernance, iGreen, iHealth, iSocial, and iManagement [4]. The combination of green campus and smart campus brings out the concept of smart green campus. The involvement of all campus communities in employing the existing resources and technology effectively and environmentally is an essential point of this concept that has a significant impact on campus sustainability [5]. In 2007 a study at Tongji University successfully applied green campus employing green concepts, ecological technologies, energy, and resource development practices that became a model of cost-effective campus expenditure [6]. Another study on proactive environmental management creates a healthy learning environment, build environmental awareness, and reduce campus operational costs [7]. Sustainable green transport is at the heart of the green transportation concept [8]. In 2014, IPB University proclaimed the green campus concept embracing five main programs: Green Transportation, Green Movement, Green Space, Green Building, and Green Energy. In 2015 the gas-fueled green transportation buses became the main program in priority. A total of five buses at Dramaga campus serves five different routes to support the mobility of the campus community in the 267-ha area. Along the bus routes, there are 19 stops, with an average of 1,918 passengers per day. Each green transportation bus has a departure schedule, bus stop arrival, and a mobile phone. However, in its implementation, schedules and routes were often ignored by bus drivers and passengers who demanded a different course. Therefore, real-time campus bus locations on its routes or bus stops were the main focus of this study.

Many researchers conducted studies on Android-based vehicle tracking systems using a global positioning system (GPS). One study provided current and offline information on bus, bus number, and bus routes, including estimate time needed to reach the destination [9]. Another study provided campus bus locations, bus details, driver details, stop location, and bus routes for students using Google Maps [10]. However, both studies demanding its user to login and entered the bus license plate. Further research gave an enhanced level of reliability in the navigation solutions of GPS-denied environments [11]. Another study sent their data in the form of latitude and longitude coordinates through short messages on the user mobile wherein the coordinates will be further plotted in the Android app automatically [12]. GPS satellites continuously transmit digital radio signals that carry satellite location data and time toward the corresponding recipient. The GPS composes of three segments, user segment (cell phones and personal digital assistants-PDAs), space segment (collection of satellites), and control segment [13]. The atomic clock that has punctuality of one-millionth of a second equipped the GPS satellites. Based on this information, the receiving station knows how long it takes to send signals to the recipient on earth. The longer time to get to the receiver, the bigger the position of the satellite from the receiving station [14].

WebSocket first emerged in HTML5, allowing web server and web browser real-time communication. The server responds according to the request sent by the client. When a connection does establish, the initial HTTP-based communication changes into full-duplex communication [15]. A client is a subscriber when the client connects the server through a web-socket connection and receives data from the server while the link lasts. This connection is an ideal solution for real-time data exchange [16]. Google Maps is a server-side, which means that maps are stored on Google servers but use by users. The Google Maps API is one of the Google services that gives access to Google maps and modifies the map as needed on the client side [17]. After the map loads, interactions and modifications are happening. Adjustments can be the addition of markers, lines, or polygons.

This study developed real-time tracking apps for green transportation movement as an effort to demonstrate smart green campus concepts using an Android-based smartphone. The developed apps are passenger app for the campus community and driver app for green transportation driver. The apps development applies for mobile devices because around 63.1 million Indonesians are smartphone usage [18]. The apps development utilizes Global Positioning System (GPS) technology and Google Maps as a viewer of the bus routes. Also, WebSocket, especially SocketIO, facilitates a real-time communication system. The novelty in this study is that in addition to the passenger app, we also developed the driver app. Each mobile app has no requirement to login and key in the license plate because all passengers and driver's data are available in the database. Lastly, the tracking of every active green transportation is displayed on a single page, meaning passengers do not need to swap among pages.

2. Methods

The study records the year 2018 longitude and latitude data of the Dramaga campus (**Table 1**). The data include bus stops locations, real-time bus positions, bus routes, passengers' locations at specified bus stops, bus information, number of passengers at each bus stop, and map of campus. Real-time coordinates of longitude and latitude were gathered using Google Maps and GPS features of the Android smartphone. The green transportation buses and prospective passengers on the specified bus stop location coordinate data were further transmitted to the server and used by the driver and passenger apps accordingly. Bus stop latitude and longitude positions were measured using a smartphone's GPS and Google Maps. Dramaga campus has 15 actives out of 19 available bus stops at an area of 267 ha. The departure route starts consecutively from bus stop 1 - Graha Widya Wisuda to bus stop 15 – FEM/FEMA (**Figure 1**). The driver app will capture data on the number of passengers waiting at the bus stop location in real-time mode. There are seven green transportation buses maneuvers at the campus. Five buses are in operation at 06.00 - 18.00 hours and two buses are set aside for backups.

Table 1 Longitude and Latitude of Campus Bus Stop

Bus Stop Location	**Longitude**	**Latitude**
1. Graha Widya Wisuda	106.730421	-6.560520
2. Berlin	106.730901	-6.558936
3. FMIPA	106.731424	-6.557861
4. Asrama Putri	106.731583	-6.556086
5. Tanoto Forest building	106.730604	-6.555764
6. Menwa	106.729613	-6.555749
7. FAHUTAN	106.727786	-6.555795
8. International Dormitory	106.726983	-6.556112
9. Alhurriyyah Mosque	106.725555	-6.556588
10. GORLAM	106.724446	-6.556368
11. FPIK	106.723648	-6.556632
12. FKH	106.721843	-6.556785
13. Techno park	106.726804	-6.557457
14. LSI	106.726857	-6.558434
15. FEM/FEMA	106.728191	-6.559750

Figure 1 Campus map of green transportation route and its 19 bus stops.

The V-model development method is used to guide the development of passenger and driver mobile apps. Each step of the development stage followed by the interrelated test scenarios and its test attainment (**figure 2.a**). Here, the green transportation managers examined the test scenarios as an assurance that the system follows the development plan. The mobile apps development started with the analysis stage, where we collect qualitative data through interviewing the green transportation manager, intended to gather needs analysis of the green transportation stakeholders' requirements, and the creation of an acceptance test scenario. The results of this phase were to use case diagrams, use case descriptions, activity diagrams, and class diagrams. The design stage utilized mockups design that follows the class diagrams of the analysis stage. During the mockups design, we created unit testing, integration testing, and system testing scenarios. The implementation phase or coding comprised of two parts that are client-side and the server-side. Client-side coding was done using Android Studio and Java programming language. Server-side was done using the JavaScript programming language equipped with Express Js framework for the development of the back-end application programming interface (API).

Further, the unit testing, integration testing, system testing, and acceptance testing were carried out using developed test scenarios of the previous stages. Unit testing aimed at verifying the smallest unit of software design. Integration testing confirmed the relationship between one function to other functions to produce the desired output without any errors. System testing examined the server in the real environment using various types and versions of Android smartphones available at the current market. Finally, acceptance testing aimed to determine whether the application developed was following the original plan.

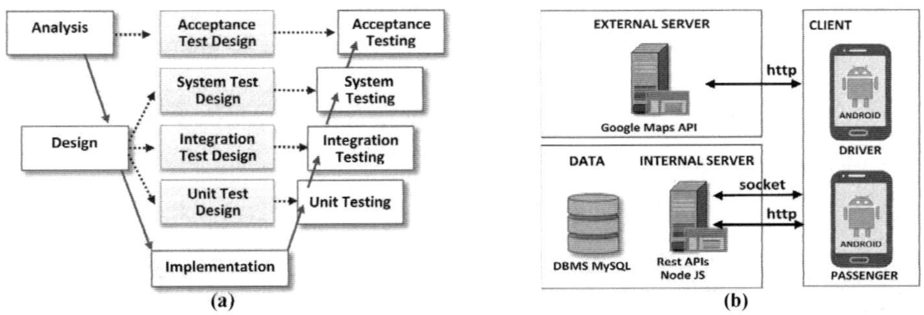

Figure 2 (a) The V-Model stages method and its related testing [19] and **(b)** Mobile apps architecture consisting of Google Maps API within the external server, data and internal server for Rest API and Node Js, and Android driver and passenger clients.

The application architecture consists of three parts, namely the client, server (internal and external), and data (**figure 2.b**). On the client-side, we built two separate apps, each for the passenger and the driver. The client-side was made using Android Studio with Java programming language. On data and internal server-side, we utilize MySQL DBMS, develop back-end applications using JS Node and JS Express framework, and employ the SocketIO library. The Google Maps API resides on the external server is used by the client to load maps and objects on maps such as bus stops to send and obtain bus and clients location. Communication between the client and the internal server consists of two arrangements of connection, i.e., through the socket and the REST API. The REST API is used to connect the client and the database contained in the server. The server sends the client request data to the REST API. The use of Sockets is a bridge for data exchange, especially location data in real-time between the clients and the server.

3. Results and Discussion

3.1 Passenger and Driver Apps Development Stages

The number of drivers during daily duty is seven people; each driver drives any bus to the nearest bus pool entrance. Each green transportation bus comes with an Android-based smartphone, meaning that the smartphone attaches to the bus, not to the driver. In this study, using passenger app, the passenger views bus position, and maneuver views information of the bus and the driver, view bus route information, and send their waiting notification using personal identification number (PIN) location. Meanwhile, using the driver app, the drivers choose the bus at the nearest bus pool entrance, activate driving status, view the digital map of other active buses, view the number of passengers waiting at a specific bus stop, and renew bus smartphone number. The entire design illustrated in a use case diagram, as depicted in **figure 3.a**. Following that, we created class diagrams to explain how the system works (**figure 3.b**). Also, we conducted analyses of the workings or activities of the functions contained within the use case diagram by creating activity diagrams and examined acceptance test scenarios. Approved test scenarios for the passenger side applied for real-time bus monitoring and pin location notification sent by the passenger. To enable notification, passengers should be within 25 meters radius of a specific bus stop distance. Approved test scenarios for the driver side implemented during login, bus selection, status activation, bus contacts update, and logout activities.

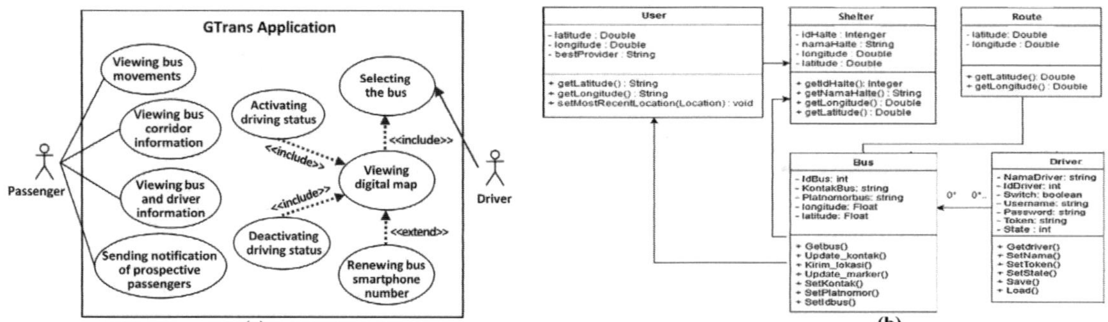

(a) (b)

Figure 3 (a) Use case diagram with passenger and driver actors and **(b)** Class diagram of green transportation tracking app.

The design stage presented in the form of mockups utilizing extensible markup language (XML) of the Android Studio, which gave an overview of the appearances and functions of the app. During the design stage, we also created test scenarios for unit tests, integration tests, and system tests of passenger and driver prototypes. Based on the analysis and stakeholder tests approval, passenger app consisted of the main page, the help page, the driver and bus information page, splash screen pages, and a pop-up message notification in the middle of the page regarding their PIN pointed to their current location (**figure 4.a**). Meanwhile, the driver app comprises of three activities: login activity, select bus activity, and maps activity (**figure 4.b**). Approved system testing scenario for both apps apply on a various Android operating system such as Jelly Bean, Android KitKat, Android Lollipop, Android Marshmallow, and Android Nougat.

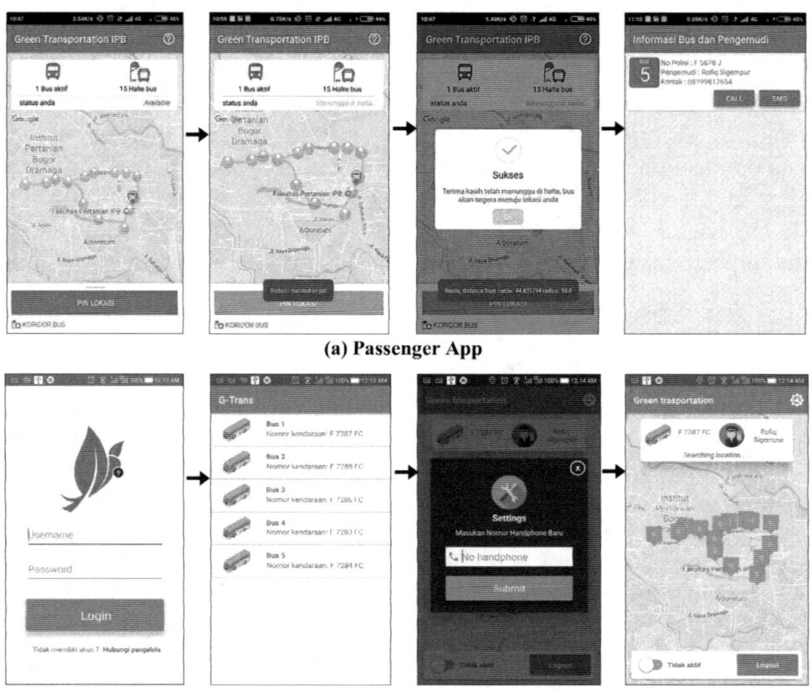

(a) Passenger App

(b) Driver App

Figure 4 XML design for **(a)** use of passenger app at a bus stop, send PIN notification, receive pop up incoming bus message, receive bus and driver data; **(b)** Driver key in driver's username and password, select nearest entrance bus number, key in bus's smartphone number, and activate the bus to maneuver.

The implementation stage of this study was carried out for the driver and the passenger apps. Concerning coding for the passenger app, we provide a socket in the server to detect the connection requested by the client. Here we make client socket to be associated with the socket server. Hence, the socket on the client serves to establish the initialization of connection to the socket on the server. Following this, real-time location data exchange is obtained by initializing the event with the method called "lokasi" on the socket server or client (**figure 5.a**). Passenger app encompasses several functions, i.e., function for a passenger to receive bus location longitudinal data (**figure 5.b**), function to visualize bus movement on passenger map, and passenger notification waiting at a specific bus stop. Function on visualization bus movement utilize a particular marker bus appeared on the app/screen. However, if the marker bus is not established, then the marker bus is displayed. Also, passenger notification employed a socket on the server that received the data and added a counter to the number of prospective passengers based on the bus stop id.

Concerning coding for the driver app, we created the map function. This function comprises of three activities, i.e., route and bus stops information, update bus contact number, and change driver status. Route data and bus stops are visualized using Google Maps API. Here, we make a WebSocket connection to receive information on the number of passengers waiting at each bus stop (**figure 5.c**) and the location of other active buses to be further projected into a digital map. The passenger app employs REST API so that clients receive data from the database. This process is performed on the bus and driver information viewing function. The implementation utilizes the GET method by combining data from the driver, bus, and driver bus tables. The resulted data in the "status" attribute in the driver bus table is valued "1". The output results are data in JSON format with two objects, namely results and data (**figure 5.d**). Also, we make popup input forms to store new contact numbers and a switch button to change the status of the driver. When driver status is active, the app pops up active status and send driver location to the server via a WebSocket connection — location data sending occurs when the smartphone finds a new location. Location data send via a WebSocket connection is immediately displayed and not stored in the database, so that server performance is faster. When the driver's status is deactivated, then the status was inactive, and hence, the app stops sending the location.

```
public Socket mSocket;
{
    try {
        mSocket= IO.socket(SERVER)}
    catch(URISyntaxException e){
        Log.e("error","on create :
"+e.toString());}
}
a
@Override
protected void onCreate(Bundle
savedInstanceState) {
    super.onCreate(savedInstanceState);
    setContentView(R.layout.activity_maps);

    mSocket.connect();
    mSocket.on("lokasi",handlingLocationBus);
```

(a)

```
private Emitter.Listener handlingLocationBus = new
Emitter.Listener() {
    @Override
    public void call(final Object... args) {
        JSONObject data = (JSONObject)args[0];
        String id = new String();
        String longitude= new String();
        String latitude = new String();
        try {
            id= data.getString("id");
            longitude= data.getString("longitude");
            latitude = data.getString("latitude");
        } catch (JSONException e) {
            e.printStackTrace();
        }
        final String dataID = id;final String
dataLongitude = longitude;
        final String dataLatitude = latitude;
        runOnUiThread(new Runnable() {
            @Override
            public void run() {
      movingBus(dataID,dataLongitude,dataLatitude);
            } }); }};
```

(b)

```
private Emitter.Listener handling_halte_awal = new Emitter.Listener() {
    @Override
    public void call(final Object... args) {

        JSONArray array = (JSONArray) args[0];
        String paket_id = new String();
        int ix=0;
        for (; ix < array.length();ix++){
            try {
                final String count = ""+array.getInt(ix);
                final String ix_string = ""+ix;
                runOnUiThread(new Runnable() {
                    @Override
                    public void run() {
                        update_halte(ix_string,count);

                    }
                });
            } catch (JSONException e) {
                e.printStackTrace();
            }
```

(c)

```
{
    "results": 200,
    "data": [
        {
            "kontak_bus": "081223991535",
            "id_driver": 1,
            "nama_driver": "Rofiq Sigempur",
            "id_bus": 1,
            "nomor_bus": 1,
            "platnomor_bus": "F 7287 FC"
        }
]}
```

(d)

Figure 5 (a) Client socket initialization to enable real time location data exchange; **(b)** passenger receive screen visualization of driver's location through the JavaScript Object Notation (JSON); **(c)** segment code to display the number of passengers at the bus stop; and **(d)** REST API output results in JSON format.

3.2 Implementation Tests

During the unit test, we tested the output correctness of each function enclosed within the developed passenger and driver apps using the black box method. All test results successfully passed. On integration tests, we carried out with successful results using the black box method. Integration tests aimed to assess and ensure functions within the two developed apps to have interacted correctly. On system tests, we utilized requirement testing methods to ensure that the developed apps work well on the real environment using different Android-based operating systems. Here, the developed passenger and driver apps are successfully functioning on five different Android-based mobile devices and operating systems (**table 2**).

Table 2 Successful compatibility test using different mobile devices and operating systems

No	Mobile Device	Operating System	Screen Size (inch)	Resolution (pixels)	Test Result
1.	Samsung Galaxy Mega J5	Android Jelly Bean	6	720x1,280	Successful
2.	Lenovo A6000	Android KitKat	5	720x1,280	Successful
3.	Xiaomi Redmi 3	Android Lollipop	4,7	720x1,280	Successful
4.	Asus 2 Zenfone Laser	Android Marshmallow	5,5	720x1,280	Successful
5.	Samsung J7 Prime	Android Nougat	5,5	1,080x1,920	Successful

The passenger acceptance tests were carried out using alpha testing techniques. Here, stakeholders tested the apps by executing each test scenario agreed previously during test design. The stakeholder acted as a passenger and tested real-time monitoring function of the green transportation bus and notification of potential passengers sent using the location pin. Remind that passengers' location cannot exceed the 25-meter radius of a specific bus stop. Further, stakeholders acted as the bus driver logged in and chose a particular bus to drive, activated driving status, maneuvered to verify longitudinal location changed and sent to the passenger app, and updated the bus contacts and logged out. At the end of the test, stakeholders agreed and accepted the passenger and the driver apps developed.

3.3 Study Review

Tens of thousands of undergraduate and graduate students, professors, and visitors are present in the campus doing activities that spread over classrooms, in between class schedules, research and education labs, and offices over a reasonably large area. Some sites are within walking distance, but other far distant sites need green transportation, especially for moving class over a limited duration in-between class. Green transportation passengers are very dependent on the presence and timeliness of the bus arriving at the bus stop in order to get to the destination promptly. The passenger app is beneficial for green transportation users to find out the arrival of the bus at a bus stop. In this case, passengers can track the distance of the bus from where they are waiting at the bus stop. The driver app is beneficial for drivers to select any bus parked near the bus garage entrance. It is thus speeding up the work process. In the driver app, there is a notification of the number of passengers waiting at a bus stop. In this way, the bus driver is triggered to go to the target bus stop immediately. More and more passengers waiting at a bus stop will encourage bus drivers to maneuver to the specified bus stop instantaneously.

The limitation of this study is that it only covers a small part of the many elements contained in the smart campus. More elements for the smart campus are available; these are smart inventory, smart parking, automated street light, global systems for mobile communication-based alerts, air quality or noise level, light intensity monitoring system, garbage or waste overflow alert system, and garden automation water flows [20]. Nevertheless, the utilization of GPS technology with Google Maps is expected to provide significant implications in institute/college for the classy environment and less pollution as an endorsement of campus activities that are in line with the smart green campus approach. The concept of smart green campus links green transportation and the Internet of Things (IoT) of GPS features of the Android-based mobile devices. We believed that the majority of campus pupils have at least one mobile device. Green transportation, i.e., gas-fuelled buses, becomes essential to conserve the environment if we compare to gasoline-fuelled buses effects.

4. Conclusion

With the help of green transportation apps, the campus community can navigate the bus movement maneuvering surrounding campus bus stops within the 267-ha area. This navigation is supported by the back-end application that enables the server to receive/send real-time data from the client-side (passenger and driver). The campus community can send PINs location to notify their position to the nearest bus in operation. Multiple driver app features are available, which allow the driver to choose the available bus to ride, find out the number of passengers waiting at the bus stop, and sending its location to passengers. The successful use of green transportation apps on five different Android-based operating systems proven that the study help improves the utilization of campus transportation efficiently and hence supporting the green campus concept.

References

[1] Xue R., Wang J., Chen L., Using the IOT to construct a ubiquitous learning environment, Proceedings of 2011 Second International Conference on Mechanic Automation and Control Engineering. p. 7878-7880, Jul. 2011.
[2] Kwok L.F., A vision for the development of i-campus, Smart Learning Environments-a Springer Open Journal, 2(2):1-3.
[3] Kar A.K., Gupta M.P., How to Make a Smart Campus – Smart Campus Programme in IIT, Report, Indian Institute of Technology Delhi, 2015,
[4] Ng J W.P., Azarmi N., Leida M., Saffre F., Afzal A., and Yoo P.D., The intelligent campus (iCampus): End-to-end learning lifecycle of a knowledge ecosystem, Proc. 6th International Conference on Intelligent Environments, Kuala Lumpur Malaysia, p. 332–337, Dec. 2010.
[5] Aion N., Helmandollar L., Wang M. dan Ng J.W.P., Intelligent campus (iCampus) impact study, Proc. of the 2012 IEEE/WIC/ACM International Conference on Web Intelligence and Intelligent Agent Technology Workshops, Macau, China; p. 291–295, Dec. 2012.
[6] Tan H., Chen S., Shi Q., Wang L., Development of green campus in China. Journal of Cleaner Production, (64), p. 646-653. Feb. 2014
[7] Sahoo M.K., Mishra S.K., Green Campus – A Competitive Advantage and Sustainability for Management Institutions. DRIEM Business Review. 1(1):74-80, 2014
[8] Andriani D.M., Yuliastuti N., Assessment of transportation systems that lead to green transportation in the city of Surakarta (Penilaian sistem transportasi yang mengarah pada green transportation di Kota Surakarta), Journal of Regional and City Development (Jurnal Pembangunan Wilayah dan Kota, 9(2), p. 183-193, 2013.
[9] Jemilda G., Bala R.K., Johnson B., Linga G., Mobile application for college bus tracking, International Journal of Computer Science and Mobile Computing, 4(3), p. 500-507. Mar. 2015.
[10] Supriya S., Pooja S., Monika Z., Roshni J., Sonekar S.V., Real-time college bus tracking application for Android smartphone, International Journal of Engineering and Computer Science, 6(2), p. 20281-20284. 2015.
[11] Namdev D. and Pal Saroha G., Android Based Integrated Vehicle Navigation System, International Journal of Electronics and Communication Engineering and Technology (IJECET), 8(4), p. 1–6, July-August 2017.
[12] Saini J., Agarwal M., Gupta A., Manjula R., Android App Based Vehicle Tracking Using GPS And GSM, International Journal of Scientific & Technology Research, 6(09), p. 53-58, Sep. 2017.
[13] Moses K.B., Dada O.A., Akinwomni F.C., GSM Base Stations Location Monitoring using Geographical Information System, International Journal Computer Network and Information Security, (8), p. 39-45.
[14] Abidin H.Z., Determination of Position Using GPS and its Application (Penentuan Posisi Dengan Menggunakan GPS dan Aplikasinya) Jakarta(ID): PT Pradyan Paramita, 2007

[15] Skvorc D, M Horvat, S Srbljic. 2014. Performance Evaluation of WebSocket Protocol for Implementation of Full-Duplex Web Streams. Information and Communication Technology, Electronics and Microelectronics; 2014 May 26-30; Opatija, Croatia. p. 1003-1008.
[16] Wessels A., Mike P., Jahrain J., Syed R., Remote Data Visualization through WebSockets. 2011 Eighth International Conference on Information Technology: New Generations; p. 1050-1051, Apr. 2011.
[17] Sunyoto A., Overview: Google Maps API V3, Undergraduate dissertation, College of Information and Computer Management (Sekolah Tinggi Manajemen Informatika dan Komputer), Yogyakarta, 2010.
[18] Indonesian Internet Service Provider Association (Asosiasi Penyelenggara Jasa Internet Indonesia), 2016 Indonesian Internet User Profile (Profil Pengguna Internet Indonesia 2016).
[19] Dennis A., Wixom, Barbara H., System Analysis and Design, ed. 5, New Jersey (US): Wiley Inc. 2012.
[20] Shivaraj Kumar TH, Sriraksha TA, and Noor U Saba. An IoT Based Secured Smart e-Campus. International Journal of Humanities and Social Science Invention. Vol 6(3), p.88-93

IOP Conf. Series: Materials Science and Engineering **874** (2020) 012021 doi:10.1088/1757-899X/874/1/012021

Big Data Analytics Technology Capability and Data-Driven Decision Making in Malaysian Higher Education Institutions: A Conceptual Framework

Mohamed Azlan Ashaari, Azlan Amran, Noor Hazlina Ahmad, Hazrieffendy Bakri, Shahrizal Nazri

Universiti Sains Malaysia, Penang, Malaysia,

Abstract. In the era of big data, many organizations are aiming to become more data-driven and increase their decision-making efficiency. Nevertheless, there is an insufficient investigation that explores the antecedents of data-driven decision making (DDDM) in the context of Malaysian Higher Education Institutions (HEIs). Therefore, the study examines existing literature and utilises the Big Data Analytics Technology Capability (BDATC) dimensions such as connectivity, compatibility and modularity to conceptualize a DDDM framework. The study utilises Resource-Based Theory (RBT) to highlight the key dimensions for BDATC and how these dimensions are being integrate in order to establish effective DDDM for excellent performance. The antecedents of DDDM is a relatively new approach that is documented in literature, where literature seems to be diversified in terms of offering theoretical and conceptual frameworks, together with a model that Higher Education Institutions (HEIs) can utilize. The study chooses HEIs that undergo Malaysian Research Assessment Instrument (MyRA) as the focus of investigation. The data was collected from the key informants of Malaysian HEIs. In conclusion, the contribution of the study is to highlight the influence of BDATC for DDDM to attain better performance of HEIs in Malaysia.

1. Introduction

Big data analytics (BDA) and The Fourth Industrial Revolution (4IR) have tremendously driven the global industry to establish an innovative strategy to stay competitive. These scenarios significantly impacted global industry players as they are required to apply new ways of doing things with technology embedded within their business processes [1]. Big data is being denoted to large datasets with sizes beyond the ability of conventional software instruments to capture, curate, manage and process the data within a specified elapsed time [2]. Meanwhile, the Big Data Analytics Capability (BDAC) is being described as an organization's ability to mobilize, deploy and utilize BDA resources efficiently; and to perform effective decision-making for strategy alignment and boosting organizational performance [3]. Despite the importance of big data analytics (BDA) and data-driven decision making (DDDM) [4], more effort needs to be conducted to understand the link between BDATC and DDDM as an emerging subject of an investigation which might bring newness to this phenomenon. Consequently, not very much is known about how BDATC can improve DDDM. As many organizations are still striving to figure out how to use data analytics, the absence of such understanding limits the ability of an organization to effectively leverage BDA for value creation. While waiting for how BDA influences organizational decision-making is better understood, realizing business value from BDA will remain a question [4].

As reported by researchers, the current practices of data analytics in Malaysia's higher education are mostly in the traditional way where the majority of data are scattered all over desktops, offices and set in different arrangements, making them difficult to capture, consolidate and merge [5]. Thus, to practically utilize these data, managing and representing the information while securing accurate data across all databases, is a major reason for implementing big data analytics in higher education. On the other hand, data management is also an issue in actualizing big data analytics in Malaysian HEIs, as information is collected from various sources [6]. This is due to the fact that data are managed by various divisions within the institution with different configurations, and since most divisions would not corporate, gathering all of the data would be a major problem. Therefore, the conceptualization of DDDM in a higher education setting would be the important factor for implementing BDATC towards effective DDDM, and it is presumably critical to attaining superior performance.

Therefore, the subsequent section presents the literature review and related studies. Then, followed by the conceptual framework and the formulation of the hypotheses. The final section was the conclusion.

2. Literature review

Presently, more organizations want to operate smarter, more responsive, and more well-organized businesses by using accurate data to support efficient and effective decision-making [7]. This scenario is generally known as data-driven decision making (DDDM), which emphasizes making decisions based on the analysis of data rather than purely on intuition, political influence and power. DDDM is specifically beneficial and imperative when a business owns large datasets that are interconnected and include time-series data reflecting past, current and future performances [7]. Top management should comprehend that data-driven decision making (DDDM) takes priority in their day-to-day business operation [8].

2.1 The Resource-Based Theory (RBT) as the underpinning theory

The RBT has been extensively known as one of the most popular and effective theories to describe how organizations reach and maintain competitive advantage because of the resources they possess or hold under their supervision [9]. Based on the fundamental concept of RBT, an organization is seen as a collection of invaluable tangible and intangible resources, which can be brought together to create a competitive advantage [10]. The foundation of RBT describes resources as rare, inimitable and non-substitutable organizational exclusive assets that empower an organization to employ a value-creating strategy to produce income [11]. This notion was subsequently separated to differentiate between possession of strategic resource and capability-building, two different mechanisms that are significant to RBT. Possession of strategic resource includes activities of finding and buying or managing resources that are seen as being of strategic value, while capability-building is related to the instrumentation and supervision of these resources into strategically beneficial assets [12]. In addition, resources are tradable and nonspecific organization assets, and capabilities are non-tradable organization-specific capacities that integrate, deploy and use other resources within the organization [13]. Furthermore, there is researcher described the difference between possession of strategic resource and capability-building in their foundational work and illustrated possession of strategic resource is a major feature since it not only benefits the organization by way of obtaining strategic resources but is also imperative for the economic outcome of the organization by evading possibly exhausted or non-valuable resources [12].

2.2 Measuring Malaysian Higher Education Institutions performance based on Malaysian Research Assessment Instrument (MyRA)

Nowadays, the higher education sector in Malaysia is under increasing pressure to transform and be relevant to regional, national, and international demands. Unfortunately, despite huge investments by the Malaysian Government in the higher education sector, the organizational performance of HEIs, in general, is very much lacking. Based on the U21 report, Malaysia ranked 11th highest out of 50 countries, just behind Singapore for government spending on higher education, investment and R&D [14]. However, this substantial investment did not translate into successful outcomes, as Malaysia ranked 39th in terms of research output, institution rankings, enrolment, and employability. In contrast, China, at the 43rd ranking, allocated the least government expenditure on higher education, yet managed to achieve better returns on their investment, as shown by their ranking as the 21st in that particular category. This clearly shows the lack of efficiency in strategic planning on the part of the Malaysian Government's investment in higher education. As a result, Malaysian HEIs are constantly searching for actionable insights for DDDM to generate strategies they can use to meet these new demands such as scoring better results in Malaysian Research Assessment or popularly known as MyRA [15]. MyRA is an important performance indicator for Malaysian HEIs, particularly in research output [16].

MyRA is underpinned by a number of important key areas of measurement. The measurement consists of eight sections (A-H) which are: A) General information, B) Quantity and quality of researchers, C) Quantity and quality of research, D) Quantity of postgraduates, E) Innovations and intellectual property, F) Professional services and gifts, G) networking and linkages, and H) Services. All universities are required to attempt a MyRA self-assessment using the web-based system that takes input data of each criterion and computes the marks obtained, which shows an indicator of a RU qualification. MyRA assessment was developed to inculcate the culture of excellence in research and development in higher education which requires strategic planning to improve performance by ensuring that the best decision is made [15]. In strategic planning, it is important to have comprehensive understanding on the factors and its connection affecting each key area of measurement in order to design relevant policies to improve its performance. However, the web-based system only allows university management to keep track of values and the calculation. Generally, it is a static model that does not attribute to strategic planning or give information on the relationship between variables. Based on that argument, implementing BDAC is capable of capturing the dynamic, complex and non-linear behaviour of a system, and supporting DDDM for strategic planning [17]. Therefore, the focus of this study is to examine whether Malaysian HEIs possess resources to implement BDAC and how BDAC influences DDDM to attain excellent performance, as well as to investigate the relationships between various elements and their interrelationships that generate performance improvement over time.

Therefore, DDDM has the potential to enable institutions to thoroughly examine their present challenges, identify ways to address them as well as predict possible future outcomes. However, because the role of BDATC and DDDM is a new phenomenon in higher education, its conceptual relevance, as well as the opportunities and limitations it might bring, is still unknown [18]. The study describes the conceptual underpinning of DDDM research and presents possible opportunities as well as limitations associated with unlocking the value of BDAC in higher education.

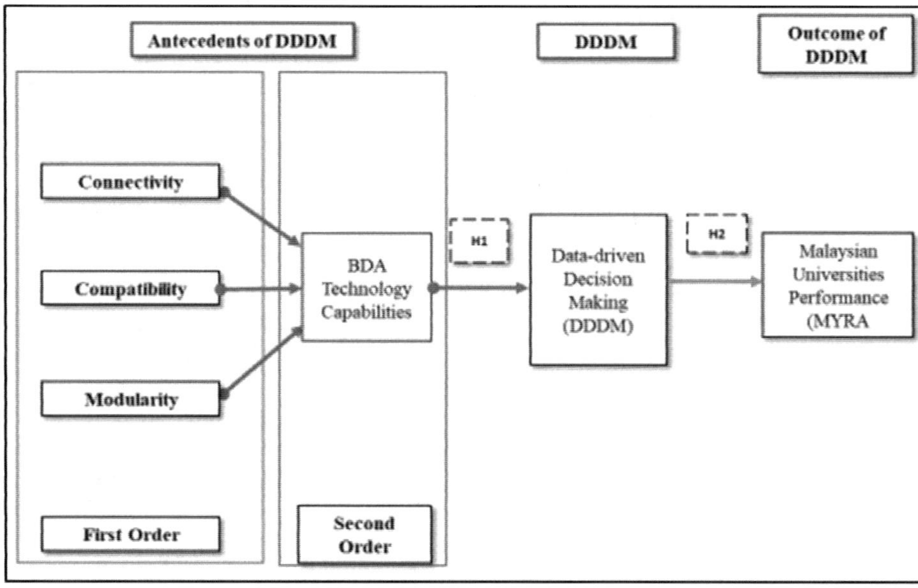

Figure 1. Proposed conceptual framework.

3. Conceptual framework and formulation of hypothesis

In order to further progress, more effort needs to be initiated in order to explore the components of DDDM capabilities and its outcomes. Thus, the study reviews the past literature review on BDAC in order to examine its influence on DDDM. The study then summarizes three approaches on how BDAC can influence DDDM, which is the Big Data Analytics Technology Capabilities (BDATC). Then, via applying the Resource-Based Theory (RBT), the identification of BDA resources to develop BDATC for effective DDDM for the purpose of improving the performance of Malaysian HEIs would be achieved. Based on Figure 1, the research proposed BDATC as the antecedents of DDDM and the performance of HEIs as the outcome of DDDM. Furthermore, several researchers highlight that technology dimensions are important in a big data environment, which helps each other improve DDDM in order to enhance firm performance [19–22]. These dimensions of BDATC and their relationships with DDDM are supported by scholar who highlighted that technology management capability is essential to explore and manage a variety of data [23].

The study hypothesized that BDATC has a positive impact on DDDM in Malaysian HEIs. The analysis that examined the influence of BDATC on DDDM suggested that connectivity, compatibility, and modularity were important dimensions for DDDM in order to achieve better HEIs performance. As literature in BDATC has devoted, organizations should be willing to allocate investments in BDATC to the gain success of their organizations [23,24]. Organisations that concentrate on BDATC for DDDM will influence and shape organizations in their business activities and will push their competencies toward superior performance. In Malaysian HEIs, the ability of infrastructure and technical software to quickly improve, deploy and support necessary system components [25] is crucial for DDDM in order to develop an effective strategic plan to improve performance of HEIs. Thus, in a dynamic, competitive and unpredictable market nowadays, the tendency of HEIs being exposed to technological change and demands due to the changing landscape of higher education and the impact of globalization is intense.

The outcome from the study will outline the influence of connectivity, compatibility and modularity capability towards DDDM for an effective strategic plan. For example, uncertainties or changes could be triggered by globalization and thus technologies will provide HEIs with a time frame to prepare an efficient response in confronting the changes. Therefore, in the context of HEIs, the antecedents of DDDM have been deliberated on and discussed in the literature review according to the three (3) dimensions of BDATC such as 1) BDA infrastructure (Connectivity), 2) BDA process integration (Compatibility) and 3) BDA standardization and flexibility (Modularity). Therefore, based on the abovementioned figure 1, the study hypotheses are being proposed as follows;

H1: BDATC has a significant positive effect on DDDM.
H2: DDDM has a significant positive effect on Malaysian HEIs performance.

4. Conclusion

In a nutshell, founded on the research framework established in the study, there are three significant BDATC resources, which are connectivity, compatibility and modularity that can directly influence DDDM and have a significant relationship with DDDM and the performance of HEIs in the context of Malaysian HEIs. These three significant BDATC resources that are essential to HEIs were proven through experimental investigation. The study discovered that in order for HEIs to attain greater performance, these three BDA resources must be put in place. By pursuing RBT, the study has leverage on the proposing theory by constructing a framework for scrutinising the antecedents and outcome of DDDM towards achieving greater performance from HEIs. These three BDATC resources are vital for Malaysian HEIs to distinguish them from their rivals. An invaluable discovery with experimental evidence has verified and recommended that BDATC dimensions such as connectivity, compatibility and modularity should be the antecedents of DDDM for Malaysian HEIs to concentrate on and improve in order to attain superior performance for HEIs. The impact of combining both BDA resources together has formed effective DDDM, thus affecting the performance of Malaysian HEIs in particular. Over and above, the study also examined the effect of DDDM on the performance of HEIs. DDDM was exhibited based on BDATC. Therefore, the purpose of the theoretical framework constructed upon the study is to harmonise the relationship between the BDA resources and BDATC needed for DDDM in a challenging and volatile landscape. The extraction of BDA resources and BDATC was planned to determine their effects of DDDM towards the performance of HEIs.

In summary, the study took another step forward by recommending the examination of RBT in the context of BDATC and DDDM, and also the investigation of BDATC in the context of DDDM from two different viewpoints: i) a developing country which in this case is Malaysia and ii) the higher education sector. The study has added to the body of knowledge by offering three new discoveries as deliberated before, related to BDATC dimensions and the performance of HEIs that will have a huge effect towards the performance of Malaysian HEIs and the necessity of having an effective DDDM when it comes to the relationship between BDATC and the performance of HEIs. In conclusion, the contribution of the study is to highlight the influence of BDATC for DDDM to attain better performance of HEIs in Malaysia as well as assisting the policy-makers in Malaysian higher education industry to understand the role of BDATC in influencing the DDDM towards achieving the vision and mission set by Ministry of Education (MOE), Malaysia.

5. References

[1] Akter S, Wamba S F, Gunasekaran A, Dubey R and Childe S J 2016 How to improve firm performance using big data analytics capability and business strategy alignment? *Int. J. Prod. Econ.* **182**

[2] Bharadwaj A, El Sawy O A, Pavlou P A and Venkatraman N 2013 Digital business strategy: Toward a next generation of insights *MIS Q. Manag. Inf. Syst.*

[3] Cao G and Duan Y 2014 A path model linking business analytics, data-driven culture, and competitive advantage *ECIS 2014 Proceedings - 22nd European Conference on Information Systems*

[4] Cao G, Duan Y and Li G 2015 Linking Business Analytics to Decision Making Effectiveness: A Path Model Analysis *IEEE Trans. Eng. Manag.* **62** 384–95

[5] Hanapiyah Z M, Wan Hanafi W N and Daud S 2018 Issues, Challenges and Opportunities of Big Data Management in Higher Education Institutions in Malaysia *Indian J. Sci. Technol.* **11** 1–6

[6] Rahmat A, Ahmad A and Ta'a A 2016 Business Intelligence Readiness for Higher Learning Institution (IHL): Preliminary Study and Research Model *Br. J. Appl. Sci. Technol.* **15** 1–10
[7] Hall D J, Jia L and Song J 2015 The conceptualization of data-driven decision making capability *Am. Conf. Inf. Syst.*
[8] Aksoy L 2013 How do you measure what you can't define?: The current state of loyalty measurement and management *J. Serv. Manag.*
[9] Barney J B 2001 Resource-based theories of competitive advantage: A ten-year retrospective on the resource-based view *J. Manage.*
[10] Mikalef P, Pappas I O, Krogstie J and Giannakos M 2018 Big data analytics capabilities: a systematic literature review and research agenda *Inf. Syst. E-bus. Manag.* **16** 547–78
[11] Barney J 1991 Firm Resources and Sustained Competitive Advantage *J. Manage.*
[12] Makadok R 1999 Interfirm differences in scale economies and the evolution of market shares *Strateg. Manag. J.*
[13] Amit R and Schoemaker P J H 1993 Strategic assets and organizational rent *Strateg. Manag. J.*
[14] Universitas 21 2017 US21 Ranking of National Higher Education Systems 2017 36
[15] Ong M Y, Muniandy B, Ong S L, Tang K N and Phua K 2013 User acceptance of key performance indicators management systems in a higher education institution in Malaysia : A pilot study *Int. Online J. Educ. Sci.* **5(1)** 22–31
[16] Hock-Eam L, Taib F M, Adiana N, Abdullah H and Siew Hwa Y 2016 How Efficient Are Malaysian Public Universities? a Comparative Analysis Using Data Envelopment Analysis *Asian Acad. Manag. J.* **21** 75–97
[17] Abidin N Z, Zaibidi N Z and Karim K N 2017 Strategic planning for MyRA performance: A causal loop diagram approach *AIP Conf. Proc.* **1891**
[18] Menon M E, Terkla D G and Gibbs P 2014 *Using data to improve higher education: Research, policy and practice*
[19] Wamba S F, Gunasekaran A, Akter S, Ren S J fan, Dubey R and Childe S J 2017 Big data analytics and firm performance: Effects of dynamic capabilities *J. Bus. Res.* **70** 356–65
[20] Davenport T H, Barth P and Bean R 2012 How "big data" is different *MIT Sloan Manag. Rev.*
[21] Elgendy N and Elragal A 2016 Big Data Analytics in Support of the Decision Making Process *Procedia Comput. Sci.* **100** 1071–84
[22] Mikalef P, Krogstie J, Mikalef P, Danielsen F and Olsen D H 2017 Big Data Analytics Capability : Antecedents and Business Value Completed Research Paper *Twenty First Pacific Asia Conf. Inf. Syst.*
[23] Akter S, Wamba S F, Gunasekaran A, Dubey R and Childe S J 2016 How to improve firm performance using big data analytics capability and business strategy alignment? *Int. J. Prod. Econ.* **182** 113–31
[24] Tulasi B 2013 Significance of Big Data and Analytics in Higher Education *Int. J. Comput. Appl.* **68** 21–3
[25] Adrian C, Abdullah R, Atan R and Jusoh Y Y 2018 Conceptual Model Development of Big Data Analytics Implementation Assessment Effect on Decision-Making *Int. J. Interact. Multimed. Artif. Intell.* **5** 101

IoT Business Model Development for Food Safety Monitoring System in the Poultry Slaughterhouse

Achmad Fadillah, and Dikky Indrawan*

IPB University, Bogor, Indonesia

*rdikky@apps.ipb.ac.id

Abstract. Internet of Things (IoT) has changed the industrial direction and open a new opportunity for the food business. The food safety business took this technological leap to increase food safety in the industry. The technological leap focuses on the improvement of traceability system in order to guarantee the appropriate level of protection for biological and chemical hazards in the poultry slaughtering. Consumers' concerns about food safety are the key to new technological businesses by using the internet platform and reengineer the computer system. The different business model for IoT provides a system building for new ventures. This paper aims to describe a new business based on a new internet technology development. An internet-based business on food safety notification for the slaughterhouse with regards to food safety hazards is the main business backbone. The system development will be based on a sensor and link with the android system that integrated conveyor system to detect the possible biological and chemical hazards. The future new business model of IoT system is integrated with HACCP sensors.

1. Introduction

The food business opportunity and direction were changed by the Internet of Things (IoT) [1]. The food safety business took this technological leap to increase food safety in the industry [2]. The technological leap is giving a new opportunity for the future development of the industry. One of the opportunities is an improvement of the traceability system in order to guarantee the appropriate level of protection for biological and chemical hazards in the poultry slaughtering. Since many consumers concern with their food safety and food sources, which previous study prove the consumer switch their buying habit due to food unsafety incident [3]. Consumers' concerns about food safety are the key to new technological businesses by using the internet platform and reengineer computer systems [2]. The different business model for IoT provides a system building for new ventures. The Internet of Things (IoT) in poultry slaughterhouse is the network of physical devices of slaughter plant appliances that work via electronic equipment. The network uses software to connect, collect, and connect data using sensors, actuators, and connectivity [4]. Each stage with food safety hazards is identified and embedded in the computing system within the Internet infrastructure. The conceptual idea of using IoT as a strategic business needs to be evaluated. Therefore, this study focuses on the use of IoT to safeguarding food safety and control the appropriate level of protection in the poultry slaughterhouse. This paper aims to describe a new business internet based on new internet technology development.

2. Materials and methods

The paper proposed an initial concept of IoT in food safety monitoring as a business model. The proposed model was developed by the basics of 3D scanning, image analysis, comparison, and classification. The model developed by two approached. First, this paper used a qualitative approach to identify business strategy. We identified the food safety monitoring system as a strategy to develop business model types by Porter's generic strategies [5, 6]. Second, adapted business model canvas to develop the value proposition for business model [7, 8]. The nine elements of the business model canvas analyzed the food safety monitoring system for a new business. These nine elements are business model canvas (BMC) are customer segments, value propositions, channels, customer relationships, revenue streams, key resources, key activities, key partnerships, and cost structure. BMC was mapped to describe food safety notification for the slaughterhouse with regards to food safety hazards.

3. Results and Discussion

A food safety notification for the slaughterhouse is the main business idea. The backbone of this idea with regards to food safety hazards is to notify the food unsafety event during slaughtering to the company. The concept of system development will be based on a sensor and link with the android system that integrated conveyor system to detect the possible biological and chemical hazards. The scanning process will give an input for image comparison analysis. A classification of microbial pathogens will process the data from the comparison stage. The sensor improvement focuses on the detection of nanomaterials by electrochemical or optical sensors.

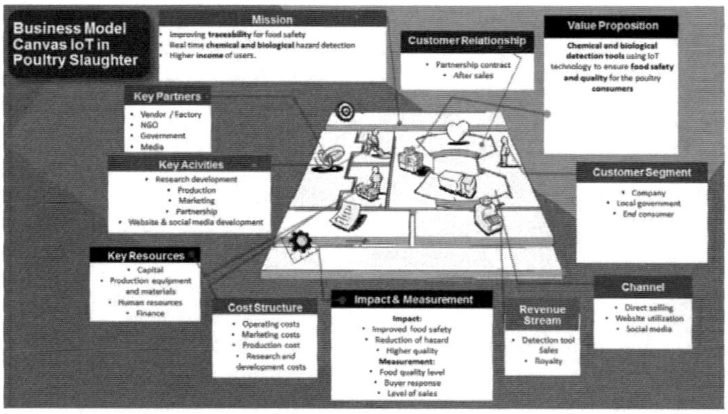

Figure 1. Business canvas model for Food safety monitoring system in the Poultry Slaughterhouse

The study found that the business model should develop a focus strategy from Porter as a strategy to be competitive. A narrow focus supports the company on a segmentation strategy or even a more niche strategy. Since a competitive advantage will be gaining from this market by product innovation and/or brand marketing. We found that the competitive advantage provided by the focus strategy is differentiation focus to develop a food safety monitoring system. The main value proposed is chemical and biological detection tools based on food safety notification for the slaughterhouse. The food monitoring system by utilizing IoT in supporting to improve food safety and traceability. BMC proposed companies as the customer segment. It will be easier to maintain companies compare to the retail market. However, the end-consumers could be developed as the customers as well, since many small-medium enterprises working in the slaughtering business. This market development needs customer relationship programs to keep and increase consumer loyalty. Such loyalty program that could be offered by BMC were partnership contract and after-sales program. It can be followed by using other marketing channels, namely direct selling, website utilization, and social media optimization. The IoT business model for food safety monitoring was proposed to be more focused on the core competencies of research, design, development sensor, and nanomaterial detection technology [1]. This business model optimized innovative technology by collaborating with key partners to enter the new market. By focusing on the specific segment, IoT development for food safety monitoring could be reduced, and limited capital investment. Another benefit of the focus strategy by Porter was the ability to drive the demand. The strategy classifies as product-driven that will be able to pull the consumers and adapt to the economic situation.

The BMC featured key activities in research development to support the innovation on production, marketing channels via a partnership with many websites and social media [7]. The BMC proposed the value of a better way to integrate sensor technology with nanomaterials and pathogen into food safety grids and technically based on more effective hazard prediction [4]. It will be improving the diagnosis and treatment of poultry slaughtering with connected devices for better assessment in slaughterhouse. The application will be given to real-time dashboards in smart devices that will ultimately deliver real-time insight to the company and safeguarding millions of customers' health. The developed feature could give more trust toward poultry consumption in their choice of market channels and guarantee their food safety [3]. In the end, technology can protect the traditional and modern poultry market channels, and it will reduce the possibility of an unhealthy poultry market to supply the market with sick poultry [9].

4. Conclusion

The paper presented an initial study for IoT in the food monitoring business by using BMC. The study found that the offered value from the BMC was a sensor to detect the possible biological and chemical hazards. The focus strategy will give a competitive advantage in the IoT business that drives technology improvement and product innovation. The future new business model of IoT system in the poultry slaughterhouse is integrated with HACCP sensors

References

[1] G. Witjaksono, A. A. S. Rabih, N. bt Yahya, and S. Alva, "IOT for Agriculture: Food Quality and Safety," in *IOP Conference Series: Materials Science and Engineering*, 2018, vol. 343, no. 1: IOP Publishing, p. 012023.

[2] M. Archana, S. Uma, and T. R. Babu, "Monitoring and Controlling of Poultry Farm Using IoT," *International Journal of Innovative Research in Computer and Communication Engineering,* vol. 6, no. 4, pp. 3573-3579, 2018.

[3] D. Indrawan, G. Tacken, and H. Hogeveen, "What drives the choice of poultry market channel and the change of purchase behavior due to highly pathogenic avian influenza outbreaks?," *Poultry science,* vol. 97, no. 10, pp. 3652-3660, 2018.

[4] C. Verdouw, H. Sundmaeker, B. Tekinerdogan, D. Conzon, and T. Montanaro, "Architecture framework of IoT-based food and farm systems: A multiple case study," *Computers and Electronics in Agriculture,* vol. 165, p. 104939, 2019.

[5] G. G. Dess and P. S. Davis, "Porter's (1980) generic strategies as determinants of strategic group membership and organizational performance," *Academy of Management journal,* vol. 27, no. 3, pp. 467-488, 1984.

[6] R. Tanwar, "Porter's generic competitive strategies," *Journal of business and management,* vol. 15, no. 1, pp. 11-17, 2013.

[7] A. Osterwalder and Y. Pigneur, "Designing business models and similar strategic objects: the contribution of IS," *Journal of the Association for information systems,* vol. 14, no. 5, p. 3, 2012.

[8] A. Osterwalder, Y. Pigneur, G. Bernarda, and A. Smith, *Value proposition design: How to create products and services customers want*. John Wiley & Sons, 2014.

[9] D. Indrawan, K. M. Rich, P. van Horne, A. Daryanto, and H. Hogeveen, "Linking supply chain governance and biosecurity in the context of HPAI control in Western Java: a value chain perspective," *Frontiers in veterinary science,* vol. 5, 2018.

Prototype Application of Online Mental - Health Screening Design Among Junior and High School Through Strength and Difficulties Measure

E Sirait*, B Utomo and K Siregar
Universitas Indonesia, Depok, Indonesia

*efridasirait@yahoo.com

Abstract. Mental disorders are a major cause of disability and death throughout the world every year. Indonesia has low mental health and is mostly experienced by adolescents. Indeed, recognizing the signs and symptoms of stress in adolescents early is a preventive measure before the condition worsens and causes many losses. Mental health service programs have so far been limited, both in schools and in health care centres. Facing such conditions, this study proposes a mobile application model that can be an early detection tool for adolescent mental health and information sharing between adolescents and teachers or health workers. The screening results will be followed up by the health service if a problem is found. This research builds a prototype of the mobile application to the modelling stage. The model is produced from the system requirements analysis, the model presented here is in the form of; Context Diagram, Entity Relational Diagram (ERD), Relational Diagram Table (TRD), Flow Chart and User Interface Design. Early mental health detection uses the Strength and Difficulties Questionnaire (SDQ). Data collection is done by interview, observation and study of literature. This information system can be used by primary health care and implemented in schools, making it easier for adolescent mental health screening.
Keywords: Mental Health, Screening, Mobile Application, Junior and High School, Strength and Difficulties Questionnaire (SDQ)

1. Introduction

People with mental health are those who are mentally healthy. They are able to accept themselves as a whole, have stable emotions, good social community, and able to take responsibility for themselves and their environment, and have expectations that are in line with their potential. While people who are not mentally healthy are those who experience mental health disorders. They are anxiety, depression, dissociative disorders, somatoform, suicide thought, and personality problem [1].

Adolescence is a unique period. Characterized by emotional behaviour and related neural circuitry. Cortolimbic circuitry will develop dynamic changes, which play a role in both healthy and disruptive cognitive control of emotion [2]. During development, adolescent brains have the most elastic or plastic than childhood or adulthood and are able to adapt extraordinary to social, physical, sexual, and intellectual challenges [3].

Mothers of anxious children, regardless of their own anxiety status, were found to lack autonomy and less warm of act. Both the behaviour of parents and adolescents influence each other, so the style parenting for children with anxiety can be the impact of adolescent behaviour, not just causative factors [4].

IOP Conf. Series: Materials Science and Engineering **874** (2020) 012023 doi:10.1088/1757-899X/874/1/012023

Schools is a place where teenagers spend a lot of time every day. Almost all teenagers go to school and faced with cognitive and social demands. School is an important place for adolescent mental health, for it is place to gain knowledge and inspiration for some people, while for others it may be a place of despair and failure [5].

The World Health Organization reported that 10-20% of children and adolescents around the world have mental health problems [6]. It is estimated that 50% of all mental disorders are experienced at the age Of 14 years and 75% at the age of 18 years [7].Most common disorders in children and adolescents is a generalized anxiety disorder and depression [8];[9].

According to the Royal Society for Public Health and Young Health Movement, the prevalence of anxiety and depression has increased by 70% in the last 25 years in young people [10].

Mental disorders are a major cause of disability [11] and contribute around 8 million deaths worldwide each year. This is a major problem in public health [12] .People with mental disorders can die 10-25 years younger than the general population [13] .Only a few people receive care with mental disorders [14] and often the quality of care is below service standards [15]. Public health workers must be able to develop, implement and disseminate programs and policies to prevent mental disorders, increase access and ensure quality care and improve quality of life in people with mental disorders [16].

The prevalence of stress in adolescents tends to increase [17]. About 30% of adolescents have reported problems with feeling sad or depressed to disrupt their daily activities [18].

Base on Riskesdas 2018, the prevalence of mental- emotional disorders in people aged> 15 years increased from 6% (Riskesdas 2013) to 9.8% [19]. The total population of Indonesia in 2019 reached 268 million, with a total age group of 0-19 years totalling 47,352,000 inhabitants, which is about 17.6% of the total population of Indonesia [20] .In the future, adolescents will be the main drivers of economic and social change, so it needs to be the focus of attention and a strategic centre of intervention in human resource development.

Based on the results of the study of Rofingatul Mubasyiroh, et al, in 2015 showed that there were 60.17% of junior- high-school students in Indonesia experiencing symptoms of mental -emotional disorders. With details of symptoms; feeling lonely by 44.54%, feeling anxious by 40.75% and ever wanting to commit suicide by 7.33%. The highest mental social symptoms based on gender are 64.8% Female students with the highest age category being 13-15 years old at 62.3% [21].

The increase in mental health problems is proportional to the increase in mental health services. The reach of mental health services must reach rural communities, not just in big cities. This is included in the distribution of health services. The number of mental health workers is still very limited, the people themselves are less concerned with their mental health conditions so that they cannot be detected and not handled properly [22].

Adolescent health care is more challenging than children and adult, for their physical, intellectual, and emotional development is developing rapidly [23]. The presence of primary care providers an important opportunity to screen teenagers with mental health problems and refer them to those who need care. About 70% of teenagers have contacted doctors in the last 12 months [24];[25] .

According to WHO, a mental health information system is a "system for collecting, processing, analysing, disseminating and using information about mental health services and other people's mental health needs [26]. This can help develop a system of care that is fair and appropriate in certain contexts, can help improve physician communication, adherence to best practice norms and reduce errors in clinical care. In addition, mental health service providers can identify patterns in presenting complaints and treatment needs, which in turn informs the delivery of services and practices.

The priority program at primary health care is physical health. Mental health is one of the efforts to develop health in community health centres. There are no operational standards that are applied to adolescent mental health screening. Mental health monitoring activities for adolescents only during school visits that are in line with school activity units. Therefore, adolescent mental health conditions cannot be monitored.

The use of mobile phones has become a major need at this time, especially among teenagers to communicate with parents and friends. The internet is also a major need in all circles. Seeing this condition and potential, this research forms an application model that can detect adolescent mental health quickly and precisely in large populations. This application is used to share information with teenagers, teachers and health workers. Therefore, people with mental health problems can be found immediately to be given care or treatment to prevent mental health. Students with problematic data will be integrated into the web of health workers. So, good collaboration between teachers and health workers in realizing good student mental health.

2. Method

To answer the research objectives above, the method to be used is prototyping. However, the stages used in this research are in the form of literature studies which are then continued with modeling as the basis for prototyping web- and android-based teen mental health applications. This research produces modeling that focuses on the early stage System Requirement Analysis, Context Diagram (CD), Entity Relational Diagram (ERD), Relational Diagram Table (TRD), Flow Diagram and Interface Design [27]. The application that will be built contains screening of adolescent mental health in schools using a standardized questionnaire from Strength and Difficulties Questionnaire (SDQ) [28] .This modeling presents a general description of the application to be built.

3. Results and Discussion

The application model is designed to make it easier for users to understand the information system to be built. Base on the application, users can get information quickly, cheaply and precisely. This condition is obtained from qualitative study results from qualitative study by interview, observation and study literature. The following is a system requirements analysis for mental-health.

3.1. System Requirements Analysis

Table 1. System Requirements Analysis

	Current Conditions	Needs
Input	There is no valid standard operational of public health services to screen adolescent mental health at school. In general, physical health examine are the main program, while mental health is rarely conducted particularly in schools.	The mental health screening application design can be used personally to find out early adolescent mental health in preventing mental disorders and disabilities. Data obtained from students answers through the system and managed by the teacher
Process	Adolescents have emotional levels that are unstable and hard to express it. Because its emotional, relation among peers, teachers and community become unproductive. Significant changes are marked by low achievement of academic and lack of interest.	Development of an early mental health detection application is complemented by communication between students, teachers, health services and parents through the feature or content provided, like video, post share and article.
Output	There are no accurate reports or records regarding the mental health conditions of adolescents. School activities by the health unit are not monitored. Lack of health workers who are experts in mental health may cause inappropriate management	Application development will be equipped with a report format for mental health status. There is intervention by health worker to solve the student's problem with mental health by reported of teachers. And, rise the relation among students-teachers-parents and health workers.

IOP Conf. Series: Materials Science and Engineering **874** (2020) 012023 doi:10.1088/1757-899X/874/1/012023

3.2. Context Diagram
The Context diagram pays attention to system boundaries, interactions between external entities and the system. Can be seen in Figure 1.

3.3. Entity Relational Diagram
The ERD describes the relationships between entities in the system. Each entity has attributes and primary keys as database in the application. Can be seen in Figure 2

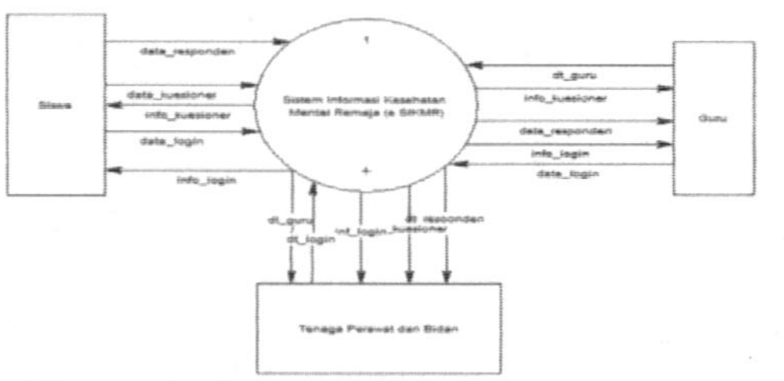

Figure 1. Context Diagram

Figure 2. Entity Relational Diagram (ERD)

3.4. Table Relational Diagram (TRD)
The TRD describes the relationship between entities. The relation between school and students is *one-to-one*, the relation between respondents/user and questionnaire is *one-to-many*, the relation between the teacher and the questionnaire is *one-to-many*, the relation between the nurses / midwives and the teacher is *one -to-many*, the relation between nurses / midwives with reports is *one-to-many*. Can be seen in Figure 3.

3.5. Flow Chart
The Flow chart is a description of the processes that exist in the system. Can be seen in Figure 4.

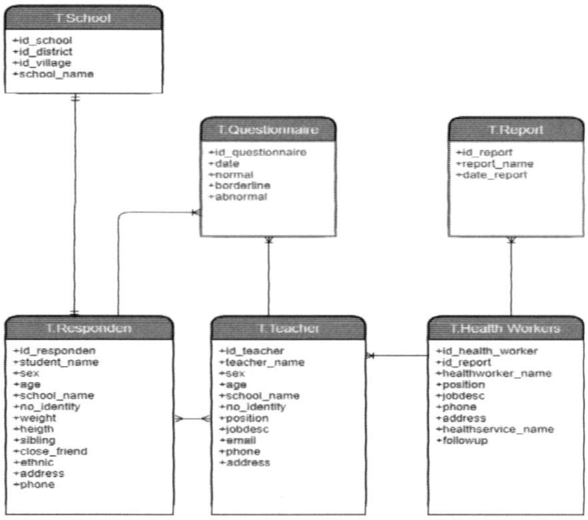

Figure 3. Table Relational Diagram (TRD)

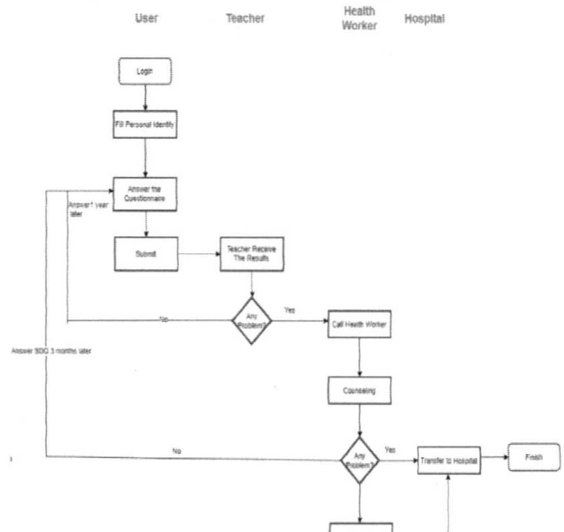

Figure 4. Flow Chart

3.6. Design Interface
There are two design interface will be built, by mobile application and web base. Mobile application is used by students and teachers, while web base is used by health workers. The screening process will be followed up by health workers who are responsible for the school as a work environment.

3.7. Mobile Application

Figure 5. Main Page

Figure 6. Login Page

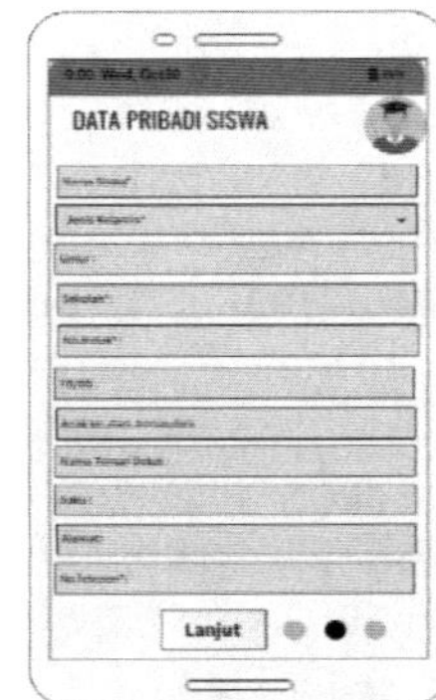

Figure 7. Student Data Page

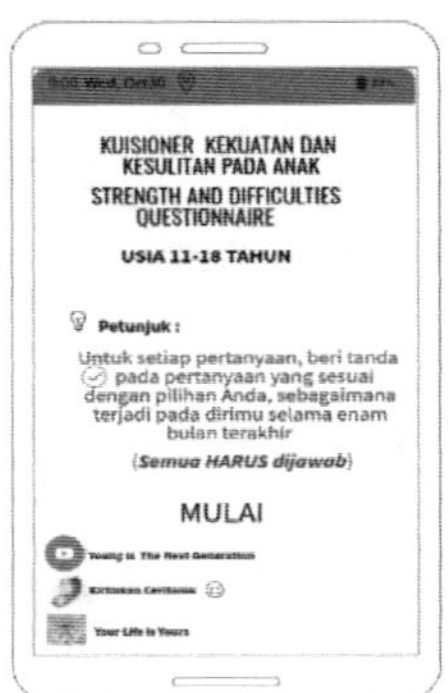

Figure.8 Questionnaire

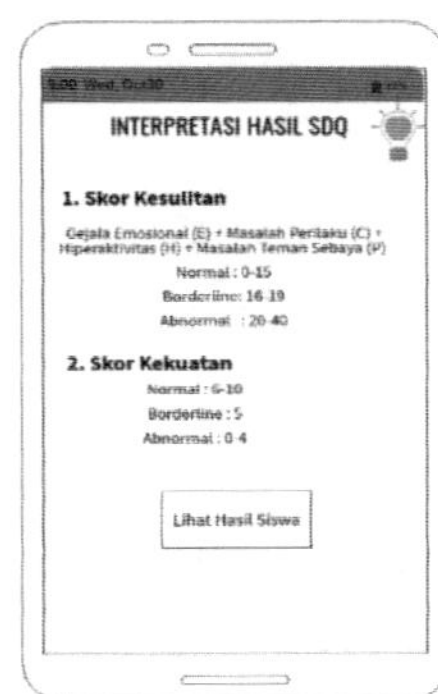

Figure 9. Page Students Scores

Figure10. Recapitulation of student data

3.8. Web (Health Worker)

Design interface for health workers (midwives and nurses) using the web managed by health workers themselves. The following interface design drawings below.

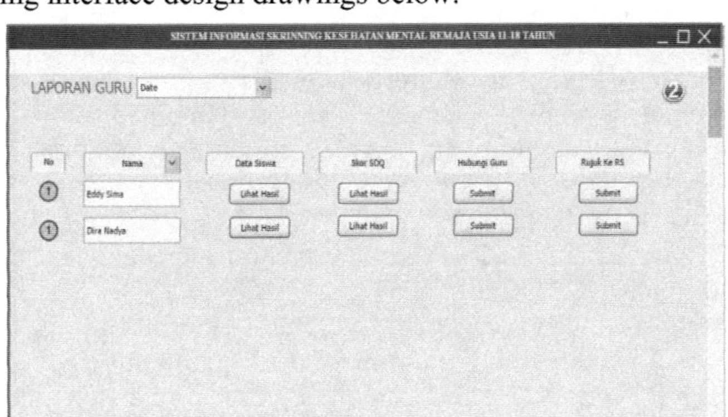

Figure 11. Display of Mental Health Information Systems web based

Teen mental health screening plan made in the form of android and the web. Online mental health screening activities are carried out to get information quickly and accurately with a large population and a short time. Each primary health care has a working area in the school in the school activity unit for junior and senior high schools in obtaining health services, both physical and mental health.

IOP Conf. Series: Materials Science and Engineering **874** (2020) 012023 doi:10.1088/1757-899X/874/1/012023

Health workers and teachers will work together in conducting mental health screening to students who are the responsibility of the teacher. Students will be directed to login the application and answer all questions according to what is considered correct. The results of the questions will be automatically recapitulated in the teacher's account if the student is already submitting in theirs account. The teacher can see student results without the students concerned being known. The system will interpret the results. If the student scores are normal then the submit button does not appear in the application, if the score is borderline and abnormal, a submit button will appear which will be integrated directly into the web at the health centre managed by nurses or midwife.

Health workers will counsel students with problems, by first communicating with teachers and parents of students. The last thing health workers can do for abnormal students is to refer to the psychiatric hospital for appropriate treatment.

Knowing the mental state of students, making teachers understand the condition of students in dealing with it as well as parents.

This application will be provided with interesting and positive content in the form of videos that can be opened at any time and up to date, youth articles and post shares that can be used to share stories and conditions experienced.

4. Conclusion

Adolescent mental health information systems produced in this study are at the modelling stage, for web-based and android applications. This research needs to be followed by an assessment to ascertain what the needs of potential users.

This application is planned to be used in primary health services, schools, for adolescent mental health screening. This application will have content that is interesting for teens, such as teen mental health videos, teen articles and sharing posts, then features for intensive communication between teachers, students, health workers, peers and parents.

References

[1] Kusumawati Farida. Buku Ajar Keperawatan Jiwa. Jakarta: Salemba Medika; 2010.

[2] Gee DG, Caballero C. Predicting Mental Health in Adolescence: Frontoinsular Circuitry, Emotion in Daily Life, and Risk for Depression. Biol Psychiatry Cogn Neurosci Neuroimaging [Internet]. 2019;4(8):684–5. Available from: https://doi.org/10.1016/j.bpsc.2019.06.004

[3] Paus T, Keshavan M, Giedd J. Paus T, Keshavan M, Giedd JN. Why do many psychiatric disorders emerge during adolescence? Nat Rev Neurosci 9: 947-957. Nat Rev Neurosci. 2009 Jan 1;9:947–57.

[4] Moore PS, Whaley SE, Sigman M. Interactions between mothers and children: Impacts of maternal and child anxiety. J Abnorm Psychol [Internet]. 2004 Aug [cited 2020 Feb 26];113(3):471–6. Available from: http://www.ncbi.nlm.nih.gov/pubmed/15311992

[5] Ogden T, Hagen KA. Adolescent Mental Health : Prevention and Intervention. Second. London and New York: Taylor & Francis Ltd.; 2005.

[6] World Health Organization. Maternal, newborn, child and adolescent health. WHO. 2017;

[7] Kessler RC, Amminger GP, Aguilar-Gaxiola S, Alonso J, Lee S, Ustün TB. Age of onset of mental disorders: a review of recent literature. Curr Opin Psychiatry [Internet]. 2007 Jul;20(4):359–64. Available from: https://pubmed.ncbi.nlm.nih.gov/17551351

[8] Morgan C, Webb RT, Carr MJ, Kontopantelis E, Green J, Chew-Graham CA, et al. Incidence, clinical management, and mortality risk following self harm among children and adolescents: Cohort study in primary care. BMJ. 2017;359.

[9] Mental Health Foundation. Children and young people [Internet]. 2018 [cited 2020 Feb 26]. Available from: https://www.mentalhealth.org.uk/a-to-z/c/children-and-young-people

[10] Royal Society for Public Health. Status of Mind: Social media and young people' s mental health and wellbeing. R Soc Public Heal. 2017;(May):1–32.

[11] Whiteford HA, Degenhardt L, Rehm J, Baxter AJ, Ferrari AJ, Erskine HE, et al. Global burden

of disease attributable to mental and substance use disorders: findings from the Global Burden of Disease Study 2010. Lancet [Internet]. 2013 Nov 9 [cited 2019 Nov 3];382(9904):1575–86. Available from: https://www.sciencedirect.com/science/article/pii/S0140673613616116

[12] Walker ER, McGee RE, Druss BG. Mortality in mental disorders and global disease burden implications: a systematic review and meta-analysis. JAMA psychiatry [Internet]. 2015 Apr 1;72(4):334–41. Available from: https://www.ncbi.nlm.nih.gov/pubmed/25671328

[13] Colton CW, Manderscheid RW. Congruencies in increased mortality rates, years of potential life lost, and causes of death among public mental health clients in eight states. Prev Chronic Dis. 2006;3(2):1–14.

[14] Kohn R, Saxena S, Levav I, Saraceno B. The treatment gap in mental health care. 2004;005736(03).

[15] Wang PS, Lane M, Olfson M, Pincus HA, Wells KB, Kessler RC. Twelve-month use of mental health services in the United States: Results from the National Comorbidity Survey Replication. Arch Gen Psychiatry. 2005;62(6):629–40.

[16] Reisinger Walker E, Jennie Kwon M, Delia Lang BL, Richard Levinson MM, Benjamin Druss MG. Mental Health Training in Schools of Public Health: Histor,Current Status, and Future Opportunities [Internet]. Available from: http://ceph.

[17] Anniko M, Boersma K, Tillfors M. Adolescent Stress : Prevalence , Sources of Stress & the Role of Cognitive Processes. 2017;(September).

[18] Li Q, Xue Y, Zhao L, Jia J, Feng L. Analyzing and Identifying Teens' Stressful Periods and Stressor Events from a Microblog. IEEE J Biomed Heal Informatics. 2017;21(5):1434–48.

[19] Kementerian Kesehatan Republik Indonesia. HASIL UTAMA RISKESDAS 2018. 2018;

[20] Badan Pusat Statistik (BPS). Proyeksi Penduduk Indonesia 2010-2035. Jakarta; 2013.

[21] Mubasyiroh R, Suryaputri IY, Tjandrarini DH. Determinan Gejala Mental Emosional Pelajar SMP-SMA di Indonesia Tahun 2015. Bul Penelit Kesehat. 2017;45(2):103–12.

[22] Afifah KA. Literasi Kesehatan Mental Pada Tenaga Kesehatan. Univ Muhammadiyah Surakarta. 2016;

[23] Stang J, Story M. Adolescent Growth and Development A [Internet]. 2005 [cited 2020 Feb 26]. Available from: http://www.epi.umn.edu/let/pubs/adol_book.shtm

[24] Das JK, Salam RA, Lassi ZS, Khan MN, Mahmood W, Patel V, et al. Interventions for Adolescent Mental Health: An Overview of Systematic Reviews. J Adolesc Heal [Internet]. 2016 Oct 1 [cited 2020 Feb 26];59(4):S49–60. Available from: https://www.sciencedirect.com/science/article/pii/S1054139X16301665

[25] Salam RA, Arshad A, Das JK, Khan MN, Mahmood W, Freedman SB, et al. Interventions to Prevent Unintentional Injuries Among Adolescents: A Systematic Review and Meta-Analysis. J Adolesc Heal [Internet]. 2016 Oct 1 [cited 2020 Feb 26];59(4):S76–87. Available from: https://www.sciencedirect.com/science/article/pii/S1054139X16302397

[26] WHO. Mental Health Information System: Mental Health Policy and Service Guidance Package. 2005.

[27] P.Deek F, A.M.McHugh J, M.Eljabiri O. Strategic Software Engineering: An Interdisciplinary Approach. New York: Aurbach Publications; 2005.

[28] Ortuño-Sierra J, Aritio-Solana R, Fonseca-Pedrero E. Mental health difficulties in children and adolescents: The study of the SDQ in the Spanish National Health Survey 2011–2012. Psychiatry Res. 2018;259(May 2017):236–42.

IOP Conf. Series: Materials Science and Engineering **874** (2020) 012024 doi:10.1088/1757-899X/874/1/012024

Data Mining Optimization uses C4.5 Classification and Particle Swarm Optimization (PSO) in the location selection of Student Boardinghouses

Ari Waluyo[1], Hendra Jatnika[2], Marlina Rahmi Shinta Permatasari[3], T Tuslaela[4], Indah Purnamasari[4], Agus Perdana Windarto[5*]

[1]Politeknik Dharma Patria, Indonesia
[2]Institut Teknologi PLN, Indonesia
[3]Universitas Bina Sarana Informatika, Indonesia
[4]STMIK Nusa Mandiri, Indonesia
[5]STIKOM Tunas Bangsa, Indonesia

Email: agus.perdana@amiktunasbangsa.ac.id *

Abstract. The purpose of this study is to select the location of student boarding houses using Particle Swarm Optimization (PSO) and C4.5 optimization techniques. The source of the data was obtained by observing and giving questionnaires to 150 respondents who were lodging in the Pematangsiantar-Simalungun area. from the data set of 81 records and using 5 parameters of assessment ((C1) water cleanliness, (C2) Facilities, (C3) Transportation, (C4) Security, and (C5) Conditions) obtained the results of modeling using the C4.5 + PSO algorithm has better accuracy is 97.78% compared to the C4.5 model whose accuracy is 97.53%. Thus, it is evident that the PSO applied to the weighting of the C4.5 attribute increases the value of accuracy..

1. Introduction

Ideal residence is a place that can protect and become a place of rest for us and a place to grow a healthy life spiritually and physically [1]. There are various types of residences that are permanent or temporary, temporary residences are rented houses and boarding houses. Boarding houses become one of the places to stay for students, students, workers, employees who are far from their homes. Boarding houses are an option for them, especially students, to become temporary residences while carrying out work. The choice of boarding houses is an alternative that must be thought wisely because it considers many aspects such as affordable prices, convenient location, transportation facilities and others. Many boarding houses are an obstacle for students to choose eligibility according to their wants and needs. Boarding houses that are safe, secure, comfortable become one of the dreams of students who are far from where they live. In choosing a boarding house dream there are criteria in choosing it. The number of criteria in the selection of boarding houses is an obstacle for students. So it is necessary to classify the criteria in selecting the location of student boarding houses. The method used for the selection of the location of the boarding houses of students uses Datamining [2]–[4] with C4.5 classification techniques [5], [6] and Particle Swarm Optimization (PSO) [7], [8]. The method often used for classification is C4.5 [5], [6]. This method is also called a very strong and well-known decision tree for classification and prediction [9]. The ability of this C4.5 method can produce decision trees that are easily interpreted [10], have an acceptable level of accuracy [11], are efficient in handling discrete type attributes and can handle discrete and numeric type attributes [7]. However, the

C4.5 method has several disadvantages including overlapping often when classes and criteria used are very numerous [8] and overfitting occurs because there is noise training data, which is irrelevant data so that the tree has a long and unbalanced subtree. In this problem can be overcome by optimized using the Particle Swarm Optimization (PSO) algorithm. The PSO algorithm is used to optimize accuracy on C4.5 to get maximum results. Research using PSO optimization for prediction has been done a lot before, namely for the prediction of sea tides where PSO is used to optimize the minimum error value in the network in order to obtain the ideal neural network weights. PSO and artificial neural networks [12]–[16] have several input parameters such as, the number of input neurons, learning rate, swarm, c1, c2 min inertia, max inertia. The data used are 1000 which are divided into 700 training data and 300 testing data. The test results showed that the prediction accuracy was 91.56% using 90 swarm, learning rate 0.9 and iteration 20 times [8]. Subsequent research on the classification of credit analysis using the c4.5 algorithm and PSO [7]. From the results of experiments conducted a C4.5 algorithm model based on Particle Swarm Optimization (PSO) got the best results at 70%, while the C4.5 algorithm model without Particle Swarm Optimization (PSO) was only 68.6%. Based on this, it is expected that the research results can classify the location of student boarding houses.

2. Methodology

2.1. Data Mining

Data mining is in the form of observational analysis of data sets to find unexpected relationships and to summarize data in new ways that can be understood and useful for data owners. There are several settlement techniques used in data mining, including: clustering, classification, estimation and association [2].

2.2 Classification

Data classification is a process that finds the same properties in a set of objects in a database and classifies them into different classes according to the specified classification model [17]. The purpose of classification is to find a model of a training set that distinguishes attributes into appropriate categories or classes, the model is then used to classify attributes whose classes have not been known before. Classification techniques are divided into several techniques including ID3, CART, and C4.5 [9].

2.3 Decision Tree C4.5

C4.5 method is to change the tree generated in several rules. The number of rules is equal to the number of paths that might be built from the root to the leaf node [6]. In general, C4.5 algorithm to build a decision tree with the following general steps:

a) Select the attribute as the root
b) Create a branch for each value
c) Divide cases in branches
d) Repeat the process for each branch until all cases in the branch have the same class [4], [5].

2.4 Research Method

The research data were obtained by conducting direct observations and giving questionnaires to students as many as 150 respondents who were randomly assigned. Data of 150 respondents consisted of student data from five private tertiary institutions in the area of Simalungun Regency and Pematangsiantar City. Respondents are those who live in boarding-houses. The results of observation data and questionnaires were pre-processed using Microsoft Excel software. From the results of the questionnaire obtained several criteria for the selection of the location of student boarding houses, among others: (C1) water cleanliness, (C2) Facilities, (C3) Transportation, (C4) Security, and (C5) Conditions. For water hygiene criteria (C1) have sub criteria {Good, Enough, Bad}, facility criteria (C2) have sub criteria {Luxury, Simple, Standard}, transportation criteria (C3) have sub criteria {Far, Medium, Near}, security criteria (C4) have sub criteria {Strict, Normal, Free} and condition criteria (C5) have sub criteria {Eligible, Not Eligible}.

IOP Conf. Series: Materials Science and Engineering **874** (2020) 012024 doi:10.1088/1757-899X/874/1/012024

3. Results and Discussion

The following is an example dataset used as research material (the dataset is randomly generated as many as 81 data samples / training).

Table 1. Learning Dataset

No	Water cleanliness (C1)	Facilities (C2)	Transportation (C3)	Security (C4)	Conditions (C5)
1	Good	Luxury	Far	Strict	Not Eligible
2	Good	Luxury	Far	Normal	Not Eligible
3	Good	Luxury	Far	Free	Not Eligible
4	Good	Simple	Far	Strict	Not Eligible
5	Good	Simple	Far	Normal	Not Eligible
6	Good	Simple	Far	Free	Not Eligible
7	Good	Standard	Far	Strict	Not Eligible
8	Good	Standard	Far	Normal	Not Eligible
9	Good	Standard	Far	Free	Not Eligible
10	Good	Luxury	Medium	Strict	Eligible
11	Good	Luxury	Medium	Normal	Eligible
12	Good	Luxury	Medium	Free	Not Eligible
13	Good	Simple	Medium	Strict	Eligible
14	Good	Simple	Medium	Normal	Eligible
15	Good	Simple	Medium	Free	Not Eligible
16	Good	Standard	Medium	Strict	Eligible
17	Good	Standard	Medium	Normal	Eligible
18	Good	Standard	Medium	Free	Not Eligible
19	Good	Luxury	Near	Strict	Eligible
20	Good	Luxury	Near	Normal	Eligible
21	Good	Luxury	Near	Free	Not Eligible
22	Good	Simple	Near	Strict	Eligible
23	Good	Simple	Near	Normal	Eligible
24	Good	Simple	Near	Free	Not Eligible
25	Good	Standard	Near	Strict	Not Eligible
26	Good	Standard	Near	Normal	Eligible
27	Good	Standard	Near	Free	Not Eligible
28	Enough	Luxury	Far	Strict	Not Eligible
29	Enough	Luxury	Far	Normal	Not Eligible
30	Enough	Luxury	Far	Free	Not Eligible
31	Enough	Simple	Far	Strict	Not Eligible
32	Enough	Simple	Far	Normal	Not Eligible
33	Enough	Simple	Far	Free	Not Eligible
34	Enough	Standard	Far	Strict	Not Eligible
35	Enough	Standard	Far	Normal	Not Eligible
36	Enough	Standard	Far	Free	Not Eligible
37	Enough	Luxury	Medium	Strict	Eligible
38	Enough	Luxury	Medium	Normal	Eligible
39	Enough	Luxury	Medium	Free	Eligible
40	Enough	Simple	Medium	Strict	Eligible
41	Enough	Simple	Medium	Normal	Eligible
42	Enough	Simple	Medium	Free	Not Eligible
43	Enough	Standard	Medium	Strict	Eligible
44	Enough	Standard	Medium	Normal	Eligible
45	Enough	Standard	Medium	Free	Not Eligible
46	Enough	Luxury	Near	Strict	Eligible
47	Enough	Luxury	Near	Normal	Eligible
48	Enough	Luxury	Near	Free	Eligible
49	Enough	Simple	Near	Strict	Eligible
50	Enough	Simple	Near	Normal	Eligible
51	Enough	Simple	Near	Free	Eligible
52	Enough	Standard	Near	Strict	Eligible
53	Enough	Standard	Near	Normal	Eligible
54	Enough	Standard	Near	Free	Eligible
55	Bad	Luxury	Far	Strict	Not Eligible
56	Bad	Luxury	Far	Normal	Not Eligible
57	Bad	Luxury	Far	Free	Not Eligible

58	Bad	Simple	Far	Strict	Not Eligible
59	Bad	Simple	Far	Normal	Not Eligible
60	Bad	Simple	Far	Free	Not Eligible
61	Bad	Standard	Far	Strict	Not Eligible
62	Bad	Standard	Far	Normal	Not Eligible
63	Bad	Standard	Far	Free	Not Eligible
64	Bad	Luxury	Medium	Strict	Eligible
65	Bad	Luxury	Medium	Normal	Eligible
66	Bad	Luxury	Medium	Free	Not Eligible
67	Bad	Simple	Medium	Strict	Eligible
68	Bad	Simple	Medium	Normal	Eligible
69	Bad	Simple	Medium	Free	Not Eligible
70	Bad	Standard	Medium	Strict	Eligible
71	Bad	Standard	Medium	Normal	Eligible
72	Bad	Standard	Medium	Free	Not Eligible
73	Bad	Luxury	Near	Strict	Eligible
74	Bad	Luxury	Near	Normal	Eligible
75	Bad	Luxury	Near	Free	Not Eligible
76	Bad	Simple	Near	Strict	Eligible
77	Bad	Simple	Near	Normal	Eligible
78	Bad	Simple	Near	Free	Not Eligible
79	Bad	Standard	Near	Strict	Eligible
80	Bad	Standard	Near	Normal	Eligible
81	Bad	Standard	Near	Free	Not Eligible

The Dataset Learning Summary data in table 1 will then be processed to obtain a decision tree using the help of RapidMiner 5.3 software.

3.1 The results of the C4.5 method with the RapidMiner software

The Dataset Learning Summary data in table 1 will then be processed to obtain a decision tree using the help of RapidMiner 5.3 software. Following is the C4.5 model using the RapidMiner software as shown in the following image:

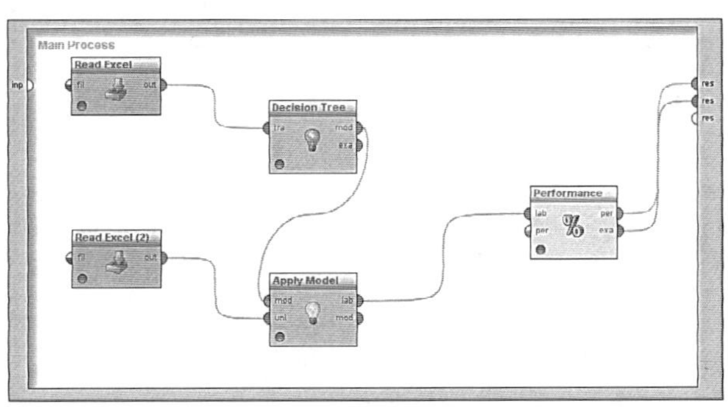

Figure 1. C4.5 model using RapidMiner software

In figure 1 it can be explained that the parameters used in the C4.5 method include: criterion: gain_ratio; minimum size for split: 4; minimum leaf size: 2; minimum gain: 0.1; minimum depth: 20; confidence: 0.25. From this model the accuracy of the results is obtained as shown below:

accuracy: 97.53%

	true Not Eligible	true Eligible	class precision
pred. Not Eligible	41	1	97.62%
pred. Eligible	1	38	97.44%
class recall	97.62%	97.44%	

Figure 2. Model C4.5 accuracy

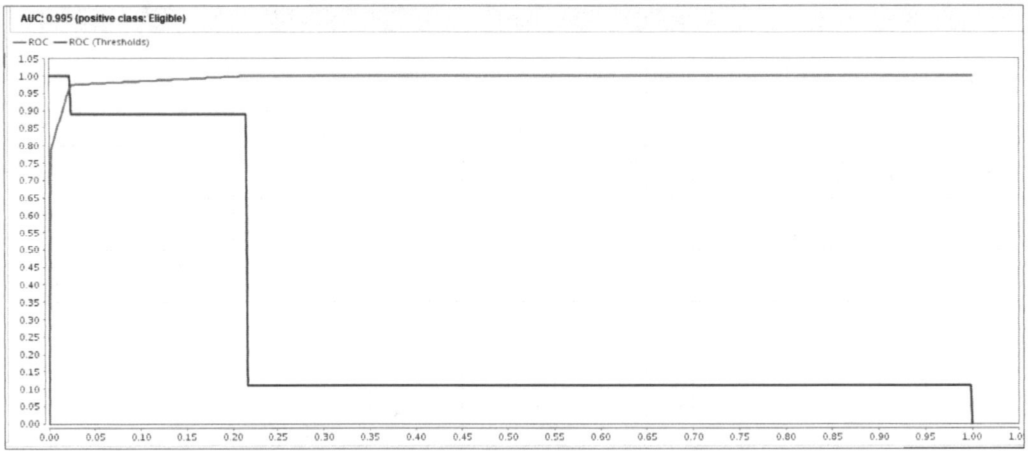

Figure 3. Model C4.5 AUC (Area Under Curve)

```
PerformanceVector
accuracy: 97.53%
ConfusionMatrix:
True:  Not Eligible     Eligible
Not Eligible:    41     1
Eligible:        1      38
precision: 97.44% (positive class: Eligible)
ConfusionMatrix:
True:  Not Eligible     Eligible
Not Eligible:    41     1
Eligible:        1      38
recall: 97.44% (positive class: Eligible)
ConfusionMatrix:
True:  Not Eligible     Eligible
Not Eligible:    41     1
Eligible:        1      38
AUC (optimistic): 0.999 (positive class: Eligible)
AUC: 0.995 (positive class: Eligible)
AUC (pessimistic): 0.990 (positive class: Eligible)
```

3.2 The results of the C4.5 + Particle Swarm Optimization (PSO) method with the RapidMiner software

Following is the C4.5 + Particle Swarm Optimization (PSO) model using the RapidMiner software as shown in the following image:

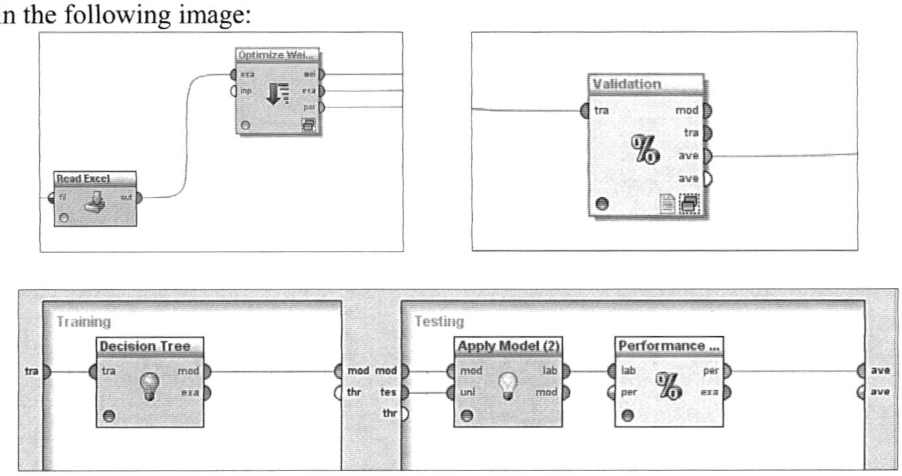

Figure 4. C4.5 + Particle Swarm Optimization (PSO) model using RapidMiner software

Using the same parameters for the C4.5 model (figure 1), optimization with Particle Swarm Optimization (PSO) uses different population sizes and maximum number of generations as shown in the following table:

Table 2. The results of the experiment used population size and maximum number of generations in the PSO + C4.5 method

No	PSO parameter	Accuracy	AUC (Area Under Curve)
1	Posize=10; generate=30	97.64%	0.836
2	Posize=20; generate=30	97.64%	0.962
3	Posize=10; generate=40	97.64%	0.971
4	Posize=15; generate=40	97.64%	0.883
5	Posize=30; generate=50	97.64%	0.825
6	Posize=50; generate=50	97.78%	0.912

Based on table 2, the selection of student boarding locations obtained results that the accuracy value of C4.5 + PSO (97.78%) is better than C4.5 without PSO (97.53%). Up around 0.25%. While the accuracy of classification using AUC, C4.5 method without PSO (0.995) is better than C4.5 + PSO method (0.912). So based on these results, the C4.5 + PSO method can improve the results of prediction accuracy compared to the C4.5 method without PSO. The results of the optimization of C4.5 + PSO states that of the 4 criteria (Water cleanliness (C1), Facilities (C2), Transportation (C3), Security (C4)) used, Water cleanliness (C1) is the most influential alternative with weight = 1.0 and Security (C4) with a weight = 0.5380 as shown in the following image:

accuracy: 97.64% +/- 4.73% (mikro: 97.53%)

	true Not Eligible	true Eligible	class precision
pred. Not Eligible	41	1	97.62%
pred. Eligible	1	38	97.44%
class recall	97.62%	97.44%	

Figure 5. C4.5 + Particle Swarm Optimization (PSO) model of accuracy

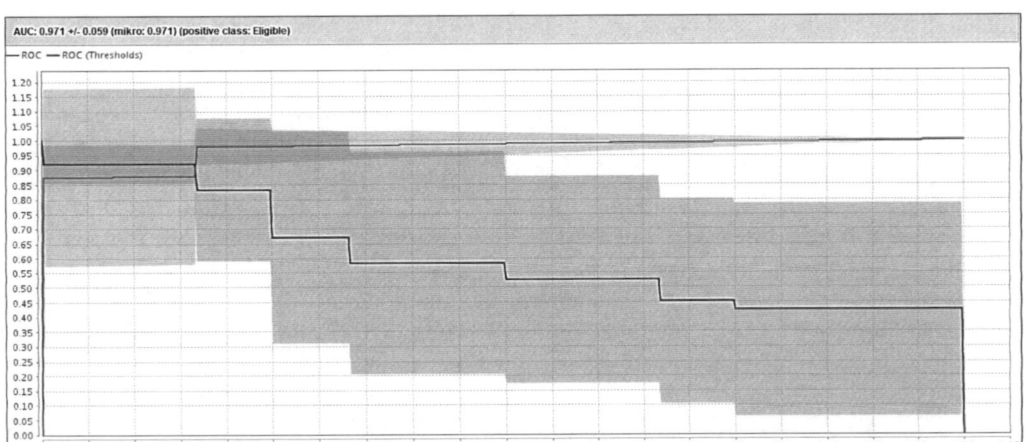

Figure 6. C4.5 + Particle Swarm Optimization (PSO) model of AUC (Area Under Curve)

```
PerformanceVector
accuracy: 97.64% +/- 4.73% (mikro: 97.53%)
ConfusionMatrix:
True:   Not Eligible        Eligible
Not Eligible:     41         1
Eligible:         1          38
precision: 97.50% +/- 7.50% (mikro: 97.44%) (positive class: Eligible)
ConfusionMatrix:
True:   Not Eligible        Eligible
```

```
Not Eligible:    41        1
Eligible:        1         38
recall: 97.50% +/- 7.50% (mikro: 97.44%) (positive class:
Eligible)
ConfusionMatrix:
True:  Not Eligible     Eligible
Not Eligible:    41        1
Eligible:        1         38
AUC (optimistic): 1.000 +/- 0.000 (mikro: 1.000) (positive
class: Eligible)
AUC: 0.971 +/- 0.059 (mikro: 0.971) (positive class: Eligible)
AUC (pessimistic): 0.958 +/- 0.085 (mikro: 0.958) (positive
class: Eligible
```

attribute	
Water cleanliness (C1)	1
Facilities (C2)	0
Transportation (C3)	0.486
Security (C4)	0.539

Figure 7. The results of the evaluation criteria using C4.5 + PSO Method

4. Conclusion

In this study modeling was carried out using the C4.5 and C4.5 + PSO algorithms with the focus of the research being the application of the PSO algorithm in weighting the attributes of the C4.5 data mining classification technique using the help of RapidMiner 5.3 software. Model validation uses 10 fold cross-validation and model evaluation uses confusion matrix and ROC curves. The results showed that the C4.5 + PSO model had a better accuracy of 97.78% compared to the C4.5 model whose accuracy was 97.53%. Thus, it is evident that the PSO applied to the weighting of the C4.5 attribute increases the value of accuracy.

References

[1] D. L. Fithri, "Model Data Mining Dalam Penentuan Kelayakan Pemilihan Tempat Tinggal Menggunakan Metode Naive Bayes," *J. SIMETRIS*, vol. 7, no. 2, pp. 725–730, 2016.

[2] A. P. Windarto *et al.*, "Analysis of the K-Means Algorithm on Clean Water Customers Based on the Province," *J. Phys. Conf. Ser.*, vol. 1255, no. 1, 2019, doi: 10.1088/1742-6596/1255/1/012001.

[3] Sudirman, A. P. Windarto, and A. Wanto, "Data mining tools | rapidminer: K-means method on clustering of rice crops by province as efforts to stabilize food crops in Indonesia," *IOP Conf. Ser. Mater. Sci. Eng.*, vol. 420, no. 1, 2018, doi: 10.1088/1757-899X/420/1/012089.

[4] D. Hartama, A. Perdana Windarto, and A. Wanto, "The Application of Data Mining in Determining Patterns of Interest of High School Graduates," *J. Phys. Conf. Ser.*, vol. 1339, no. 1, 2019, doi: 10.1088/1742-6596/1339/1/012042.

[5] W. Katrina, H. J. Damanik, F. Parhusip, D. Hartama, A. P. Windarto, and A. Wanto, "C.45 Classification Rules Model for Determining Students Level of Understanding of the Subject," *J. Phys. Conf. Ser.*, vol. 1255, no. 012005, pp. 1–7, 2019, doi: 10.1088/1742-6596/1255/1/012005.

[6] M. Widyastuti, A. G. Fepdiani Simanjuntak, D. Hartama, A. P. Windarto, and A. Wanto, "Classification Model C.45 on Determining the Quality of Custumer Service in Bank BTN Pematangsiantar Branch," *J. Phys. Conf. Ser.*, vol. 1255, no. 012002, pp. 1–6, 2019, doi: 10.1088/1742-6596/1255/1/012002.

[7] S. Saprudin, "Penerapan Particle Swarm Optimization (PSO) untuk Klasifikasi dan Analisis Kredit dengan Menggunakan Algoritma C4.5," *J. Inform. Univ. Pamulang*, vol. 2, no. 4, p. 214, 2017, doi: 10.32493/informatika.v2i4.1488.

[8] N. Nikentari, H. Kurniawan, N. Ritha, D. Kurniawan, U. Maritim, and R. Ali, "Particle Swarm Optimization Untuk Prediksi Pasang Surut Air Optimization of Backpropagation Artificial Neural Network With Particle Swarm Optimization To Predict Tide Level," *J. Teknol. Inf. dan*

Ilmu Komput., vol. 5, no. 5, pp. 605–612, 2018, doi: 10.25126/jtiik2018551055.

[9] K. Adhatrao, A. Gaykar, A. Dhawan, R. Jha, and V. Honrao, "Predicting Students' Performance Using Id3 and C4.5 Classification Algorithms," *Int. J. Data Min. Knowl. Manag. Process*, vol. 3, no. 5, pp. 39–52, 2013, doi: 10.5121/ijdkp.2013.3504.

[10] H. Siahaan, H. Mawengkang, S. Efendi, A. Wanto, and A. P. Windarto, "Application of Classification Method C4 . 5 on Selection of Exemplary Teachers," in *IOP Conference Series*, 2018, pp. 1–6.

[11] I. S. Damanik, A. P. Windarto, A. Wanto, Poningsih, S. R. Andani, and W. Saputra, "Decision Tree Optimization in C4.5 Algorithm Using Genetic Algorithm," *J. Phys. Conf. Ser.*, vol. 1255, no. 012012, pp. 1–7, 2019, doi: 10.1088/1742-6596/1255/1/012012.

[12] Sumijan, A. P. Windarto, A. Muhammad, and Budiharjo, "Implementation of Neural Networks in Predicting the Understanding Level of Students Subject," *Int. J. Softw. Eng. Its Appl.*, vol. 10, no. 10, pp. 189–204, 2016.

[13] Budiharjo, T. Soemartono, A. P. Windarto, and T. Herawan, "Predicting School Participation in Indonesia using Back-Propagation Algorithm Model," *Int. J. Control Autom.*, vol. 11, no. 11, pp. 57–68, 2018.

[14] Budiharjo, T. Soemartono, A. P. Windarto, and T. Herawan, "Predicting tuition fee payment problem using backpropagation neural network model," *Int. J. Adv. Sci. Technol.*, vol. 120, pp. 85–96, 2018, doi: 10.14257/ijast.2018.120.07.

[15] A. P. Windarto, M. R. Lubis, and Solikhun, "Implementasi Jst Pada Prediksi Total Laba Rugi Komprehensif Bank Umum Konvensional Dengan Backpropagation," *J. Teknol. Inf. dan Ilmu Komput.*, vol. 5, no. 4, pp. 411–418, 2018, doi: 10.25126/jtiik.201854767.

[16] A. P. Windarto, M. R. Lubis, and Solikhun, "Model Arsitektur Neural Network Dengan Backpropogation Pada Prediksi Total Laba Rugi Komprehensif Bank Umum Konvensional," *Kumpul. J. Ilmu Komput.*, vol. 05, no. 02, pp. 147–158, 2018.

[17] L. N. Rani, "Klasifikasi Nasabah Menggunakan Algoritma C4.5 Sebagai Dasar Pemberian Kredit," *J. KomTekInfo Fak. Ilmu Komput.*, vol. 2, no. 2, pp. 33–38, 2015.

IOP Conf. Series: Materials Science and Engineering **874** (2020) 012025 doi:10.1088/1757-899X/874/1/012025

Evaluation of PHP Framework Measured Using Object-Oriented Metrics with the Analytic Hierarchy Process

R Setiyawan [1], W Gata [2], Elbiansyah [3], D Ardiansyah [4], D Yuliandari [5], D Wijayanti [6], F P Harmono [7], Y Komalasari [8], M Lase [9], and M H Fakhriza [10]

[1] Master of Computer Science, Postgraduate Program, *STMIK Nusa Mandiri*, Indonesia
[2] Master of Computer Science, Postgraduate Program, *STMIK Nusa Mandiri*, Indonesia
[3] Master of Computer Science, Postgraduate Program, *STMIK Nusa Mandiri*, Indonesia
[4] Master of Computer Science, Postgraduate Program, *STMIK Nusa Mandiri*, Indonesia
[5] Information System, *Universitas Bina Sarana Informatika*, Indonesia
[6] Information System, *Universitas Bina Sarana Informatika*, Indonesia
[7] Information System, *Universitas Bina Sarana Informatika*, Indonesia
[8] Information Technology, *Universitas Bina Sarana Informatika*, Indonesia
[9] Information Technology, Postgraduate Program, *STMIK Nusa Mandiri*, Indonesia
[10] Information Technology, Postgraduate Program, *STMIK Nusa Mandiri*, Indonesia

E-mail: 14002193@nusamandiri.ac.id

Abstract. The existence of various PHP frameworks may become one alternative choice in developing a web application that PT XYZ wants to evaluate its Wisanggeni internal PHP framework in order to measure how far the design quality of Wisanggeni framework is working. An evaluation on PHP framework using object-oriented metric parameter and design quality is then conducted in this research. The results of measurement conducted using object-oriented metric parameters for each framework are evaluated with the Analytic Hierarchy Process (AHP) method to determine the best framework model design quality used as a reference for the development of Wisanggeni. This study also uses a combination of PHPDepend and PHPMetrics devices to measure the values of object-oriented metrics in each PHP framework. The research final result is that each framework has one value representing each framework's quality in which the quality of Wisanggeni framework is almost to the same with the Laravel and CodeIgniter framework .

Keywords: framework, PHP, AHP, object-oriented, metric, quality design.

1. Introduction

Based on the research conducted by Standish Group International, there were 7400 developed information technology projects. The result showed that only 16.2% was completed in accordance with the planning, 31.1% experienced delays, and 52.7% experienced cost overrun up to 189%. Those IT projects failed due to the ineffectiveness of time, cost, and function [4]. Famous Indonesian companies, such as Kompas.com, Okezone.com, Urbanesia.com, bejubel.com as well as PT XYZ use their frameworks to save the time in developing their IT projects. PT XYZ has an internal framework to develop its system called Wisanggeni PHP framework. The use of this framework is [15] to create the programming standards as one option to reduce the problems. In the past few years, there were various PHP frameworks offering several conveniences with various features related to the software development using the concept of object orientation which encourages PT XYZ to evaluate its Wisanggeni framework. Based on Skvorc, a survey on the Site points using a framework was conducted by the system developers in Indonesia. The frameworks preferred by the system developers are the CodeIgniter and Laravel framework that those become highly familiar among the system developers which can save the time in understanding processes related to how those frameworks are working.

The concept of object orientation has also encouraged the developers to make applications using different implementation techniques and patterns [8]. Various patterns of object orientation implementation also lead to different opinions in viewing the software quality. The software quality measurement conducted [11] based on the software's object design aspects is one field which is recently still developed. Some previous researches have produced the software quality measurement devices translated into several parameters, such as Metrics for Object-Oriented Software Engineering [13] and Metrics for Object-Oriented Design [12]. However, with these various parameter values, it seems still to be very difficult to determine which software has the better quality. Thus, we need a general method to combine these overall values into a value interpreting the software quality to the others.

Because the application domain also influences the design and implementation structure , a quality comparison should be made on a number of applications in the same domain. In making a comparison, [7] there are several parameters used based on the existing quality. Thus, the Analytic Hierarchy Process [9] method is selected as the multicriteria comparison method which generates rankings from several alternatives based on certain criteria generally used in all software types, such as desktop applications, web applications, class libraries, or APIs [8].

Several previous studies have been conducted with similar objectives regarding to the quality of software design, by Rosenberg and Hyatt [2], discussing the Software Assurance Technology Center (SATC) in NASA Goddard Space Flight Center which is approaching to the metrics by first identifying the attributes associated with the object-oriented development. Researches on SATC have produced a combination of traditional metrics and metrics [3] that measure the unique structures of object-oriented development. Each metric [16] has various weights which may affect some software's quality. The research conducted by Misra [1] has proposed that the need for a criteria evaluation for a clear object-oriented metric [10] which is in software is by evaluating its measurement theory. The method to empirically assess a good object-oriented quality design [5] is simple and may support a reasonable estimation for several factors, such as functionability, efficiency, understandability, and maintainability [14] which determine the assessment measurement for an object-oriented programming. Measurements should include most features needed for evaluation and validation of object-oriented metrics [1]. Hermawan and Mursanto [8] have conducted another approach in ranking the Java-based ERP software with CKJM 1.8-based ranking device which produces a parameter for collecting the metrics intended to measure the software quality based on six object-oriented metrics proposed by Chidamber and Kemerer [13] considering the design quality perspectives. Latelay and Azhari [6] have evaluated the software parameter quality with the six object-oriented metrics proposed by Chidamber and Kemerer [13]. The calculation of both high and low values obtained from those six metrics may affect the quality of factors related to each metric. From those multicriteria problems, one effort provided as a solution is by implementing the AHP method.

Table 1 shows similar researches conducted on metric-related software. Based on the literature study of previous researches that the evaluation of PHP framework design quality comparison conducted with AHP method has never been made before. The closest one is with that conducted by Hermawan and Mursanto [8], yet the difference in this study is that the PHP framework object is conducted on the problems which have been described before. Thus this research aims at evaluating and ranking the PHP Framework based on OO (Object-Oriented) Metrics.

Table 1. Research on Metrics-Related Software

No	Previous Research	Result
1	Rosenberg & Hyaat	Produces a combination of traditional metrics and metrics measuring the unique structure of object-oriented development [2].
2	Jamali	Identifies several OO metrics used in analysing the object-oriented programming development [10].
3	Misra	Produces evaluation criteria on the object-oriented programming paradigm [1].
4	Latelay and Azhari	Suggests the use of six Chidamber and Kemerer metrics to evaluate the software quality [6].
5	Hermawan dan Mursanto	Generates general ranking methods using Java software with a combination of AHP concepts, OO metrics, and software quality [8].

2. Methodology

The proposed settlement process will be divided into several stages for the research analysis process as explained in Figure 1.

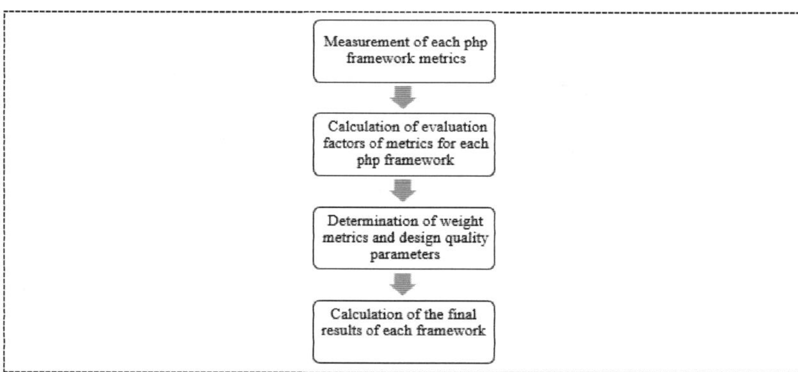

Figure 1. Evaluation Design.

The final result expected from this method is that each framework has one value interpreting the software design quality as shown in Figure 2. Thus, the value may determine the software design's PHP framework quality.

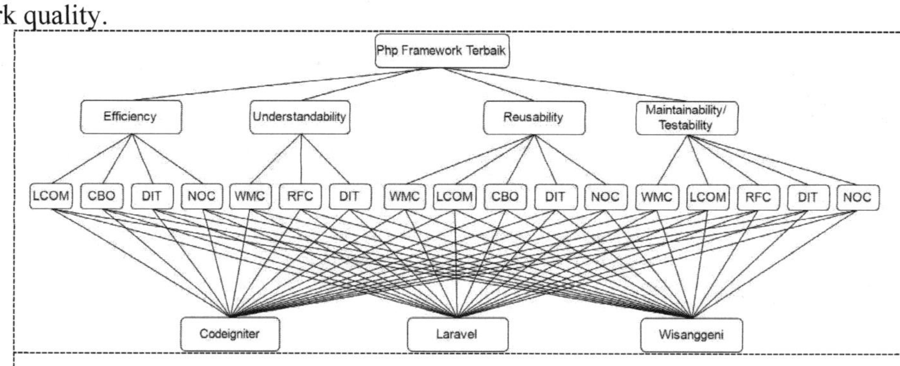

Figure 2. Hierarchical Structure of PHP Framework Evaluation

2.1. Metric Measurement of each PHP framework
At this stage, the OO measurement on the same metrics is conducted at each PHP framework. Many software or devices may be used to measure the metrics. Each device has both advantages and disadvantages. However, the calculations which support this study are those conducted with PHPDepend and PHPMetrics. Both types of matric software are specified for the PHP programming languages. The combination of two PHPDepend and PHPMetrics devices may support the metric calculation of the Chidamber Kemerer object-oriented PHP source code.

2.2. Calculating the factors evaluating the metric values for each PHP framework
Evaluation of metric values is conducted using the measurement values. The calculation is made using the AHP method to obtain the evaluation values of metric factors for each PHP framework. The stages to obtain the evaluation values of metric factors for each PHP framework are presented in Figure 3.

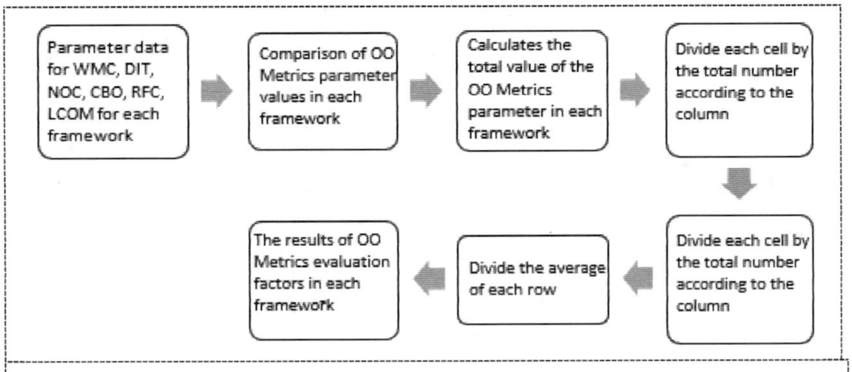

Figure 3. Metric Value Evaluation Stages.

2.3. Determining the metric weights and design quality parameters
The evaluation of design quality parameters is influenced by the metrics as presented in Table 1. To get an evaluation factor for each design quality parameter, the evaluation value of PHP framework with the metric weights is influenced by the design quality as shown in Table 2.

Table 2. Calculation of Design Quality Evaluation Factors.

Software Quality	PHP Framework
Efficiency	$(EF)_{DIT} \times (MW)_{DIT} + (EF)_{CBO} \times (MW)_{CBO} + (EF)_{NOC} \times (MW)_{NOC} + (EF)_{LCOM} \times (MW)_{LCOM}$
Understandability	$(EF)_{WMC} \times (MW)_{WMC} + (EF)_{DIT} \times (MW)_{DIT} + (EF)_{RFC} \times (MW)_{RFC}$
Reusability	$(EF)_{WMC} \times (MW)_{WMC} + (EF)_{DIT} \times (MW)_{DIT} + (EF)_{CBO} \times (MW)_{CBO} + (EF)_{NOC} \times (MW)_{NOC} + (EF)_{LCOM} \times (MW)_{LCOM}$
Maintainability/Testability	$(EF)_{WMC} \times (MW)_{WMC} + (EF)_{DIT} \times (MW)_{DIT} + (EF)_{NOC} \times (MW)_{NOC} + (EF)_{RFC} \times (MW)_{RFC} + (EF)_{LCOM} \times (MW)_{LCOM}$

Evaluation Factor (EF) and Metric Weight (MW) are adjusted to the values in the selected PHP framework.

2.4. Each framework's final result calculation of

Each framework's final result calculation multiplies the weight quality (WQ) with the calculation result of evaluation factor. Evaluation factor (EF) is adjusted with the related items, including the evaluated three PHP frameworks.

Table 3. Final results of each framework

PHP Framework	Software Quality
CodeIgniter PHP Framework, Laravel, Wisanggeni	$(EF)Efficiency \times (WQ)Efficiency + (EF)Understandability \times (WQ)Understandability + (EF)Reusability \times (WQ)Reusability + (EF)Maintainability \times (WQ)Maintainability$

3. Result

After measuring the metrics' PHP framework compare each metric and calculate the design quality with the AHP method.

3.1. The measurement results of the metric value's PHP framework

The measurement of metric values' PHP framework using a combination of PHPDepend and PHPMetrics devices containing CK metrics produce the matric values of each PHP framework.

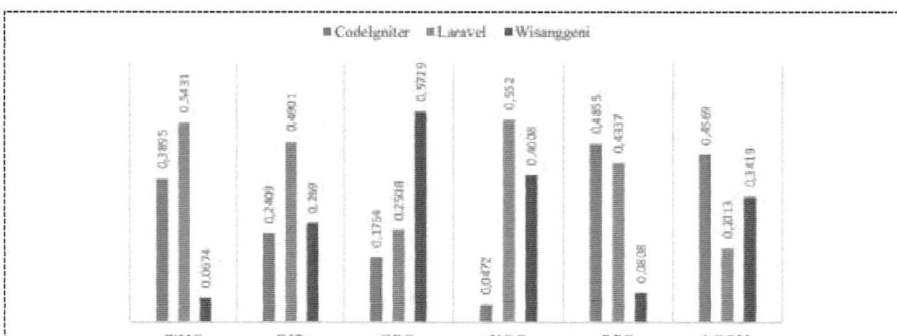

Figure 4. Measurement of metrics' PHP framework.

3.2. The results of PHP framework using AHP method

The calculation of design quality evaluation factor using AHP method is conducted with the values as presented in Figure 5.

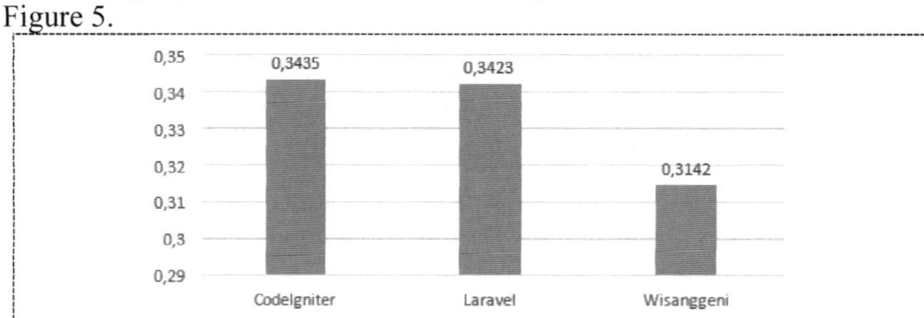

Figure 5. Final Result

4. Conclusion

This study has evaluated the Wisanggeni comparing to the CodeIgniter and Laravel PHP frameworks based on the software design quality as well as the Chidamber and Kemerer object-oriented metrics using the Analytical Hierarchy Process method. The CodeIgniter has the highest value of all by 0.3435. of The Laravel is at the second position with The value difference of only 0.0012, while the Wisanggeni has the lowest value when compared to that of both Laravel and CodeIgniter PHP framework with the value of 0.3142 as shown in Figure 5. The quality of Wisanggeni framework is almost to the same with that of both Laravel and CodeIgniter framework with a difference of only 0.02.The framework evaluation final result in this study may be used as a reference for developing the Wisanggeni PHP framework.

5. References

[1] S. Misra, "Evaluation criteria for object-oriented metrics," *Acta Polytech. Hungarica*, vol. 8, no. 5, pp. 109–136, 2011.
[2] L. Rosenberg and L. Hyatt, "Software quality metrics for object-oriented environments," *Crosstalk Journal, April*, vol. 10, no. 4, pp. 1–6, 1997.
[3] F. BritoeAbreu, M. Goulao, and R. Esteves, "Toward the Design Quality Evaluation of Object Oriented Software Systems," *Proc. 5th Int. Conf. Softw. Qual.*, no. October, 1995.
[4] D. Rosmala, M. Ichwan, and M. I. Gandalisha, "Komparasi Framework MVC (Codeigniter, dan CakePHP) pada Aplikasi Berbasis Web (Studi Kasus: Sistem Informasi Perwalian di Jurusan Informatika Institut Teknologi Nasional)," *J. Inform.*, vol. 2, no. 2, pp. 22–30, 2011.
[5] R. A. Khan, K. Mustafa, and S. I. Ahson, "An Empirical Validation of Object Oriented Design Quality Metrics," *J. King Saud Univ. - Comput. Inf. Sci.*, vol. 19, pp. 1–16, 2013.
[6] K. Letelay and A. SN, "Evaluasi Kualitas Perangkat Lunak Dengan Metrics Berorientasi Objek," vol. 2012, no. semnasIF, pp. 139–145, 2012.
[7] R. Anggrainingsih, B. O. P. Johannanda, A. P. Kuswara, D. Wahyuningsih, and T. Rejekiningsih, "Comparison of maintainability and flexibility on open source LMS," in *Proceedings - 2016 International Seminar on Application of Technology for Information and Communication, ISEMANTIC 2016*, 2017.
[8] P. Mursanto and E. Hermawan, "PEMERINGKATAN SOFTWARE APLIKASI BERDASARKAN PROPERTI KUALITAS DISAIN DAN METRICS FOR OBJECT ORIENTED SOFTWARE MENGGUNAKAN ANALYTIC HIERARCHY PROCESS," *J. Sist. Inf.*, 2012.
[9] T. L. Saaty, "Decision making — the Analytic Hierarchy and Network Processes (AHP/ANP)," *J. Syst. Sci. Syst. Eng.*, 2004.
[10] S. M. Jamali, "Object Oriented Metrics (A Survey Approach) Seyyed Mohsen Jamali Department of Computer Engineering Sharif University of Technology January 2006 Tehran Iran," *Design*, 2006.
[11] D. Jamwal, "Analysis of Software Quality Models for Organizations," 2010.
[12] Muktamyee Sarker, "An overview of Object Oriented Design Metrics Master Thesis," *Computer (Long. Beach. Calif).*, 2005.
[13] S. R. Chidamber and C. F. Kemerer, "A Metrics Suite for Object Oriented Design," *IEEE Trans. Softw. Eng.*, 1994.
[14] J. a. McCall, P. K. Richards, and G. F. Walters, "Factors in Software Quality," *at'l Tech. Inf. Serv.*, 1977.
[15] M. Fayad, D. C. Schmidt, and R. E. Johnson, *Implementing Application Frameworks*. 1999.
[16] N. E. Fenton and S. L. Pfleeger, "Software metrics: a rigorous and practical approach. 1997," *Brooks/Cole Pub Co*, 1993.

Acknowledgments
This research was conducted at *STMIK Nusa Mandiri*. The research support was also provided by PT XYZ, including the Wisanggeni framework which was used as a case study in this research.

IOP Conf. Series: Materials Science and Engineering **874** (2020) 012026 doi:10.1088/1757-899X/874/1/012026

Study of Crack Connections in Materials Composite Based on Polymer

Tjahjanti P H[1,2], Firdaus R[1], Iswanto[1], Ahnan M F[1]

[1] Department of Mechanical Engineering, Universitas Muhammadiyah Sidoarjo, Indonesia

[1] prantasi@gmail.com, [2] prantasiharmi@umsida.ac.id

Abstract. The main problem in composite materials based on polymer arises when receiving dynamic loads resulting in initial cracks that continue to propagate. When crack on the metal, it can be connected by welding. Also, when a crack on concrete can be reconnected by grouting. While when crack in composite materials based on polymer, there is no special way to deal with cracks so far. The purpose of this research is to reconnect the composite materials based on polymer which has cracked and do the impact test (ASTM D6110-04 standard) is carried out to determine the impact strength and the amount of energy to break, tensile strength test, and observe the structure using Scanning Electron Microscopy (SEM). The highest of impact strength and the amount of energy to break, and tensile strength in sample A is 50% resin + 1% catalyst + 19% calcium carbonate + 25% fine fiber + 5% roving fiber, with an impact strength value of 23.89 Joules / mm^2 and a large amount of energy to break 130.67 Joules, and value of tensile strength is 7.22 kgf/mm^2. While observed microstructure by using SEM shows the crack that happened is brittle.

1. Introduction

Composite materials are everywhere in our daily lives. Many products on the market contain composite materials or structures, such as plywood, golf clubs, canoes, and coated knives. Although the term composite is very broad and refers to the combination of two or more materials at the macroscopic level, this article refers to a composite as a fiber-reinforced plastic (FRP) or polymer matrix composite (PMC). Reinforcement typically is fibers and the matrix is a thermoset resin. Polymer matrix is most commonly referred to as resin. In about 90 percent of these composites fiberglass is used for reinforcement, and either polyester or vinyl ester is used as a matrix. The open molding technique is used for 65 percent of the composites manufactured, whereas the remaining 35 percent of the composites are manufactured through closed molding [1].

The use of natural fibres (e.g., straw, flax, hemp, banana and jute) as material reinforcement was investigated with considerable interest in last decades [2-5]. The most common concerns about the use of these fibres regard their coupling with a polymeric matrix, which needs to be compatible with the cellulose contained in the fibres [6]. A number of thermoplastic and thermoset matrices were used with this aim [7-8]. However, the best results were obtained so far with polyester and some phenolic resins [9-11]. Another issue widely investigated is the weathering behaviour of natural fibre reinforced composites, including the study of the influence of water sorption on the mechanical properties of the laminate [12].Top of Form Nowadays a lot of material that the object of research is some kind of plastic such as polyethylene, polypropylene, LDPE, PVC and others. With manipulate the properties of the base material is capable of making such materials into raw materials that are superior to each of the constituent elements. With advances in the field of composites, it opens opportunities to open new materials, the survey includes a large volume commodity plastic, cheap plastic and plastic for engineering as well as modern plastics. Thermoplastic that dominate the market (for example: PE, PVC, and PP) plastic type of development entirely new very expensive.

The results of research by [13] Bismarck A., et al, 2002 and Jamasri [14] (2005-2006) have used kenaf fiber and palm fiber waste as an amplifier for executive train (K1) and economy train (K3) panel at PT INKA Madiun. Santoso et al [15] (2006) also concluded that the optimization of increased bending and shear strength can be done by providing variations in the addition of random kenaf fibers. Based on the description of the results of research and application of the composite green panel that has been observed by the author, it shows that the prospect of utilizing kenaf natural fibers in the future will be certainly very likely to be used as the main structural panel, both those suffering from dynamic or static loads. The majority of panel plate failures that suffer from dynamic loads are initial cracks that continue to propagate. Therefore, testing and analysis of crack propagation on a green composite panel plate is important to investigate. The purpose of this research is make to joint when composite materials base on polymer if there is a crack/broken.

2. Experimental Studies

2.1 Materials Research

Materials that use are: resin (epoxy), catalyst, potassium carbonate/$CaCO_3$ powder, met/mes (soft fiber) and roving (hard fiber), and composition of each ingredient are made listed in **Table 1**.

Table 1. Composition of samples

Materials	Composition (%)			Note (unit)
Resin (Epoxy)	50	50	50	5000 ml
Catalyst	1	1	1	0.5 ml
$CaCO_3$ powder	19	19	19	150 gram
Met/mes (soft fiber)	25	20	15	(30 x 30) cm
Roving (hard fiber)	5	10	15	
Code	A	B	C	
Code D is origin				

2.2 Tensile Test

Preparation of test substance adjusted to the standard ASTM D 638 type IV (Standard Test Method for Tensile Properties of Plastics) with test conditions as follows:

- Temperature test chamber = 23°C
- Humidity test room = 50% relative for more than 40 hours.
- Speed drag = 10 mm / min

The size of the test materials shown in **Table 2** and shapes of test materials such as **Figure 1**. Machine examiners use TM 113 Universal 30 KN

2.3 Impact Test

The making of impact test specimens refers to the ASTM D6110-4 standard (**Figure 2**) with the specimens to be tested in the form of rods with a cross section of a square given a notch V by means of the machining process on the side.

Table 2. The size of the test material

Notation	Size (mm)	Tolerance
T = thick test materials	3	-
W = width at the narrow part	6	± 0,5
L = length at the narrow part	33	± 0,5
Wo = width of test materials	19	± 6
Lo = length of test materials	115	-
G = length limit of measurement for the	25	± 0,13

Notation	Size (mm)	Tolerance
test materials		
R = inner radius	14	± 1
Ro = outer radius	25	± 1
D = the distance between the clamp	64	± 5

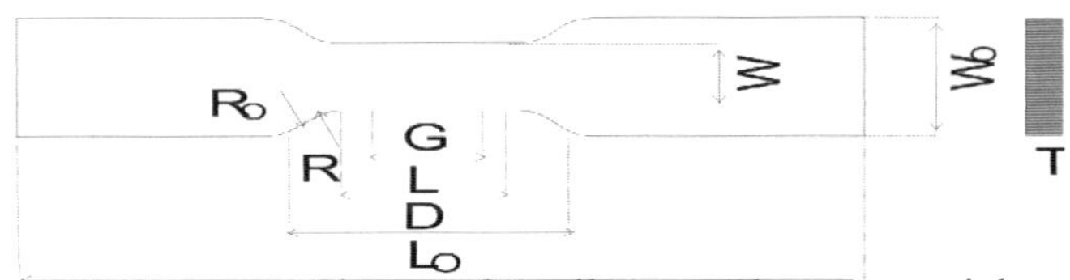

Figure 1. Shapes of tensile strength test materials

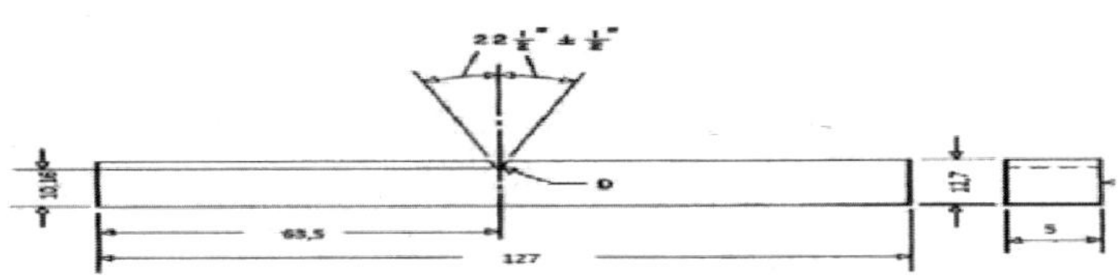

Figure 2. Sample of impact test (ASTM D6110-4 standard)

3. Result and Discussion

3.1 Result of Tensile Test

Figure 3a and **3b** shows the composite material that has been made for tensile testing, and and the sample of material has been broken during tensile testing (**Figure 3c**). Result of tensile strength found in the **Table 3**.

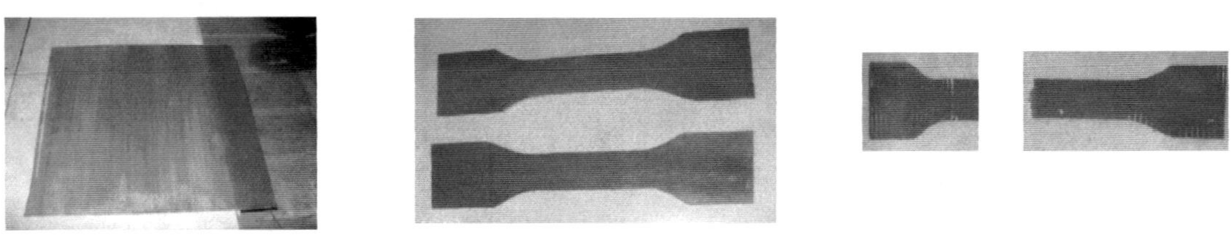

(a) (b) (c)

Figure 3. Composite material for tensile test

(a) Composite material that has been made
(b) Samples for tensile strength test
(c) Material has been broken during tensile testing

Table 3. Value of tensile strength

No.	Samples	Tensile Strength (kgf/mm^2)	Strain/Elongation
1.	A	7,22	0.92
2.	B	6,35	0.67
3.	C	5,67	0.46
4.	D (origin)	7.06	0.99

From table 3 shows that a tensile test sample code A that is sample A is 50% resin + 1% catalyst + 19% calcium carbonate + 25% fine fiber + 5% roving fiber, showed the highest tensile strength (7.22 kgf/mm2) and strain/elongation value is 0.92. While sample D (origin) have value of tensile strength and strain not much different value (difference 2.22% for tensile strength and 7.07% for elongation).

Comparison between tensile strength and elongation before fracture in tensile testing of materials PE50 with PE50 after patched, show prices are not much different value (difference 16% for tensile strength and 7% for strain/elongation)..

3.2 Result of Impact Test

The highest of impact strength and the amount of energy to break in sample A is (50% resin + 1% catalyst + 19% calcium carbonate + 25% fine fiber + 5% roving fiber) is 23.89 Joules/mm^2, and 130.67 Joules. **Figure 4a** shows the composite material that has been made for impact testing, and the sample of material has been broken during impact testing (**Figure 4b**).

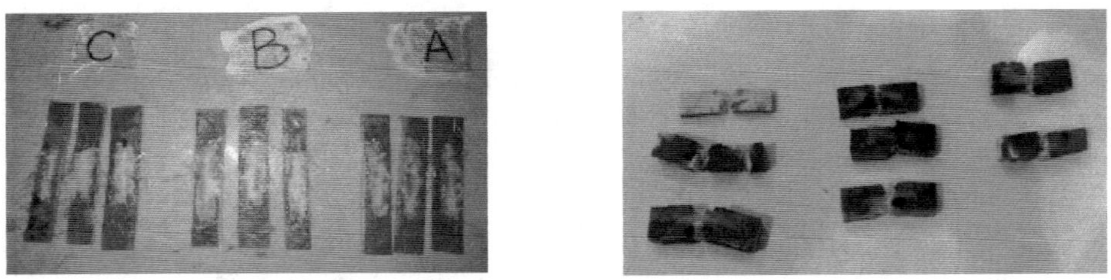

(a) (b)

Figure 4. Composite material for impact test

(a) Samples for impact test
(b) Material has been broken during impact testing

The results of the impact test data (impact strength and amount of energy to break), are complete in Table 4.

Tabel 4. Value of impact data

Samples	Amount of energy to break (Joule)	Impact strength (J/mm^2)
A	130.67	23.89
B	120.54	20.54
C	118.97	15.33
D (origin)	129.34	23.31

3.3 Result Image of Scanning Electron Microscopy (SEM)

The microstructure of composite material sample A (50% resin + 1% catalyst + 19% calcium carbonate + 25% fine fiber + 5% roving fiber) was observed using Scanning Electron Microscope (SEM) each is displayed in **Figure 5a, 5b,** and **5c.** Showed from figures that resin and fiber not united properly (fiber does not support matrix) and this material is brittle. Next research to try joining/connecting crack connections in materials composite based on polymer with 'plastic welding'

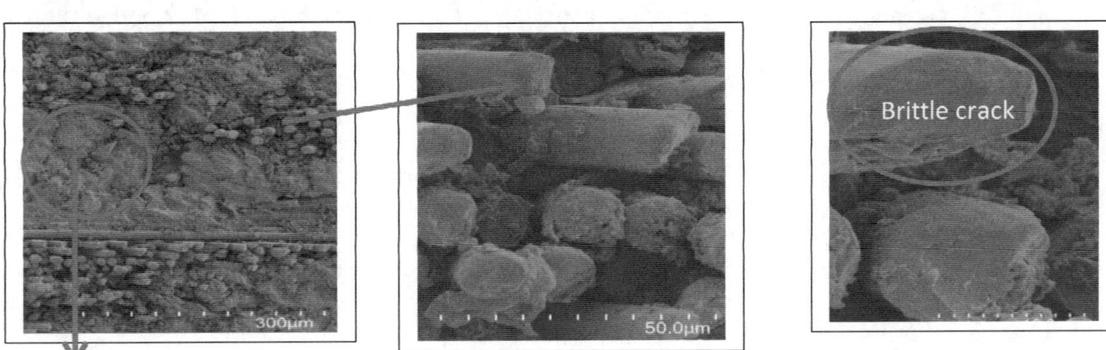

Resin and fiber not united properly
(fiber does not support matrix)

4. Conclusion

Composite material based on polymer has been successfully obtained from 50% resin + 1% catalyst + 19% calcium carbonate + 25% fine fiber + 5% roving fiber but joining/connecting with this method in this research has not produced optimum results, because between matrix and reinforcement it is not connected properly.

Acknowledgment

The work/research was financially supported by Universitas Muhammadiyah Sidoarjo and RISTEK DIKTI (grant-in-aid Penelitian Terapan Unggulan Perguruan Tinggi (PTUPT 2019) is acknowledged.

References

[1]. Jitendra S. Tate, Adekunle T. Akinola, and Dmitri Kabakov., " Bio-based Nanocomposites: An Alternative to Traditional Composites", The Journal of Technology Studies, Volume 35, Number 1, (2009).
[2]. Shah, A.N., Lakkad, S.C., "Mechanical properties of jute reinforced plastics", Fibre Science and Technology 15 (1981) 41-46
[3]. Roe, P.J., and Ansell, M.P., "Jute reinforced polyester composites", Journal of Materials Science 20 (1985) 4015-4020.
[4]. Lilholt, H. & Bjerre, A. B., "Composites based on jute-fibres and polypropylene matrix, their fabrication and characterisation", Proceedings of the 18th Risø International Symposium on Materials Science: Polymeric Composites - Expanding the Limits, Risø National Laboratory, Denmark (1997), 411-423.
[5]. Mohanty, A.K. and Misra, M., "Studies on Jute composites - a literature review", Polymer Plastics Technology and Engineering, 34 (1995), 729-792.
[6]. Sebe G., Cetin N.S., Hill C.A.S., M. Hughes., "RTM hemp fibre-reinforced polyester composites", Applied Composite Materials 7 (2000), 341-349.
[7]. Chawla, K.K., Bastos, A.C., "The mechanical properties of jute fibre and polyester/jute composites", Mechanical Behaviour of Materials 3 (1979), 191-196
[8]. Harikumar K.R., Kuruvilla J., Sabu T., "Jute sack cloth reinforced polypropylene composites: mechanical and sorption studies" Journal of Reinforced Plastics and Composites 18 (1999), 346-371.
[9]. Ray, D., Sarkar, B.K., Rana, A.K. and Bose N.R. "The mechanical properties of vinylester resin matrix composites reinforced with alkali-treated jute fibres" Composites A 32 (2001), 119-128.
[10]. Mishra, H.K.; Dash, B.N.,; Tripathy, S.S; Padhi, B.N. "Study on mechanical performance of jute-epoxy composites" Polymer-Plastics Technology and Engineering 39 (2000), 187-198.
[11]. Singh, B.; Gupta, M.; Verma, A., "The durability of jute fibre-reinforced phenolic composites", Composites Science and Technology, 60 (2000), pp.581-589.
[12]. Dash, B. N.; Rana, A. K.; Mishra, H. K.; Nayak, S. K.; Tripathy, S. S. "Novel low-cost jute-polyester composites. III. Weathering and thermal behavior" Journal of Applied Polymer Science, 7/8, (2000), 1671-1679
[13] Bismarck A., et al, 2002, Surface Characterization of Flax, Hemp and Cellulose Fibres: Surface Properties and the Water Uptake Behavior, Polymer Composite Vol 23, no. 5.
[14] Diharjo K., dkk., 2005, The Effect of Alkali Treatment on Tensile Properties of Random Kenaf Fiber Reinforced Polyester Composites, Part III of Doctorate Dissertation Research Result, Post Graduate Study, Indonesia: Gajah Mada University.
[15] Kusko, C.S, et al, 2004, Influnce of Stress Ratio on Fatigue Crack Propagation Behavior of Stainless Stell Welds, Welding Journal, pp.59-64, Bethlehem, Pa.
[16] Eichhorn S.J., et al., 2001, Review Current International Research into Selullosic Fibres and Composite, Journal of materials Science, pp. 2107-2131.
[17] Gay, et al., 2003, Composite Material, Desaign and Applications, Boca Raton: CRC Press.

[18] Helder F.S. et al., 2009, Influence Of Loading Sequance And Stress Ratio On Fatigue Damage Accumulatin Of A Composite Structure, Journal of Wuhan University of Tecnology-Master. Sci. Ed. April 2009.
[19] Kenaf Green Industries.2008. Business, Kenaf Users, Environment , New Project. URL: http: www.kenaf-fiber.com.
[20] Mathur & Nirbay M., 2007. Effect of Stress Ratio and frequency on fatigue crack growth rate of 2618 aluminium alloy silicon carbide matal matrix composite. Bulletin of Materials Science, Vol. 24, No. 2, pp.169-171, Indian Academy of science. 50

IOP Conf. Series: Materials Science and Engineering **874** (2020) 012027 doi:10.1088/1757-899X/874/1/012027

Crack Patterns Analysis on Structural Beam with *Slag Cement*

Rahmi Karolina*, Syahrizal, R.Inanda, Baskara Hutahean, and M.A.P Handana

Universitas Sumatera Utara, Medan, Indonesia

*rahmi.karolina@usu.ac.id

Abstract. The more widespread use of concrete and the increasing development shows that there is a greater demand for concrete so a new innovation for concrete is needed in an alternative use of basic materials. Slag cement is the result of fine grinding of steel slag which has the properties of cement which can function as an aggregate adhesive in concrete mixtures and has a fineness level that is either equal or more than type I portland cement. This research aims to compare high quality concrete made from slag cement and conventional concrete mixtures. Concrete beam samples with the size of 20 cm x 30 cm x 240 cm, with 2D10 compressive reinforcement and 3D10 tensile reinforcement. The number of samples are 6 with a variation of 3 high quality reinforced concrete with type I portland cement and 3 using slag cement. Flexture test is by analyzing the behavior of beams due to loading, crack patterns, and visual results of the beam with slag cement. From the results of the research, at the same load of 8 tons obtained a maximum deflection value of 9.52 mm on structural beams with slag cement and 13.30 mm on normal structural beams, thus structural beams with slag cement have better rigidity. The crack patterns that occur is the flextural crack patterns that occur in 1/3 of the middle span with a number of 5 cracks and a maximum crack width of 0.218 mm that occurs in structural beams with slag cement.

1. Introduction

In the world of building construction, development is an ongoing effort to improve the standard of living and welfare of the community. Many research has been done to get better and more environmentally friendly construction products. Concrete is an option as a structural material in building construction. The more widespread use of concrete and the increasing development shows that there is a greater demand for concrete so a new innovation for concrete is needed in an alternative use of basic materials. The more widespread use of concrete and the increasing development also shows that there is a greater demand for concrete, so we need a new innovation for concrete in the alternative use of basic materials. This innovation includes the use of steel slag waste as a fine aggregate substitution materials and slag that can be used for raw materials for making cement (Keputusan Menteri Negara Lingkungan Hidup Nomor: 231 Tahun 2010). Slag is waste produced from the by-products of the metal ore smelting process. Iron ore is reduced to iron and the remaining material from the slag that floats on the iron. This slag is periodically tapped as a liquid and it will be used for the manufacture of GGBS that must be quickly extinguished in large volumes of water, optimizing the properties of cement and produce granules similar to coarse sand. The granulated pulp is then dried and grinded into a fine powder. Although usually designated as "GGBS" in the UK, it can also be referred to as "GGBS" or "Cement Slag" Concrete which is basically a mixture of fine aggregate, coarse aggregate and cement. The content of the dominant chemical composition in slag contains iron oxide and silicate. Physically and the chemical composition of the slag is similar to clinker so that slag can be added in the final grinding of cement at the finish mill as an additional substitute material.

In Indonesia, there are many industries engaged in steel smelting and refining, including PT Krakatau Steel which is located in West Java that produces at least 150 tons of slag every day. Every ton of steel production produces 20 percent of slag waste. Slag byproducts produced from steel smelting and refining companies can be utilized as material with more value through waste co processing. If it's not utilized, the waste is included in the category of toxic and hazardous waste (B3). (Puslitbang Jalan dan Jembatan, 2011). In good fineness, slag cement shows the same or higher quality compared to Portland cement (type I) and has the characteristic of "Low Heat Hydration" which is producing a low hydration heat and CO2 emissions which are produced when production is very low. Therefore, it can replace portland cement function with certain mass ratio. Various substitution levels are starting from 30% - 70%. (Semen Indonesia, 2017). From the content of the chemical composition, it can be expected that slag can be reused for various applications in the field of Civil Engineering. Currently, the steel mill has developed the use of slag as a mixture for structural concrete in order to reduce the cost of construction or repairs to existing facilities and infrastructures. So it is expected that later, slag has a high economic value besides the negative impact due to the slag on environmental damage can be minimized so that the environment becomes clean, free of solid waste and constantly maintained.

2. Research Method

2.1 Concrete

Concrete is a mixture of coarse aggregate, fine aggregate, water and cement with or without additives which then hardened. In a concrete, there is a set of mechanical and chemical interactions of the forming materials. The diversity of concrete-forming materials are made from various types of cement, aggregate and also pozzolanic material, fly ash, high furnace slag, fiber, and others. (Neville and Brooks 1987). Concrete is also a mixture of Portland cement or other hydraulic cement, fine aggregate, coarse aggregate and water with or without additives that forms a solid mass. (SNI 02 2847 2002).

2.2 Reinforced Concrete Beams

To hold a great tensile strength on the bottom edge beam fibers, it needs to be given reinforcing steel so that it is called as reinforced concrete in this reinforced concrete beam section. Steel reinforcement is planted in the concrete in such a way so that the tensile force needed to hold the torque on the cross section of the crack can be held by the reinforced steel. Due to the nature of the concrete which is not strong enough against tensile, then the part of the beam which holds the tensile (under the neutral line) will be held by the reinforcement while the part that holds the pressure (above the neutral line) is still retained by concrete.

2.3 Constituent Materials of Concrete Beams

a. Portland Cement

According to ASTM C-150 1985, Portland cement is defined as a hydraulic cement produced by grinding a clinker consisting of hydraulic calcium silicate which generally contains one or more forms of Calcium Sulfate as an added material which is milled together with its main ingredient. The process of making portland cement begins by putting the main raw material from the quarry in the form of 70% limestone, 15% clay, Silica and Iron Oxide mashed until it can be stirred evenly according to the proportions. Then heated up to 800-9000C and put into a rotary clin to be burned at a temperature of 14500C. When the clinker is cooled, grinding is conducted along with the addition of gypsum until it becomes fine.

b. Slag Cement

Slag cement is the result of the addition of Granulated Blast Furnace Slag which has clinker-like properties into the final grinding of cement at the finish mill as an additional substitute material. The content of the dominant chemical composition in the slag contains Iron Oxides and Silicates. Physically and the chemical composition of slag is similar to clinker.

Blast Furnace Slag is a slag, a waste material from pig iron casting, where the process uses a furnace whose fuel is from a blown air (blast). Slag constituent materials are limestone, silica and alumina which react at temperatures of 16000c and in a form of liquid. Liquid slag floating on molten iron ore, separated and granulated. Liquid slag has a composition of about 30% to 40% SiO2 and about 40% CaO which is close to the chemical composition of Portland cement. If this liquid is cooled slowly, there will be a useless crystal as a mixture of cement and can be used as a substitute for aggregates that commonly known as slag steel. However, if the liquid is cooled quickly and suddenly, it will form a granulated glass which is very reactive for making slag cement. This granulated glass is dried and grinded with a granulator until it is fine or to the required size and can be used as a substitute for cement in the making of concrete. (Paul Nugraha, 2007).

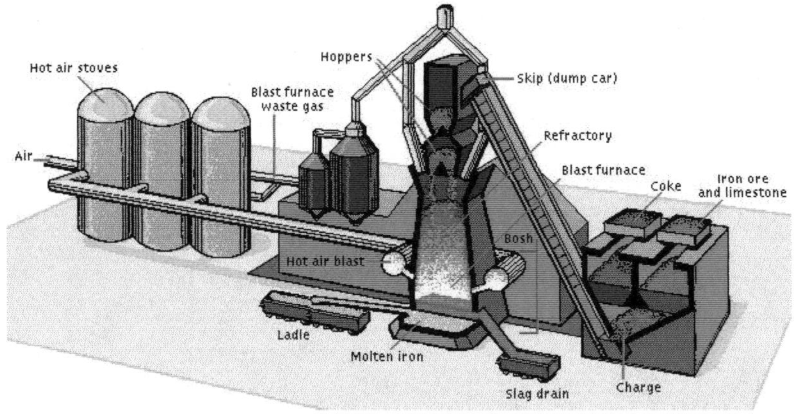

Figure 2.1. The Making Process of Slag

c. Aggregates

Based on SK.SNI T-15-1991-03, aggregates are granular materials such as sand, gravel, split, and iron furnace crust that is used together with a binder to form hydraulic or mixed cement concrete. Aggregates content in concrete ranges from 60-70% of the volume of concrete. A large proportion of aggregates in concrete indicates an important role of aggregates. Aggregates must be graded in such a way that the entire mass of concrete can function as a whole, homogeneous, and dense object in which small aggregates function as filling gaps between large aggregates. (Nawy, 1998) The selection of aggregates is important because it will affect the quality of concrete. Therefore, the aggregates used must have the following conditions:

a.The aggregate is clean.
b.Hard.
c. Free of absorption properties.
d. Does not mix with clay or mud.
e. Distribution / gradation of the aggregate size meets the requirements

d. Water

Water is needed in the making of concrete to trigger the chemical process of cement, soaking aggregates and providing ease in concrete work. Water containing hazardous compounds, which are contaminated with salt, oil, sugar, or other chemicals when used in concrete mixtures will reduce the quality of concrete and can even change the properties of the concrete produced. The water used can be in the form of fresh water (from rivers, lakes, ponds, etc.), sea water or waste water, as long as it meets the specified quality requirements (Trimulyono, 2003). The value of the ratio between the weight of water and cement for a concrete stir is called the water cement ratio (w / c). In order to make the perfect hydration process in the concrete mix, general w / c values used are from 0.40 to 0.65 depending on the quality of the concrete to be achieved. To increase workability (ease of workmanship) a higher w / c value is needed. (Dipohusodo, 1994). Generally, the strength and quality of concrete is strongly influenced by the water used.

IOP Conf. Series: Materials Science and Engineering **874** (2020) 012027 doi:10.1088/1757-899X/874/1/012027

e. Superplasticizer Master Ease 3029

Superplasticizer is greatly increases workability. Mixtures with a slump of 7.5 cm will become 20 cm. Used primarily for high quality concrete because it can reduce water by up to 30%. High flowability in concrete mixtures containing superplastizicers generally lasts around 30-60 minutes and after that it decreases rapidly. (Paul Nugraha, 2007). Superplasticizer used in this high quality concrete mixture is the type of Master Ease 3029. Master Eease 3029 is a type of high range water reduce superplasticizer produced by BASF. The Master Ease is designed to provide high rheological properties in fresh concrete thereby increasing the ease of placement and completion of concrete as well as pumping concrete for all construction activities.

f. Reinforcing Steel

Concrete-reinforcing steel is steel in the form of a circular cross-section rod used for reinforcement of concrete, which is produced from billet raw material by hot rolling. Based on the shape of the concrete, concrete-reinforcing steel can be divided into 2 (types) i.e. plain concrete-reinforcing steel and fin concrete-reinforcing steel. Plain concrete-reinforcing steel has a circular cross section with non-finned flat surfaces and fin concrete-reinforcing steel is a special-shaped concrete reinforcement whose surface has a cross and longitudinal ribs that are intended to increase adhesion and resist longitudinal motion from the bar relatively against the concrete.

2.4 Test Samples

The making of concrete samples in this research used a block mold with a size of (20 x 30 x 240) cm. There are 2 variations of materials for the samples.

Table 2.1 Beams Sample

Code	Type of Cement	Number of Samples
BN	Type I Portland Cement	3
BS	Slag Cement	3
	Total of Samples	6

2.5 Mix Design of Concrete

Mix design of concrete is intended to determine the composition or proportion of the constituent materials. The proportion of these constituent ingredients is determined through a concrete design (mix design). This is done so that the mixture proportion can meet the technical requirements economically. In determining the proportion of mixtures, the American Concrete Institute (ACI) method is used in this research based on SNI 2847: 2013.

Table 2.2 Normal Beams Mixtures

Portland Cement	Split	Sand	Water	HRWR
95,65714 kg	155,5476 kg	139,7012 kg	26784 ml	1147,886 ml

Table 2.3 Slag Beams Mixtures

Slag Cement	Split	Sand	Water	HRWR
95,65714 kg	155,5476 kg	139,7012 kg	26784 ml	1147,886 ml

2.6 Tensile Strength Test of Structural Beams

The concrete beam test refers to ASTM C-78 conducted at the Structure Laboratory of FT USU. The test steps are as follows:

1. Preparation of testing equipment: Dial Gauge, Hydraulic jack pump, hydraulic jack.
2. In the beam, a grid line with a distance of 5 cm is made first so that the crack pattern lines at the time of testing are easily seen and marked.
3. The beam is placed on top of the Static Loading Frame that has been provided. The support of the sample is located 15 cm from the right and bottom left of the sample.

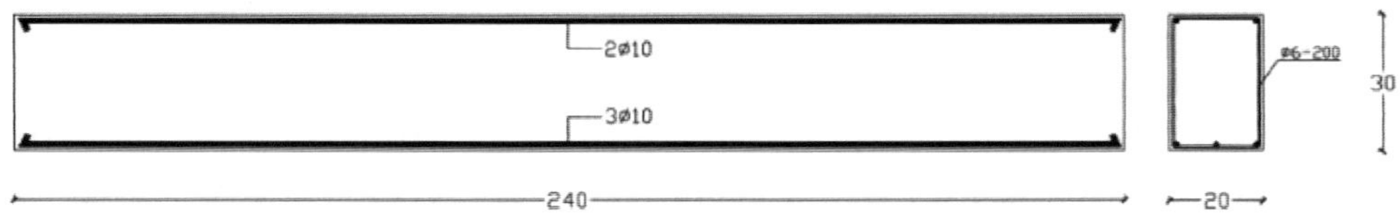

Figure 2.2 Beam Dimension

4. Dial Gauge (deflection / decrease gauge equipment) is placed under the beam with a 3-point position, i.e.1/4 L-L, center line and 1/4 L-R.
5. Profile I with a length of 70 cm is placed right in the middle of the beam, in order to divide the centralized force loading with a large of P.

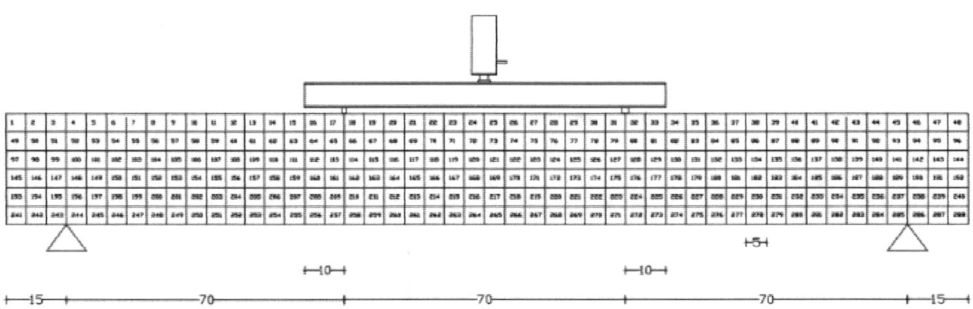

Figure 2.3 Beam Loading

6. The vertical examination of the hydraulic load and jacking equipment above the profile I was carried out in order to obtain pure vertical force.
g. After all equipment are installed, the testing phase is carried out by pumping the hydraulic jack with a constant load multiples. Load multiples that occur is read through the manometer. Hydraulic jack is pumped, the end of the hydraulic jack will press the top profile I frame and automatically the base of the hydraulic jack presses the profile I steel which is given 2 solid iron supports as a 2 point load. Every time the load is added, the mechanical dial gauge will spin and read the deflection that occurs.
h. Load and deflection is recorded when adding a load of 0.5 tons. In addition to recording the load and deflection, each additional load is observed on the sample whether cracks occur or not.
i. Loading is stopped after the beam has a maximum crack and deflection.

Figure 2.4 Beam Loading

3. Results and Discussion

3.1 Tensile Strength Test of Rebar

The tensile strength test of steel reinforcement, by using the equipment "Tensile Machine Test" is carried out at Politeknik Negeri Medan.

Figure 3.1 Tensile Test of Rebar

Plain rebar Ø 6

- yield stress value (fy) = 337,93 N/mm2
- breakdown stress value (fu) = 518,40 N/mm2

Thread rebar D 10

- yield stress value (fy) = 364,75 N/mm2
- breakdown stress value (fu) = 527,64 N/mm2

3.2 Compressive Strength Test of Concrete

From the results of previous experiments, obtained the results of the average compressive strength test of concrete :

Normal concrete = 73,75 Mpa
Slag concrete = 74,63 Mpa

3.3 Theoretical Results of Structural Beams

Table 3.1 Results of Theoretical Calculation

Normal Beam(F'c =73,75 Mpa				Slag Beam(F'c =74,63 Mpa)			
Pu (KN)	Deflection			Pu (KN)	Deflection		
	1/4L-L	1/4L-R	CL		1/4L-L	1/4L-R	CL
	mm	mm	Mm		mm	mm	mm
81,9	8,051	8,051	9,975	83,3	8,188	8,188	10,146

3.4 Test Results of Deflection

Table 3.2 Results of Theoretical Calculation

Beban (Kg)	C-L (Balok Normal) Lendutan (mm)	C-L (Balok Slag) Lendutan (mm)	1/4L-L(Balok Normal) Lendutan (mm)	1/4L-L(Balok Slag) Lendutan (mm)	1/4L-R(Balok Normal) Lendutan (mm)	1/4L-R(Balok Slag) Lendutan (mm)
0	0,00	0,00	0,00	0,00	0,00	0,00
500	0,25	0,20	0,17	0,15	0,18	0,16
1000	0,33	0,25	0,24	0,20	0,26	0,21
1500	0,43	0,38	0,35	0,26	0,36	0,28
2000	0,85	0,74	0,50	0,43	0,54	0,45
2500	1,34	1,21	0,81	0,75	0,85	0,77
3000	1,76	1,57	1,21	0,91	1,26	0,93
3500	2,24	2,02	1,75	1,36	1,79	1,39
4000	2,77	2,49	2,27	1,70	2,32	1,80
4500	3,32	2,99	2,75	2,14	2,79	2,17
5000	3,96	3,51	3,30	2,65	3,34	2,68
5500	4,96	4,09	3,83	3,47	3,89	3,51
6000	5,65	4,76	4,51	4,19	4,56	4,20
6500	6,56	5,65	5,21	4,80	5,26	4,82
7000	7,82	6,64	6,01	5,25	6,08	5,27
7500	9,21	7,83	7,11	5,83	7,15	5,86
8000	10,30	9,52	8,39	6,37	8,42	6,40

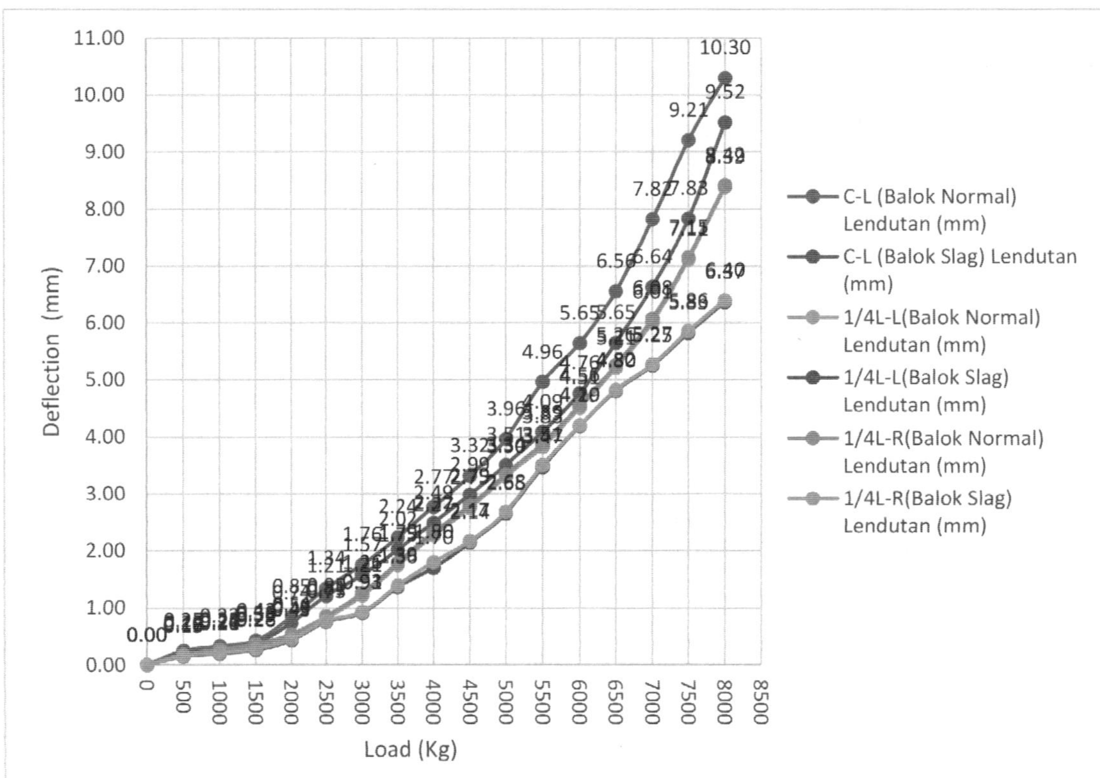

Figure 3.2 Graph of Load-Deflection Ratio (BN and BS)

IOP Conf. Series: Materials Science and Engineering **874** (2020) 012027 doi:10.1088/1757-899X/874/1/012027

3.5 Crack Patterns Occured on the Beam

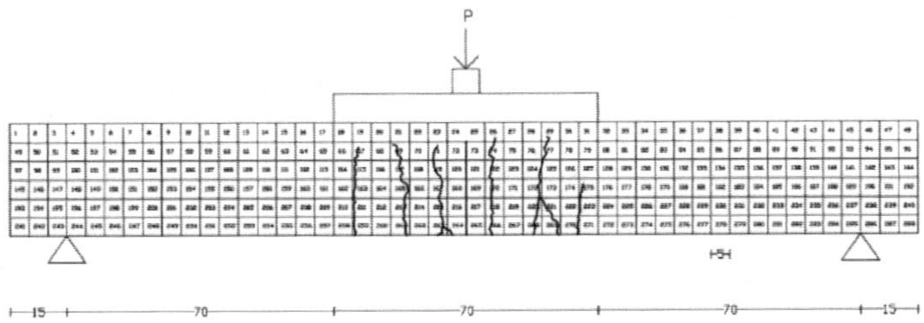

Figure 3.3 (a) Crack Pattern of Normal Structural Beam 1

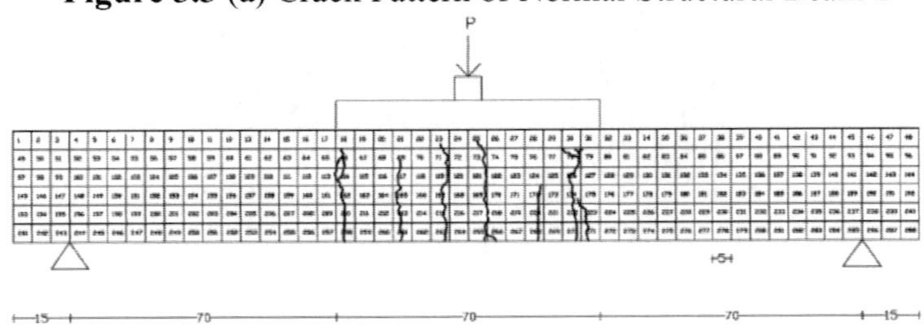

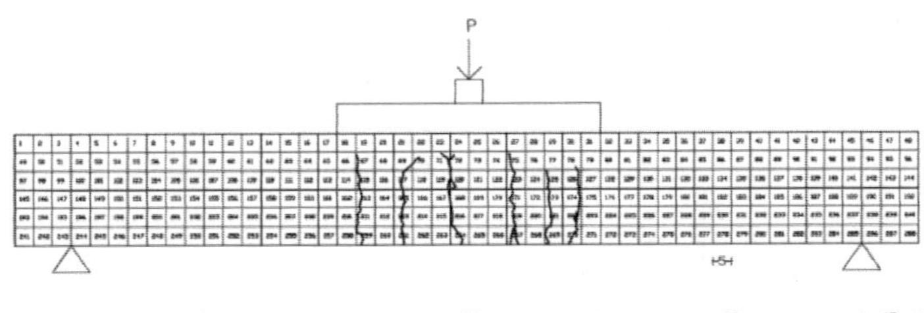

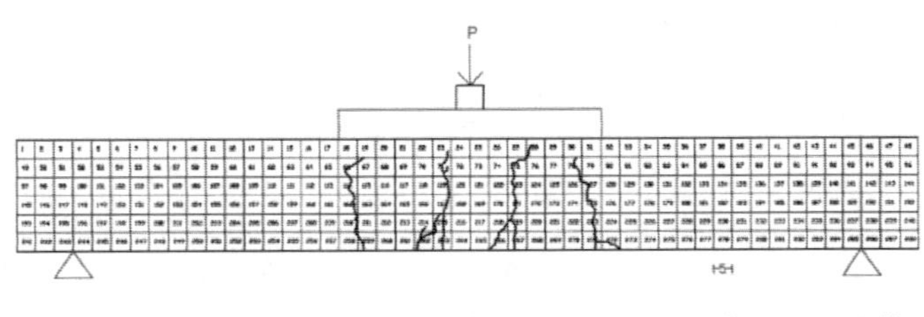

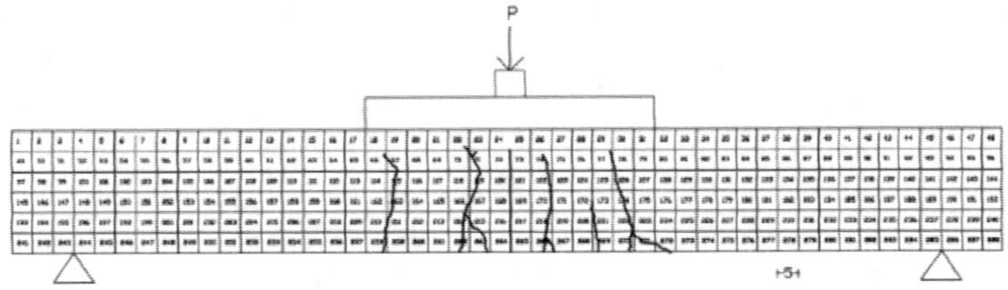

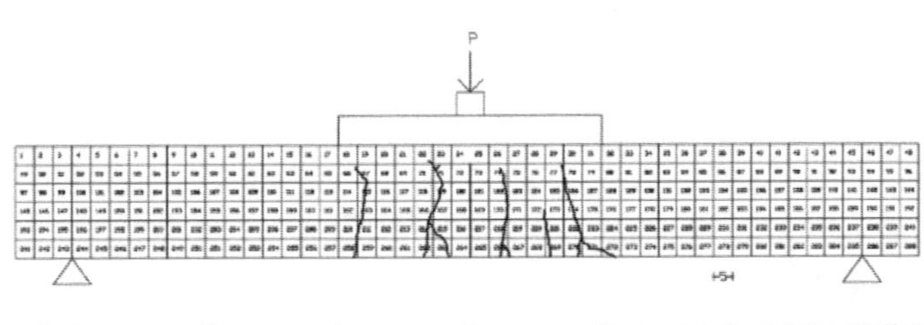

In Figure 3.3 (a), (b), (c) and figure 3.4 (a), (b), (c) it can be seen that the crack patterns that occur is almost the same, which is flexture. In this test, cracks that occur in the samples are generally begin with a subtle crack on the underside of the samples, i.e. the tensile side of the cross section, followed by flextural cracks on the side where the previous subtle cracks occurred. Flextural cracks occur in areas that have a maximum flexture torque value, that is, in the area around the center span which is exactly located in the area of two given point loads. The direction of the perpendicular crack is almost perpendicular to the axis of the beam.

3.6 Maximum Crack Width

Table 3.3 Maximum Crack Width

No	Sample	Maximum Crack Width (cm)
1	BN 1	0,219
2	BN 2	0,220
3	BN 3	0,221
4	BS 1	0,217
5	BS 2	0,216
6	BS 3	0,218

4. Conclusions

Based on the test results that have been carried out on the structural beams with slag cement, the following conclusions are obtained.

1. From the test results, structural beams using slag cement experienced initial cracks at an average load of 4.2 tons and for normal structural beams experienced initial cracks at an average load of 4 tons.
2. At the same load of 8 tons, it is obtained the maximum average deflection value in the middle of the span which is on the structural beam with a slag cement of 9.52 mm and on the normal structural beam of 10.30 mm. Thus structural beams with slag cement have better rigidity.
3. The crack patterns that occur in all samples are the flextural crack patterns. This is indicated by the direction of the vertical crack towards the main axis of the beam and occurs in the 1/3 middle span region.
4. From the test results, the number of cracks that occur in the structural beam using slag cement is less than the normal structural beam. And for the maximum average crack width that occurs on the structural beam with slag cement is 0.217 cm and on the normal structural beam is 0.220 cm.
5. Structural beam with slag cement produces brighter colors compared to the normal structural beam.

References

[1] ASTM C 1611. Standard Test Method for Slump Flow of Self Consolidating
[2] Concrete.
[3] ASTM C 642-97. Standard Test Method for Density, Absorption, and Voids in
[4] Hardened Concrete.
[5] ASTM C 78-02. Standard Test Method for Flexural Strength of Concrete (Using Sample Beam with Third-Point Loading).
[6] Badan Standarisasi Nasional.. SNI 4431 : 2011
[7] Suresh, D. & Nagaraju, K., 2015. e-ISSN: 2278-1684,p-ISSN: 2320-334X, Volume 12, Issue 4 Ver. VI (Jul. - Aug. 2015), PP 76-82. Kuppam: IOSR Journal of Mechanical and Civil Engineering (IOSR-JMCE).
[8] Turuallo, G., 2013.. s.l.:www. researchgate.net.
[9] Departemen Pekerjaan Umum, 1990. SK. SNI S -18-1990-03. Bandung: Yayasan LPMB.
[10] Wegian, M. F., 2010. Vo.3, No.4, 235-243 November 2010. s.l.:The IES Journal Part.A Civil and Structural Engineering.
[11] Pandey, R. K., Kumar, A. & Khan, M. A., 2016.Volume: 03 Issue: 02. India: International Research Journal of Engineering and Technology (IRJET)
[12] Pandiangan, J., 2016l. Medan: Universitas Sumatera Utara.
[13] Sundaramoorthy, G. & VimalBalaji, K., 2017. M.I.E.T Engineering College, Tirchirappalli: SSRG International Journal of Civil Engineering.
[14] Suresh, D. & Nagaraju, K., 2015. e-ISSN: 2278-1684,p-ISSN: 2320-334X, Volume 12, Issue 4 Ver. VI (Jul. - Aug. 2015), PP 76-82. Kuppam: IOSR Journal of Mechanical and Civil Engineering (IOSR-JMCE).
[15] Turuallo, G., 2013. s.l.:www. researchgate.net.
[16] Dipohusodo, Istimawan, 1994, Jakarta : Gramedia Pustaka Umum
[17] McCormac, C. , Jack, 2004, Edisi Kelima Jilid 1, Jakarta : Erlangga.
[18] Nawy, G. , Edward, 2008, Bandung : PT. Refika Aditama

IOP Conf. Series: Materials Science and Engineering **874** (2020) 012028 doi:10.1088/1757-899X/874/1/012028

Estimation of *Whole Blood* (WB) using Unscented Kalman Filter in Surabaya

Abdul Muhith[1*], Teguh Herlambang[1], Abdul Haris[2], Dinita Rahmalia[3]

[1]University of Nahdlatul Ulama Surabaya, Surabaya, Indonesia
[2]Universitas Panca Marga Probolinggo, Indonesia
[3]University of Islam Darul Ulum Lamongan

*abdulmuhith@unusa.ac.id

Abstract As the need for blood transfusion increases, an improved blood stock management for blood transfusion is required. In terms of quality and quantity, blood transfusion is needed by patients with various health problems. Because of that need, it is urgent that the stability of blood stock be maintained to avoid possible excessive blood stock that leads to unnecessary blood disposal. For that purpose, prediction of blood demand is required. One of the blood transfusion units is PMI, a national organization dealing with social concern and humanity field. The objective of this paper is to make estimation of blood demand for PRC and WB at PMI Surabaya using UKF method. The simulation results show that the UKF method is effective with a high accuracy and an error of less than 4%.

Keywords: *Whole Blood*, PMI Surabaya, estimation, UKF

1. Introduction

Blood is an important component in the body that carries nutrients and oxygen to all organs of the body, including vital organs such as the brain, heart, kidneys, lungs, and liver. If there is a lack of blood in the body caused by several things, the nutritional and oxygen requirements of these organs cannot be fulfilled. Tissue damage can occur quickly which leads to death. To prevent this, a blood supply from outside the body is needed. The process of transferring blood from a healthy person (donor) to a sick person / needy (recipient) is called a blood transfusion [1]. Blood transfusion has become an important part of health care. If blood transfusions are applied correctly, transfusion can save the lives of patients and can improve the health status of these patients.

Estimation of the number of requests is required in the health field. Blood is very important in human life. Blood is a liquid in the body, needed to transport oxygen needed by cells throughout the body [2]. To minimize errors, it is necessay to estimate the number of blood requests at the Indonesian Red Cross (PMI) [3]. Estimates are made because a problem can sometimes be solved using presvious information or data related to the problem [4,5,6]. One estimation method is the Extended Kalman Filter (EKF) which functions to minimize covariance error estimation in correction step [7]. In this paper, the estimation of the demand for WB blood is limited only to PMI Surabaya City. The development of the Kalman Filter method is the Unscented Kalman Filter (UKF) obtained by applying sigma point at the prediction stage. In this paper, UKF method was applied to estimate the demand for WB blood as the basis of the consideration of the blood management at PMI Surabaya City.

IOP Conf. Series: Materials Science and Engineering **874** (2020) 012028 doi:10.1088/1757-899X/874/1/012028

2. Methodology

2.1 Blood Data of Whole Blood (WB)

The data of Whole Blood (WB) is shown as follows:

Table 1. Blood Data of WB type

	Data of WB											
	Jan	**Feb**	**Mar**	**Apr**	**May**	**June**	**July**	**Aug**	**Sep**	**Oct**	**Nov**	**Dec**
2013	1018	1062	1083	1076	1222	1053	1007	1031	1039	1023	979	1052
2014	1103	953	972	812	872	906	871	1085	897	859	818	744
2015	890	865	911	797	787	687	643	801	730	686	666	723
2016	803	659	653	629	666	599	595	576	629	557	500	504
2017	579	611	675	583	637	488	647	684	581	715	749	612
2018												

2.2 Unscented Kalman Filter

Algorithm of Unscented Kalman Filter is writen as follows [7]:

- *Initiation at* $k = 0$:

$$\hat{x}_0 = E[x_0]$$
$$P_{x_0} = E[(x_0 - \hat{x}_0)(x_0 - \hat{x}_0)^T]$$
$$\hat{x}_0^a = E[x^a] = E\left[\hat{x}_0^{\,T} \;\; 0 \;\; 0\right]^T$$
$$P_0^a = E[(x_0^a - \hat{x}_0)(x_0^a - \hat{x}_0)^T] = \begin{bmatrix} P_x & 0 & 0 \\ 0 & P_v & 0 \\ 0 & 0 & P_n \end{bmatrix} \quad (1)$$

For $k = 1,2,3,\ldots,\infty$:

1) Count sigma point

$$X_{k-1}^a = \left[\hat{x}_{k-1}^a \quad \hat{x}_{k-1}^a + \gamma\sqrt{P_{k-1}} \quad \hat{x}_{k-1}^a - \gamma\sqrt{P_{k-1}}\right]$$

Dimana:

$$\gamma = \sqrt{L + \lambda}$$
$$\lambda = \alpha^2(L + \kappa) - L \quad (2)$$

2) Time-update (prediction stage)

$$X_{k|k-1}^x = f(X_{k-1}^x, X_{k-1}^v)$$
$$\hat{x}_k^- = \sum_{i=0}^{2L} W_i^{(m)} X_{i,k|k-1}^x$$
$$P_{x_k}^- = \sum_{i=0}^{2L} W_i^{(c)} \left(X_{i,k|k-1}^x - \hat{x}_k^-\right)\left(X_{i,k|k-1}^x - \hat{x}_k^-\right)^T$$
$$Z_{k|k-1} = H(X_{k|k-1}^x, X_{k-1}^n)$$
$$\hat{z}_k^- = \sum_{i=0}^{2L} W_i^{(m)} Z_{i,k|k-1} \quad (3)$$

3) Measurement update (cortrection stage):

$$P_{\bar{z}_k,\bar{z}_k} = \sum_{i=0}^{2L} W_i^{(c)} \left(Z_{i,k|k-1} - \hat{z}_k^-\right)\left(Z_{i,k|k-1} - \hat{z}_k^-\right)^T$$
$$P_{x_k,z_k} = \sum_{i=0}^{2L} W_i^{(c)} \left(X_{i,k|k-1}^x - \hat{x}_k^-\right)\left(Z_{i,k|k-1} - \hat{z}_k^-\right)^T$$
$$K_k = P_{x_k,z_k} P_{\bar{z}_k,\bar{z}_k}^{-1}$$
$$\hat{x}_k = \hat{x}_k^- + K_k(z_k - \hat{z}_k^-)$$
$$P_{x_k} = P_{x_k}^- - K_k P_{\bar{z}_k} K_k^T \quad (4)$$

IOP Conf. Series: Materials Science and Engineering **874** (2020) 012028 doi:10.1088/1757-899X/874/1/012028

3. Discussion and Analysis

From the blood data of WB types in Table 2, a mathematical function was obtained for the blood supply of WB types using Mathematica software. The Mathematica software simulation resulted in a function of WB as follows:

$$f(x) = 1025.13 - 1.3789x - 0.371\,x^2$$
$$f'(x) = -1.3789 - 0.742\,x \qquad (5)$$

from equations (2), the modified WB blood stock function model in (5) is discreted using the finite difference method and obtained as follows:

$$f_{k+1} = (-1.3789 - 0.742\ x_k)\Delta$$

After the function was got, then it was computation simulated with the Matlab software. In this paper a simulation was applied by applying the Unscented Kalman Filter (UKF) algorithm to the function of blood stock of WB type. The simulation results were evaluated by comparing the real conditions in the field with the estimation results of UKF. This simulation used $\Delta t = 0{,}1$ and 100, 200, and 300 iteration. Figure 1 is the estimation results using UKF methods using 100 iteration. Figure 2 is the simulation result of the UKF methods using 200 iteration. Figure 3 is a simulation of the UKF methods by 300 iteration.

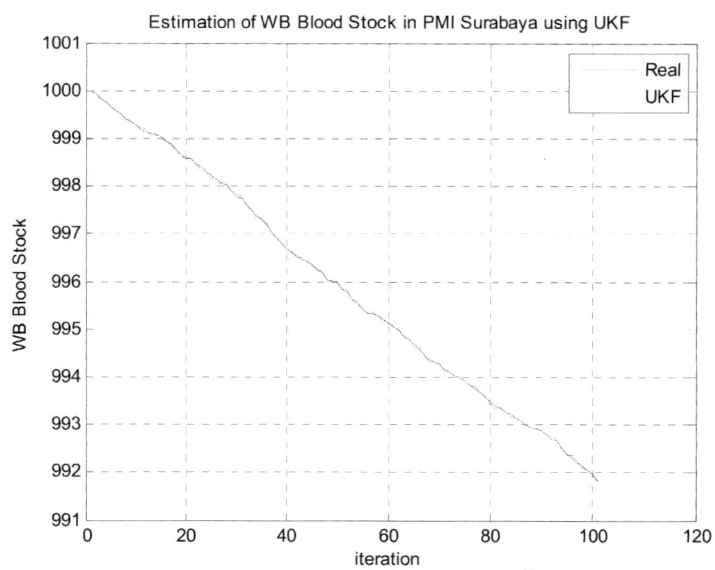

Figure 1. Estimation of WB blood using UKF methods with 100 iteration

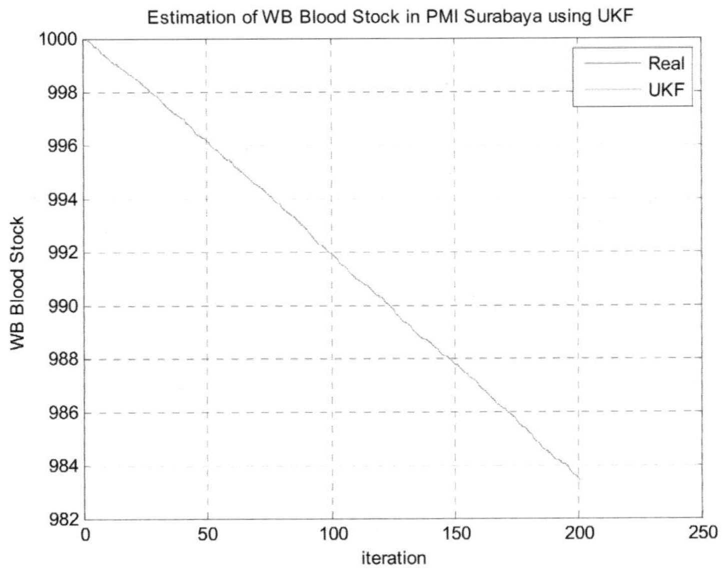

Figure 2. Estimation of WB blood using UKF methods with 200 iteration

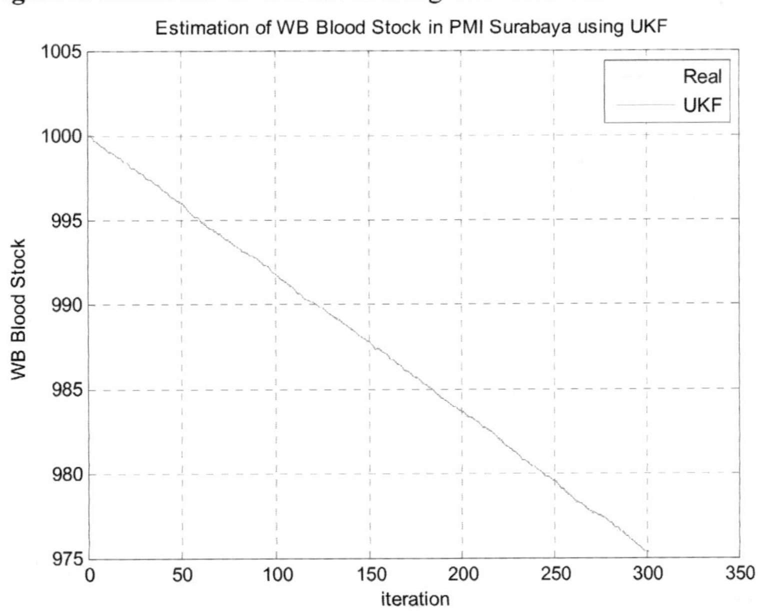

Figure 3. Estimation of WB blood using UKF methods with 300 iteration

Figures 1, 2 and 3 show that the estimation results of WB blood stock have high accuracy with errors of less than 4% as we can see in the real graphs. The accuracy of Unscented Kalman Filter methods showed no significant difference. In Figure 1 and Table 2, it appears that the Unscented Kalman Filter method using 300 iteration with RMSE of 0.00321has higher accuracy than that of 200 and 100 iteration with RMSE of 0.00407and 0.00491, but the difference is not much. Likewise, in Figures 2 and 3 with 200 and 300 iteration. In conclusion, UKF method can be used as a method estimating either WB blood stock or other blood types.

Table 2. Comparison of the RMSE values by the UKF based on 100, 200 and 300 iteration

	100 Iteration		**200 Iteration**		**300 Iteration**	
	RMSE	**Accuracy**	**RMSE**	**Accuracy**	**RMSE**	**Accuracy**
RMSE	0.00491	96,2%	0.00407	97%	0.00321	97,5%
Simulation Time	6,189 s		8,256 s		10,371 s	

In general, the methods of Unscented Kalman Filter can be used as a method to estimate Whole Blood (WB) blood stock with high accuracy. Based on the numeric simulation results above, it is likely that method can also be used to estimate other type of blood stock, so it can support the work of PMI Surabaya's blood transfusion management in blood bank

4 Conclusion

Based on the results of the simulation analysis and discussion, Unscented Kalman Filter (UKF) method can be used as a method to estimate WB blood stock with excellent accuracy and errors of less than 4%. Based on the simulation results above, it is likely that both methods can also be used to estimate othertype of blood stock, so it can support the work of PMI Surabaya's blood bank management in particular and PMI in all cities in Indonesia in general.

References

[1] Weinstein, R. 2012 Clinical Practice Guide on Red Blood Cell Transfusion. American Society of Hematology

[2] Ohsaka, A., Abe, K., Ohsawa, T., Miyake, N., Sugita, S., & Tojima, I. (2008). A komputer-assisted transfusion management sistem and changed transfusion practices contribute to appropriate management of blood components. *Transfusion, 48*(8), 1730-1738.

[3] Shanty, W., Firdaus., Herlambang, T., 2018. "Prediction of Availability of Packed Red Cells (PRC) at PMI Surabaya City Using Ensemble Kalman Filter as Management of Blood Transfusion Management", The Second Internatonal Conference on Combinatorics, Graph Teory and Network Topology, University of Jember-Indonesia, 24-25 Nov 2018, Journal of Physics: Conf. Series 1211 (2019) 012031

[4] Herlambang, T., 2017, "Design of a Navigation and Guidance System of Missile with Trajectory Estimation Using Ensemble Kalman Filter Square Root (EnKF-SR). International Conference on Computer Applications and Information Processing Technology (CAIPT)-IEEE, Bali Indonesia 8-10 Augsut 2017

[5] Karya, D.F., Puspandam, K. and Herlambang, T., 2017. "Stock Price Estimation Using Ensemble Kalman Filter Square Root Methods", The First Internatonal Conference on Combinatorics, Graph Teory and Network Topology, University of Jember-Indonesia, 25-26 Nov 2017, Journal of Physics: Conf. Series 1008 (2018) 012026.

[6] Puspandam, K., Fidita, D.F., Herlambang, T.,and Khusnah,H., 2018. "Ensemble Kalman Filter for Crude Oil Price Estimation,", The Second Internatonal Conference on Combinatorics, Graph Teory and Network Topology, University of Jember-Indonesia, 24-25 Nov 2018, Journal of Physics: Conf. Series 1211 (2019) 012031

[7] Karya, D.F., Puspandam, K., Herlambang, T., and Rahmalia D., 2018. "Development of Unscented Kalman Filter Algorithm for stock price estimation", The Second Internatonal Conference on Combinatorics, Graph Teory and Network Topology, University of Jember-Indonesia, 24-25 Nov 2018, Journal of Physics: Conf. Series 1211 (2019) 012031.

IOP Conf. Series: Materials Science and Engineering **874** (2020) 012029 doi:10.1088/1757-899X/874/1/012029

Computational algorithm PAOR for time-fractional diffusion equations

A Sunarto[1*] and J Sulaiman[2]

[1]Department Tadris Matematika, IAIN Bengkulu, Indonesia.
[2]Department Matemathics with economics, Universiti Malaysia Sabah, Malaysia.

*andang99@gmail.com

Abstract. We deal with the application of an unconditionally implicit finite difference approximation equation of the one-dimensional linear time fractional diffusion equations (TFDE's) via the Caputo's time fractional derivative. Based on this implicit approximation equation, the corresponding linear system can be generated in which its coefficient matrix is large scale and sparse. To speed up the convergence rate in solving the linear system iteratively, we construct the corresponding preconditioned linear system. Then we formulate and implement the Preconditioned AOR (PAOR) iterative method for solving the generated linear system. One example of the problem is presented to illustrate the effectiveness of PAOR method. The numerical results of this study show that the proposed iterative method is superior to PSOR and PGS, GS iterative method.

1. Introduction

Based on previous studies in [1,2,3,4] many successful mathematical models, which are based on fractional partial derivative equations, have been developed. Following to that, there are several methods used to solve these models. For example, researchers have proposed finite difference methods such as explicit and implicit [5,6,7]. Also it is pointed out that the explicit methods are conditionally stable. Therefore, we discretize the time-fractional diffusion equation (TFDE's) via the implicit finite difference discretization scheme and Caputo's fractional partial derivative of order α in order to derive a Caputo's implicit finite difference approximation equation. This approximation equation leads a tridiagonal linear system. Due to the properties of the coefficient matrix of the linear system which is sparse and large scale, iterative methods are the alternative option for efficient solutions. As far as iterative methods are concerned, it can be observed that many researchers such as Young [8], Hackbusch [9] and Saad [10] have proposed and discussed several families of iterative methods. Among the existing iterative methods, the preconditioned iterative methods Hoang-hao [11], Gunawardena [12] have been widely accepted to be one of the efficient methods for solving linear systems.

Because of the advantages of these iterative methods, the aim of this paper is to construct and investigate the effectiveness of the Preconditioned AOR (PAOR) iterative method for solving time fractional diffusion equations (TFDE's) based on the Caputo's implicit finite difference approximation equation. To investigate the effectiveness of the PAOR method, we also implement the PSOR and PGS,GS iterative methods being used a control method.

To show the effectiveness of PAOR method, let time fractional diffusion equations (TFDe's) be defined as

$$\frac{\partial^{\alpha} U(x,t)}{\partial^{\alpha}} = a(x)\frac{\partial^2 U(x,t)}{\partial x^2} + b(x)\frac{\partial U(x,t)}{\partial x} + c(x)U(x,t) \qquad (1)$$

where *a(x), b(x)* and *c(x)* are known functions or constants, whereas α is a parameter which refers to the fractional order of time derivative.

2. Preliminaries

In this section, the space-fractional diffusion equation (1) is solved. In order to find solution in Eq.(1), let us define $h = \frac{\ell}{m+1}$, where, $m=n+1$ is positive even integer. By implementing definition (2) we obtain

To constructing the Caputo's implicit finite difference approximation equation of Eq.(1), the following are some basic definitions for fractional derivative theory which are used in this paper.

Definition 1.[8] The Riemann-Liouville fractional integral operator, J^{α} of order-α is defined as

$$J^{\alpha} f(x) = \frac{1}{\Gamma(\alpha)} \int_0^x (x-t)^{\alpha} f(t) dt, \quad \alpha > 0 \ x > 0 \tag{2}$$

Definition 2.[8] The Caputo's fractional partial derivative operator, D^{α} of order -α is defined as

$$D^{\alpha} f(x) = \frac{1}{\Gamma(m-\alpha)} \int_0^x \frac{f^{(m)}(t)}{(x-t)^{\alpha-m+1}} dt, \quad \alpha > 0 \tag{3}$$

with $m-1 < \alpha \leq m$, $m \in N$, $x > 0$

To obtain the numerical solution of Eq.(1) with Dirichlet boundary conditions, firstly we derive an implicit finite difference approximation equation based on the Caputo's derivative definition and the non-local fractional derivative operator. This implicit approximation equation can be categoried as unconditionally stable scheme. A discretize approximation to the time fractional derivative in Eq. (1) by using Caputo's fractional partial derivative of order α, is defined as[10,11]

$$\frac{\partial^{\alpha} u(x,t)}{\partial t^{\alpha}} = \frac{1}{\Gamma(n-1)} \int_0^{\infty} \frac{\partial u(x-s)}{\partial t} (t-s)^{-\alpha} ds, \quad t > 0, \quad 0 < \alpha < 1 \tag{4}$$

3. Approximation For Fractional Diffusion Equation

In this paper, FSAOR, HSAOR and QSAOR iterative methods will be applied to solve linear system generated from the discretization of the problem in Eq.(1) as shown in Eq.(10). To derive the formulation of both proposed methods, let the coefficient matrix A in Eq.(10) be expressed as

From Eq. (4), the formulation of Caputo's fractional partial derivative of the first order approximation method is given as

$$D_t^{\alpha} U_{i,n} \cong \sigma_{\alpha,k} \sum_{j=1}^{n} \omega_j^{(\alpha)} \left(U_{i,n-j+1} - U_{i,n-j} \right) \tag{5}$$

and we have the following expressions

$$\sigma_{\alpha,k} = \frac{1}{\Gamma(1-\alpha)(1-\alpha)k^{\alpha}} \quad \text{and} \quad \omega_j^{(\alpha)} = j^{1-\alpha} - (j-1)^{1-\alpha}.$$

Before discretizing Eq.(1), let the solution domain of the problem be partitioned uniformly. To do this, we consider some positive integers *m* and *n* in which the grid sizes in space and time directions for the finite difference algorithm are defined as $h = \Delta x = \frac{\gamma - 0}{m}$ and $k = \Delta t = \frac{T}{n}$ respectively. Based on these grid sizes, we construct the uniformly grid network of the solution domain where the grid points in the space interval $[0, \gamma]$ are indicated as the numbers $x_i = ih$, $i = 0,1,2,\ldots, m$ and the grid points in the time interval $[0, T]$ are labeled $t_j = jk$, $j = 0,1,2,\ldots, n$. Then the values of the function $U(x,t)$ at the grid points are denoted as $U_{i,j} = U(x_i, t_j)$.

By using Eq. (5) and the implicit finite difference discretization scheme, the Caputo's implicit finite difference approximation equation of Problem (1) to the grid point centered at $(x_i, t_j) = (ih, nk)$ is given as

$$\sigma_{\alpha,k}\sum_{j=1}^{n}\omega_j^{(\alpha)}\left(U_{i,n-j+1}-U_{i,n-j}\right)=a_i\frac{1}{h^2}\left(U_{i-1,n}-2U_{i,n}+U_{i+1,n}\right)+b_i\frac{1}{2h}\left(U_{i+1,n}-U_{i-1,n}\right)+c_iU_{i,n}, \quad (6)$$

for $i=1,2...,m-1$.

Based on Eq. (6), this approximation equation is known as the fully implicit finite difference approximation equation which is consistent first order accuracy in time and second order in space. Basically, the approximation equation (6) can be rewritten based on the specified time level. For instance, we have for $n \geq 2$:

$$\sigma_{\alpha,k}\sum_{j=1}^{n}\omega_j^{(\alpha)}\left(U_{i,n-j+1}-U_{i,n-j}\right)=\left(\frac{a_i}{h^2}-\frac{b_i}{2h}\right)U_{i-1,n}+\left(c_i-\frac{2a_i}{h^2}\right)U_{i,n}+\left(\frac{a_i}{h^2}+\frac{b_i}{2h}\right)U_{i+1,n}, \quad (7a)$$

$$\therefore \sigma_{\alpha,k}\sum_{j=1}^{n}\omega_j^{(\alpha)}\left(U_{i,n-j+1}-U_{i,n-j}\right)=p_iU_{i-1,n}+q_iU_{i,n}+r_iU_{i+1,n},$$

where

$$p_i=\frac{a_i}{h^2}-\frac{b_i}{2h},\ q_i=c_i-\frac{2a_i}{h^2},\ r_i=\frac{a_i}{h^2}+\frac{b_i}{2h}.$$

Also, we get for $n = 1$,

$$-p_iU_{i-1,1}+q_i^*U_{i,1}-r_iU_{i+1,1}=f_{i,1},\quad i=1,2,...,m-1 \quad (7b)$$

where

$$\omega_j^{(\alpha)}=1,\ q_i^*=\sigma_{\alpha,k}-q_i,\ f_{i,1}=\sigma_{\alpha,k}U_{i,1}.$$

From, Eq. (7b), it can be seen that the tridiagonal linear system can be constructed in matrix form as

$$AU=f \quad (8)$$

where

$$A=\begin{bmatrix} q_1^* & -r_1 & & & & \\ -p_2 & q_2^* & -r_2 & & & \\ & -p_3 & q_3^* & -r_3 & & \\ & & \ddots & \ddots & \ddots & \\ & & & -p_{m-2} & q_{m-2}^* & -r_{m-2} \\ & & & & -p_{m-1} & q_{m-1}^* \end{bmatrix}_{(m-1)x(m-1)}$$

$$\underset{\sim}{U}=\left[U_{11}\quad U_{21}\quad U_{31}\quad \cdots\quad U_{m-2,1}\quad U_{m-1,1}\right]^T$$

$$\underset{\sim}{f}=\left[U_{11}+p_1U_{01}\quad U_{21}\quad U_{31}\quad \cdots\quad U_{m-2,1}\quad U_{m-1,1}+p_{m-1}U_{m,1}\right]^T.$$

4. Preconditioned Aor Iterative Method

In relation to the tridiagonal linear system in Eq. (8), it is clear that the characteristics of its coefficient matrix are large scale and sparse. As mentioned in Section 1, many researchers have discussed various iterative methods such as Young [8], Hackbusch [9], Saad [10], Sunarto *et.al* [13]. To obtain numerical solutions of the tridiagonal linear system (8), we consider the Preconditioned AOR (PAOR) iterative method [11, 14], which is the most known and widely using for solving any linear systems.

Before applying the PAOR iterative method, we need to transform the original linear system (8) into the preconditioned linear system

$$A^*\underset{\sim}{x}=\underset{\sim}{f}^* \quad (9)$$

where,

$$A^*=PAP^T,$$

$$\underset{\sim}{f}^*=P\underset{\sim}{f},\ \underset{\sim}{U}=P^T\underset{\sim}{x}.$$

IOP Conf. Series: Materials Science and Engineering **874** (2020) 012029 doi:10.1088/1757-899X/874/1/012029

Actually, the matrix P is called a preconditioned matrix and defined as [12]

$P = I + S$ where

$$S = \begin{bmatrix} 0 & -r_1 & 0 & 0 & 0 & 0 \\ 0 & 0 & -r_2 & 0 & 0 & 0 \\ 0 & 0 & 0 & -r_3 & 0 & 0 \\ 0 & 0 & \ddots & \ddots & \ddots & 0 \\ 0 & 0 & 0 & 0 & 0 & -r_{m-1} \\ 0 & 0 & 0 & 0 & 0 & 0 \end{bmatrix}_{(m-1)x(m-1)}$$

and the matrix I is an identical matrix. To formulate PAOR method, let the coefficient matrix A^* in (8) be expressed as summation of the three matrices

$$A^* = D - L - V \tag{10}$$

where D, L and V are diagonal, lower triangular and upper triangular matrices respectively. By using Eq. (9) and (10) , the formulation of PAOR iterative method can be defined generally as [11,14]

$$\underset{\sim}{x}^{(k+1)} = (D - \omega L)^{-1}[\beta V + (\beta - \omega)D + (1 - \beta)D]\underset{\sim}{x}^{(k)} + \beta(D - \omega L)^{-1} f \tag{11}$$

where $\underset{\sim}{x}^{(k+1)}$ represents an unknown vector at $(k+1)^{th}$ iteration. The implementation of the PAOR iterative method can be described in Algorithm 1.

Algorithm 1: PAOR method

i. Initialize $\tilde{U} \leftarrow 0$ and $\varepsilon \leftarrow 10^{-10}$.

ii. For $j = 1,2,\ldots,n$ Implement

For $i = 1,2,\ldots,m-1$ calculate

$$\underset{\sim}{x}^{(k+1)} = (D - \omega L)^{-1}[\beta V + (\beta - \omega)D + (1 - \beta)D]\underset{\sim}{x}^{(k)} + \beta(D - \omega L)^{-1} f$$

$$\underset{\sim}{U}^{(k+1)} = P^T \underset{\sim}{x}^{(k+1)}$$

Convergence test. If the convergence criterion i.e $\left\| \underset{\sim}{U}^{(k+1)} - \underset{\sim}{U}^{(k)} \right\| \leq \varepsilon = 10^{-10}$ is satisfied, go to Step (iii). Otherwise go back to Step (a).

Iii Display approximate solutions.

5. Numerical Experiment

With approximation Eq.(7), we consider one example of the time fractional diffusion equation to test the effectiveness of the Gauss-Seidel (GS), Preconditioned Gauss-Seidel (PGS), Preconditioned SOR (PSOR) and PAOR iterative methods. In order to compare the effectiveness of these two proposed iterative methods, three criteria have been considered such as number of iterations, execution time (in seconds) and maximum absolute error at three different values of $\alpha = 0.25$, $\alpha = 0.50$ and $\alpha = 0.75$. For implementation of both iterative schemes, the convergence test considered the tolerance error, which is fixed as $\varepsilon = 10^{-10}$.

Let us consider the time fractional initial boundary value problem be given as [15]

$$\frac{\partial^\alpha U(x,t)}{\partial t^\alpha} = \frac{\partial^2 U(x,t)}{\partial x^2}, \quad 0 < \alpha \leq 1, 0 \leq x \leq \gamma, \quad t > 0 \tag{12}$$

where the boundary conditions are stated in fractional terms

$$U(0,t) = \frac{2kt^\alpha}{\Gamma(\alpha+1)}, \; U(\ell,t) = \ell^2 + \frac{2kt^\alpha}{\Gamma(\alpha+1)}, \tag{13}$$

and the initial condition

$$U(x,0) = x^2 \tag{14}$$

All results of numerical experiments for Problem (12), obtained from implementation of GS, PGS, PSOR and PAOR iterative methods are recorded in Table 1 at different values of mesh sizes, m = 128, 256, 512, 1024, and 2048.

6. Conclusions

For the numerical solution of the time fractional diffusion problems, the paper presents the derivation of the Caputo's implicit finite difference approximation equations in which this approximation equation leads a linear system. From observation of all experimental results by imposing the PGS,PSOR and PAOR iterative methods, it is obvious at $\alpha = 0.25$ that number of iterations have declined approximately by 64.87-99.79% corresponds to the PAOR iterative method compared with the GS method. Again in terms of execution time, implementations of PAOR method are much faster about 4.95-98.97% than the PSOR and PGS method. It means that the PAOR method requires the least amount for number of iterations and computational time at $\alpha = 0.25$ as compared with PSOR and PGS iterative methods. Based on the accuracy of both iterative methods, it can be concluded that their numerical solutions are in good agreement.

References

[1] Sunarto, A..Implicit Finite Difference Solution For Time-Fractional Diffusion Equations Using AOR Method. *IOP Journal of Physics Conference Series*. 495, 100-107, 2014.

[2] Baleanu, D., Machado, J.A.T., Catani, C., Baleanu, M.T., & Yang, X.J.,Loacal Fractional Variational Iteration & Decomposition Method for Wave Equation On Cantor Sets Within Local Fractional Operators. *Abstract and Applied Analysis,*1-6, 2014.

[3] Ege, S.M & Misirli, E., A New Method For Solving Non Linear Fractional Differential Equations. *New Trend In Mathematicals Sciences,*5(1), 225-233. Meerschaert, M.M, &Tadjeran, C. 2004. Finite difference approximation for fractional advection-dispersion flow equations, *JCAM*, 172, 145-155, 2017.

[4] Mao, Z,. Xiao, A., Yu, Z & Shi,L., Finite Difference and Sinc-Collacation Approximations to a Class of Fractional Diffusion Wave Equations. *Journal of Applied Mathematics,* 1-11, 2014.

[5] Zhang, Y., A Finite Difference Method for Fractional Partial Differential Equation, *Applied Mathematics and Computation*, 215, 524-529, 2009.

[6] Feras, L.L., Ford, N.J., Morgado, M.L & Robelo, M.,Numerical Method for The Solution of The Time Fractional Diffusion equation. *Part 1 LNCS 8579,*117-131, 2014

[7] Al-Shibani, F.S., Md.Ismail,A.I &Abdullah, F.A., Explicit Finite Difference Methods for The Solution of The One Dimensional Time Fractional Advection Diffusion Equation. *AIP Conference Proceeding 160,* 380-385, 2014.

[8] Young, D.M. Iterative solution of large linear systems. London: Academic Press, 1971.

[9] Hackbusch, W, Iterative solution of large sparse systems of equations. New York: Springer-Verlag, 1995.

[10] Saad, Y.,Iterative method for sparse linear systems. Boston: International Thomas Publishing, 1996.

[11] Deghan, M and Hajarian, M., Modified AOR Iterative Methods To Solve Linear Systems, *Journal of Vibration and Control.* 20 (5).661-669, 2014

[12] Gunawardena, A.D, Jain, K & Synder, L., Modified Iterative Methods for Consistent Linear Systems. *Linear Algebra and Its Applications*, 152-154, 123-143, 1991.

[13] Sunarto, A and Sulaiman, J., Solving Space-Fractional Diffusion Equations By Using HSSOR Method, *Fast and East Journal,* 97(4), 159-69, 2017.

[14] Li, A., A New Preconditioned AOR Iterative Method and Comparison Theorems for Linear Systems, *IAENG International Journal of Applied Mathematics,* 42, 1-3, 2016.

[15] Ali, D., Erman, S.,Ozgur, B & Korkmaz, E., Analysis of Fractional Partial Differential Equations by Taylor Series Expansion. *Boundary Value Problem*, 68, 1-12, 2014.

Table 1. Comparison of number iterations (K), the execution time (seconds) and maximum errors for the iterative methods using example at $\alpha = 0.25, 0.50, 0.75$

		α = 0.25			**α = 0.50**			**α = 0.75**		
M	**Method**	**K**	**Time**	**Max Error**	**K**	**Time**	**Max Error**	**K**	**Time**	**Max Error**
128	**GS**	21017	37.73	9.9e-5	13601	5.92	9.8e-5	6695	2.94	1.3e-4
	PGS	7292	35.86	9.9e-5	4715	2.23	9.8e-5	2319	1.93	1.3e-4
	PSOR	281	2.24	9.9e-5	229	1.95	9.8e-5	164	1.63	1.3e-4
	PAOR	280	1.12	9.9e-5	225	1.50	9.8e-5	160	1.59	1.3e-4
256	**GS**	77231	343.63	1.0e-4	50095	42.17	9.9e-5	24732	20.70	1.3e-4
	PGS	26884	261.56	9.9e-5	17417	16.68	9.8e-5	8585	12.37	1.3e-4
	PSOR	1428	16.90	9.9e-5	1171	12.61	9.8e-5	814	8.90	1.3e-4
	PAOR	1100	12.44	9.9e-5	950	10.75	9.8e-5	713	8.13	1.3e-4
512	**GS**	281598	2747.34	1.2e-4	183181	339.85	1.0e-4	90783	166.75	1.3e-4
	PGS	98422	1916.28	1.0e-4	63298	123.01	9.9e-5	31619	62.78	1.3e-4
	PSOR	5524	113.86	9.9e-5	4520	91.37	9.8e-5	11695	61.98	1.3e-4
	PAOR	4397	92.58	9.9e-5	3754	78.34	9.8e-5	2780	59.09	1.3e-4
1024	**GS**	1017140	68285.36	1.0e-4	663971	2454.53	1.0e-5	330622	1209.39	1.4e-4
	PGS	357258	14064.44	1.4e-4	232784	1007.47	1.0e-5	115617	820.93	1.3e-4
	PSOR	20574	817.59	9.9e-5	16842	662.23	9.8e-5	11695	456.23	1.3e-4
	PAOR	16487	699.81	9.9e-5	14058	607.00	9.8e-5	10394	429.58	1.3e-4
2048	**GS**	3631638	58914.30	1.3e-4	2380946	17795.25	1.3e-4	1192528	8794.26	1.7e-4
	PGS	121156	4104.17	1.3e-4	19153.0	3239.84	13e-5	112899	1305.5	1.3e-4
	PSOR	75580	3043.59	1.3e-4	61941	2894.7	9.9e-5	43070	337.1	1.3e-4
	PAOR	56289	3002.21	1.3e-4	46535	2870.12	9.9e-5	33819	305.2	1.3e-4

The Rubber Compound Combination To Strenghthen Tire Valve of 80/90 R-14

Edi Widodo, Hani Suhami, Rachmad Firdaus, Mulyadi, A'rasy Fahruddin

Mechanical Engineering Department, Universitas Muhammadiyah Sidoarjo, Indonesia

Abstract. Tires are part of the wheel of the vehicle in direct contact with the road surface. Tires are the basis for supporting vehicle loads. The basic material of the tire is rubber so it has elastic and soft properties, strong but yet susceptible to taper and sharp objects. Minor damage to the tire will cause air leakage, and its can occur in all parts of the tire. This study discusses tire leakage caused by valve damage. Valve damage is caused by high pressure, lack of pressure, inner tire leakage or shock load collision which causes the tire valve to release. Reused tire damaged in the valve section can be repaired with modifications of the worn and damaged valve. The addition of plastic-coated compound to add strength and toughness of the valve, was applied to get the benefits of tire recycling. The added layer includes a plastic layer, compound and balloon layer. The three layers are tested to find out the layer that is able to increase the strength of the valve more significantly. Tire valve pressure testing which is coated with plastic can have greater strength than other composites.

Keywords: valve strength,compound sheet, canisir

Introduction

Tires are part of the wheel of the vehicle in direct contact with the road surface. Leaked tires can be patched using rubber tires. The tire part that has leaked is given rubber compound and heated so that the base material will blend with the tire. Tires that are damaged in the valve section will usually not be pinched and will be discarded. In this study the valve that is damaged will be given additional reinforcement material so that it can be reused. The addition of material to the valve is rarely done because it is not common in tire repair [1]. Broken valves will be difficult to repair with economic considerations, the best option is to replace them with new ones. This research attempts to refute the prevailing opinion in tire repair. The reuse of worn out valves with the addition of reinforcing material is useful in increasing tire usage. The use of used tires will reduce waste disposal of used tires that are harmful to the environment. [2]. Strengthening the valve with some alternative reinforcement material, is carried out to obtain strength values against pressure loads and fatigue loads. [3][4]

Table 1. Standard Dimension of Dumbbell Component Size On Tensile Strength Test.

Dimension	Tolerance	Die C (mm)
A	± 1	25 mm
B	Max	40 mm
C	Min	115 mm
D	±6b	32 mm
D-E	±1	13 mm
F	2	19 mm

IOP Conf. Series: Materials Science and Engineering **874** (2020) 012030 doi:10.1088/1757-899X/874/1/012030

Dimension	Tolerance	Die C (mm)
G	±1	14 mm
H	±2	25 mm
L	±2	33 mm
W	± 00.5-0.00	6 mm
Z	±1	13 mm

Table 1 Indicates the standard dimension of the dumbbell dimension in the tensile strength test in the form of compound sheets or natural rubber composites and in the form being in one test variation.[5][6]

Tensile testing process generates maximum tensile strength values that force is given (F) for the test material divided by the surface area (Ao) test material, as in the show similarities. [7]

$\partial max = F / Ao$ [8]

Information :

∂ max = Maximum voltage (MPa)

F = Style (N)

Ao = Surface area (mm2)

Methodology

The research methodology begins with a study of the literature on rubber compound material.[9] The material of rubber compound mixed with plastic will be used in the tire fitting of the motorcycle tire. From the selection of the material, the initial composition of the mixture of the rubber tire valve of the rubber compound is covered with plastic.[10]

The tensile test process shall comply with the standard (ASTM 12) [11] in which the thickness and length shall be in accordance with the standards of the process carried out by clamping the specimens in the tensile testing machine chasset and drawn in the assurance before the tensile test of the material shall be checked first and drawn slowly. Slowly straight and until the specimens from each of the 5 test specimens are broken.

Result

The tensile test process is to know the strength of materials that will be tested. This tensile test process is done in BLK Laboratory of Surabaya. Tensile test is done by using ASTM D412 standard (Rubber). The tensile test conducted on composite materials that have been processed each of the different coating materials may be a fracture in the HAZ area.[12][13] At the taken test performed on the composite material already processing each different coating material in a single fault on a tire valve of size 80/90 R-14.

The tensile test is conducted to determine the mechanical properties of the rubber composite material which is coated with different materials as the test material in this research. Tensile test result is generally the tensile strength parameter (ultimate strength) and yield strength. The strength of the material is indicated by the percentage of material extension.[14]

Figure 1. Tensile test prosess

The tensile test is conducted to determine the mechanical properties of the rubber composite material which is coated with different materials as the test material in this research. Tensile test result is generally the tensile strength parameter (ultimate strength and yield strength). The strength of the material is indicated by the percentage of material extension.[15]

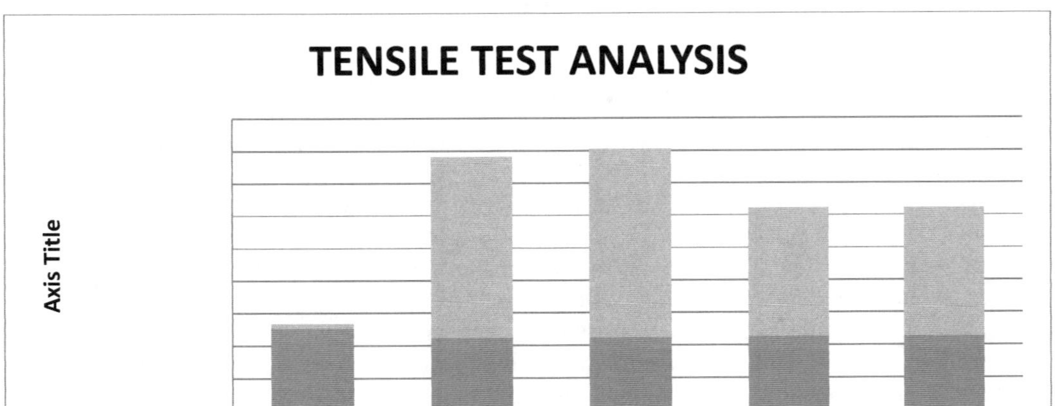

	unthreatment	nilon layers	plastic	tube	without layer
elongasion	25.69	1116.29	1158.39	785.62	785.62
peak kgf/mm2	82.98	73.34	73.51	74.19	74.19
peak kgf	414.88	366.72	367.56	370.94	370.94
length mm	10	10	10	10	10

Figure 2. Tensile test analysis

The results of the diagram diagram test Figure 4.13 shows the balloon numbers 366.72 kgf. Tire layer in kgf. Plastic layer 370.94 kgf and without layers of material 368.72 kgf.

In tensile experiments performed on the composite material each of the different lining materials can be found in HAZ, there are 2 bases to be tested.

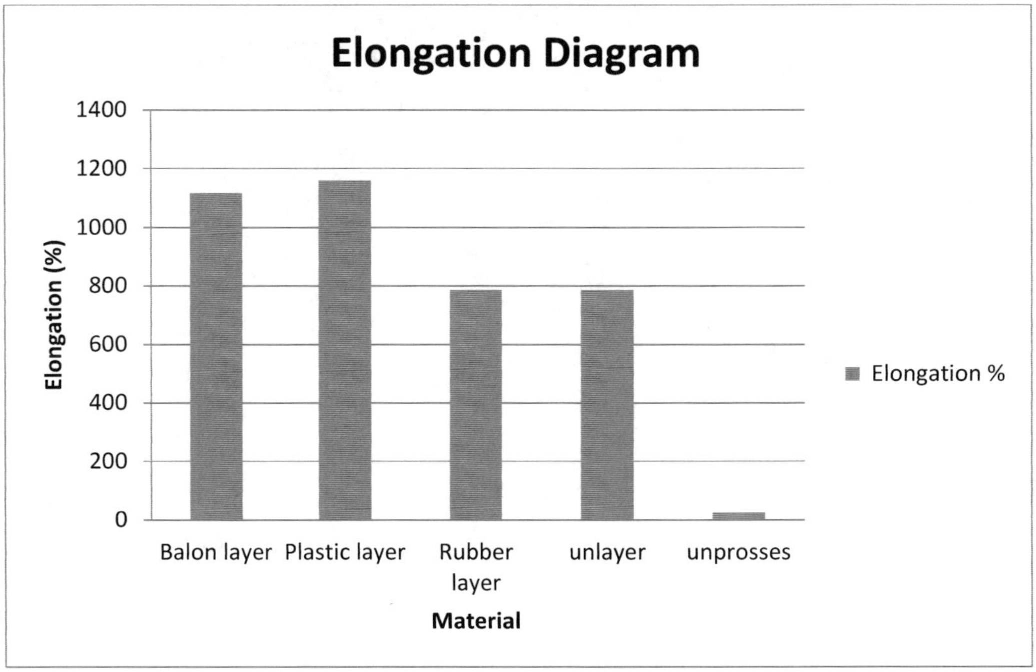

Figure 3. Elongation diagram

The result of the test diagram of the drawing diagram 2 shows the figures on the balloon 82.98 peaks kgf / mm2. The Inner tire layer = 73.34 kgf / mm2. Plastic coating = 73.51 kgf / mm. Without material layer = 74.19 kgf / mm2. Without heating process = 74.19 kgf / mm2.

The Pressure Test Result illustrated in Figure 3 shows a diagram of the balloon layer at the number line of 1116.29% elongasion. The result of the composite tire material composite test in the tire line is 785.62% of elongasion. The Pressure test of composite material of rubber compound layer in united with plastic compound give 1158.39% elongasion. Composite material without composite layer that is in the line of numbers 785.62% elongasion, and without heating process show 25.69% elongasion with this can be concluded that from 5 composite material that is tested that the material composite with layer of plastic compound is much stronger.

Calculation of Tensile Strength Test

In this case specimen data is taken as an example of tensile test results for base metal / base metal as follows:

Specimen area (Ao): 5 mm^2

Specimen length (L0): 10 mm^2

Drag Tension (Ơ): 73.51 Kgf / mm^2

Elongasion (ϵ): 1158.39%

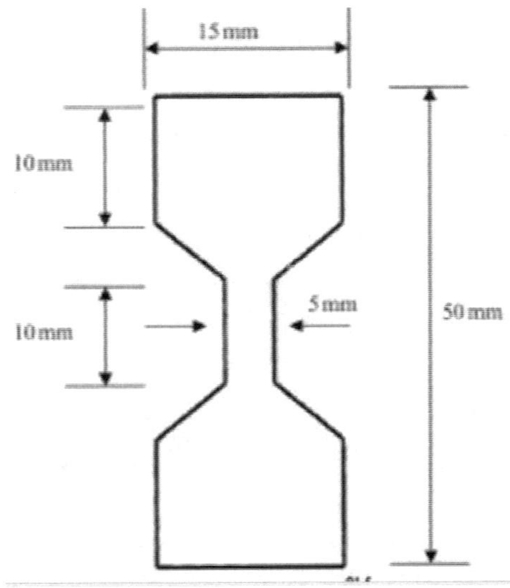

Figure 4. tensile test speciment

From the above data can be calculated to know the force / load (F) which is given by:

α = F / Ao

F = σ.Ao

= 73.51 Kgf / mm² × 5 mm²

= 367.55 kgf / mm²

While the length increase (L) is obtained by:

Є = (ΔL) / Lσ × 100

ΔL = Lσ / (100 x ε.

= 10 / (100 x 1158.39

= 8.63 mm

Table.2. Tensile test

NO	Length mm	Peak Kgf	Peak Kgf/mm2	Elongasion %
1	10.00	414.88	82.98	25.69
2	10.00	366.72	73.34	1116.29
3	10.00	367.56	73.51	1158.39
4	10.00	370.94	74.19	785.62
5	10.00	370.94	74.19	785.62
Average				1158.39

Table 2 is an overview of preliminary data from the tensile test which is the graph of elongation and pressure. Where the figure shows the magnitude of the tensile extension to know the modulus of elasticity in the test material.

Pressure. Duration Time / second
From figure diagram 4.4 is the result of test data of tire valve in size 80/90 R-14.deangan includes 4 specimen method that is:
1. The balloon layer shows in the time line number (0: 4) sec.
2. The inner tire layer shows in the timeline number (2: 5) sec.
3. The plastic layer shows in the time line number (4: 5) sec.
4. No layers show in the timeline numbers (3: 5) sec.
5. Conclusion: From the test results the tire valve pressure is coated with a stronger canal plastic in comparison with other composite materials, which at the time of its longer testing process is damaged.

Based on the results of test data and data analysis s of motor valve strength tires size 80/90 R-14 with the addition of plastic compound, its can be concluded some technical information as follows:
1. From the test The ultimate tensile strength of the material which is united between the compound and the plastic of the rubber compound (base rubber) of the resulting data shows that for 10 mm length, peak 36756 kgf, peak 73.51 kgf / mm2, elongation 1158.39 %.
2. Test tensile test results are done by using ASTM D412 standard (rubber).
3. From the compressive strength test on tire valve in size 80/90 R-14 processed with addition of plastic canon compound.data obtained by air pressure 60 bar.dengan time 0:36:73 sec.
4. The result of tensile testing is done by using standard SNI 0101. above air pressure 32 psi.

Suggestions
From tensile testing of compound material in united with canisir plastic or other composite material that the composite material is coated with the most powerful plastic and pliable because the plastic has the properties that blend with the rubber that always follow the appropriate shape of the counter in the lead. From the analysis and data obtained that each different composite material has unequal properties and strengths and from each different composite material also the power generated from the material

Acknowledgement
This research was supported by Universitas Muhammadiyah Sidoarjo. We thank our colleagues from Engineering Laboratory which provided insight and expertise that assisted this research. Although they may not agree with all of the interpretations of this paper. Thank you to all parties involved in helping to carry out this research.

References
[1] S. Fujii, "Crash Analysis Of Motorcycle Tire," *JSAE Rev.*, vol. 24, no. 4, pp. 471–475, 2003.
[2] F. A. Ghasemi, G. Payganeh, M. Rahmani, and M. R. Kalaee, "Investigating The Effects Of Waste Ground Rubber Tire," *Dig. J. Nanomater. Biostructures*, vol. 7, no. 4, pp. 1859–1868, 2012.
[3] E. T. Mohamad, N. Latifi, A. Marto, R. Moradim, and S. V. A. N. K. Abad, "Effects Of Relative Density On Shear Strength Characteristics Of Sand-Tire Chips Mixture," *Electron. J. Geotech. Eng.*, vol. 18 D, pp. 623–632, 2013.
[4] P. Gotteland, S. Lambert, and L. Balachowsky, "Strength Characteristics of Tyre Chips – Sand Mixtures," *Stud. Geotech. Mech.*, vol. XXVII, no. 1, pp. 55–66, 2005.
[5] I. O. Bakare, F. E. Okieimen, C. Pavithran, H. P. S. Abdul Khalil, and M. Brahmakumar, "Mechanical And Thermal Properties Of Sisal Fiber-Reinforced Rubber Seed Oil-Based Polyurethane Composites," *Mater. Des.*, 2010.
[6] S. Ramarad, M. Khalid, C. T. Ratnam, A. L. Chuah, and W. Rashmi, "Waste Tire Rubber In Polymer Blends: A Review On The Evolution, Properties And Future," *Progress in Materials Science*. 2015.
[7] B. S. Thomas and R. Chandra Gupta, "Properties Of High Strength Concrete Containing Scrap Tire Rubber," *J. Clean. Prod.*, 2016.
[8] B. S. Thomas, R. C. Gupta, and V. J. Panicker, "Recycling Of Waste Tire Rubber As

Aggregate In Concrete: Durability-Related Performance," *J. Clean. Prod.*, 2016.

[9] M. Rajkiewicz, M. Ślączka, and J. Czakaj, "Rubber Technology Compounding and Testing for Performance," in *Key Engineering Materials, Volume 1*, 2014.

[10] K. Sengloyluan, K. Sahakaro, W. K. Dierkes, and J. W. M. Noordermeer, "Silica-Reinforced Tire Tread Compounds Compatibilized By Using Epoxidized Natural Rubber," *Eur. Polym. J.*, 2014.

[11] ASTM International, "Standard Specification for Rubber Examination Gloves," *Astm*, 2015.

[12] ASTM-D638-14, "Standard Test Method for Tensile Properties of Plastics," *ASTM Stand.*, 2014.

[13] A. Skrobak, M. Stanek, D. Manas, M. Ovsik, V. Senkerik, and M. Reznicek, "Mechanical Properties Of Rubber Samples," in *Key Engineering Materials*, 2014.

[14] A. Tröls *et al.*, "Stretch Dependence Of The Electrical Breakdown Strength And Dielectric Constant Of Dielectric Elastomers," *Smart Mater. Struct.*, 2013.

[15] A. N. Gent and G. R. Hamed, "Materials and Compounds," in *Engineering with Rubber*, 2012.

Experimental Study of Blade Angle Effect on Two-Stage Vertical Shaft Hydrofoil Water Turbines on Power and Efficiency

A N E B Santoso, A Fahruddin, and Iswanto

Mechanical Engineering Department, Universitas Muhammadiyah Sidoarjo, Jl. Raya Gelam No.250, 61271, Sidoarjo, Indonesia

arasy.fahruddin@umsida.ac.id

Abstract. The limitations of fossil energy sources are one of the most fundamental energy problems. Therefore, it is necessary to make power plants with alternative sources that are environmentally friendly and affordable. One effort that can be done is using water energy to drive turbines. This study aims to determine the effect of blade angle on the total power and efficiency produced by the hydrofoil water turbine. In this study using a two-stage hydrofoil turbine with NACA 64-212 standard blade. The turbine blade angle is varied by 4, 6, and 8 degrees. Water discharge is varied by 40, 50, and 60 L/s. The results show that the largest total power is 3.046 Watt, achieved at a 4-degree blade angle with a water discharge of 60 L/s. And the highest efficiency is 11.22%, achieved at a 4-degree blade angle with a water discharge of 40 L/s.
Keywords: Hydrofoil; Water turbine; Two-stage; Blade angle.

1. Introduction

The issue of renewable energy innovation as an alternative to fossil energy has grown rapidly. This appears as an anticipatory step to the reduction in fossil energy sources due to massive use, the reduction in fossil energy can affect the scarcity and increase in the price of electricity [1]. On the other hand, the limitations of electrical energy are one of the most fundamental energy problems. This can be seen from the many remote areas that have not been electrified. Therefore, it is necessary to continue to develop power plants with renewable energy that are easily applied in remote areas. One effort that has been made is to utilize water energy as a turbine mover [2,3].

However, hydroelectric power plants that are often used are hydroelectric power plants with high fall and large discharge models. While hydroelectric power plants with the model of flat currents and small water discharges have not been widely used, whereas in some areas of Indonesia have considerable potential to be developed. So an effort and research need to be done to utilize the potential of a horizontal water flow energy source for electricity generation, one of which uses a vertical shaft crossflow turbine type [4,5]. This research is very important considering the potential of hydropower is spread almost throughout Indonesia and is estimated to reach 75,000 MW [6,7].

Choi, et al (2008) conducted a numerical study (CFD) to determine the effect of turbine structural configuration on the performance and internal flow characteristics of cross flow type turbines by varying the shape of the nozzle, runner inlet angle (nozzle angle), runner blade angle and number of blades. The results showed that the shape of the nozzle, runner blade angle, and number of blades greatly influenced the performance and shape of fluid flow in the turbine [8]. Mafruddin, (2016) has conducted research on blade angle variations in cross-flow turbines, and concluded that the nozzle angle and blade angle greatly affect the performance of the water turbine. The highest turbine efficiency of 77% was obtained using a 16° blade angle [9]. According to Adrian, et al (2014), the performance of kinetic turbines is very dependent on the number of blades, flow direction, turbine blade angle, flow velocity, and blade dimensions. However, in the kinetic turbine the main factor is the speed of water flow [7]. Asroful, et al (2016) conducted research on the effect of variations in the angle of the input blade of the blade on the performance of the kinetic turbine and concluded that the angle of the input of the blade of the blade affected the performance of the kinetic turbine. The smaller the angle of the input blade of the bowl, the greater the curvature of the blade. The greater the curvature of the blade of the bowl, the greater the tangential force, torque, power, and efficiency [6].

Crossflow turbines generally use horizontal shafts and curved blades so that they require directional flow and casing [10–12]. In this research, the crossflow turbine made adopts the Darrieus turbine design which is generally used for Vertical Axis Wind Turbine (VAWT) type wind turbines which have high efficiency and are capable of producing large enough torque. Darrieus turbine blade shape in the form of aero-foil or hydrofoil has the advantage of not taking into account the direction of fluid flow because of its symmetrical shape, and is able to operate at head and low speed [13]. To maximize the amount of water energy that can be captured, the crossflow turbines can be mounted in stages [14]. To obtain optimal performance, it is necessary to examine the effect of the blade angle on the Two-Stage Vertical Shaft Hydrofoil Water Turbines on Power and Efficiency.

2. Method

This research was conducted by direct experimental by making a prototype. Turbines are made in two pieces and arranged parallel to the artificial channel as shown in Figure 1. The turbine runner is bound to the channel, then the channel is dipped into the river flow. The shape of the turbine blade refers to the NACA (National Advisory Committee for Aeronautics) standard which provides standards in the design of an airfoil or hydrofoil. Hydrofoil geometry has a major influence on aerodynamic characteristics with important parameters in the form of lift (lift force generated) [15]. The type of blade used is NACA 64-212 as shown in Figure 2.

Figure 1. Testing installation.

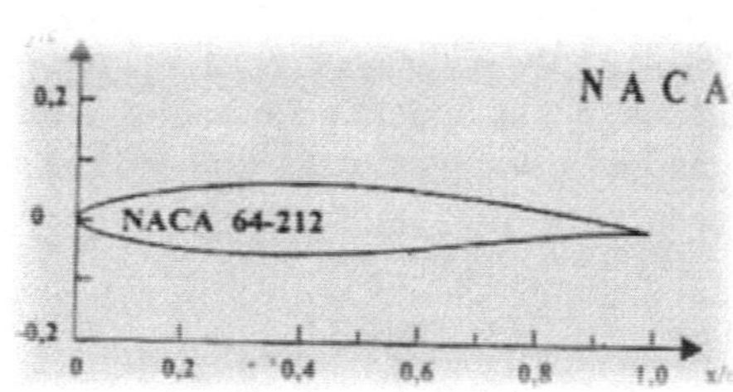

Figure 2. NACA 64-212.

The second stage turbine blade angle is varied by: 4°, 6°, and 8° as shown in Figure 3. The source of water flow is varied by 40 L/s, 50 L/s, and 60 L/s by adjusting the height of the water cap. The rotation speed of the water turbine is measured using a digital tachometer. While the output torque is measured by the braking method on a pulley using a belt that is attached to a spring balance. To calculate the power of water flow in a certain cross section, the following equation is used in the calculation [11]:

$$P_w = \frac{1}{2}.\rho.A.V^3 \quad (1)$$

Where P_w is water power, ρ is water density, V is water velocity. To calculate the turbine torque, the equation is used:

$$T = F.l \quad (2)$$

Where F = force (F_1-F_2), l = pulley radius. Whereas to calculate the turbine power the following equation is used,

$$P_t = T.\omega \quad (3)$$

Where T is turbine torque, ω is turbine angular velocity. The efficiency of a crossflow turbine is determined by the comparison between the turbine power produced and the water power, as shown in the equation [3]:

$$\eta = \frac{P_t}{P_w} x\ 100\% \quad (4)$$

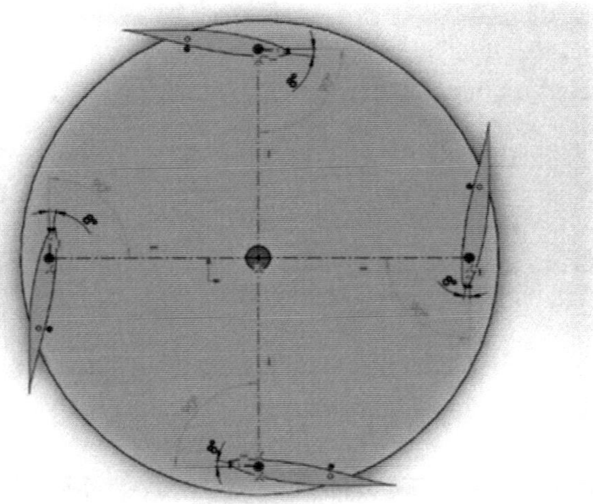

Figure 3. Turbine blade angle.

3. Result and Discussion

3.1. Effect of blade angle variations on the speed rotation of turbine

Based on Figure 4, the highest rotation is obtained at 4° blade angle variations with 60 L/s water discharge and the resulting speed is 338 rpm, while the lowest rotation is obtained at 8° angle variation with 40 L/s water discharge and the resulting speed is 225 rpm. The rotational speed of the turbine is affected by water discharge and blade angle. From Figure 4 it can be seen that the greater the flow of water and the smaller the angle of blade opening applied, the faster the rotation generated by the turbine. And conversely, the smaller the flow of water and the greater the angle opening of the blade, the slower the turbine rotational speed.

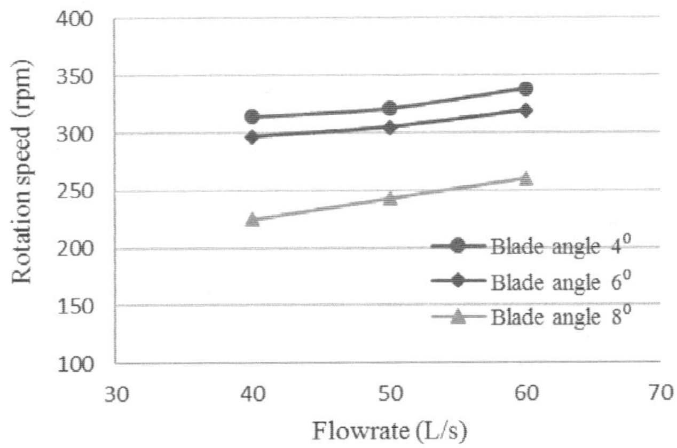

Figure 4. Graph of the effect of blade angle on the first-stage turbine speed rotation.

Based on Figure 5, it can be seen that the second highest level of turbine rotation is obtained at 4° blade angle variation with 60 L/s discharge and the resulting rotational speed is 246 rpm, while the lowest rotation is obtained at 8° angle variation with 40 L/s water discharge and speed the result is 127 rpm. As in the first level turbine, the highest rotation in the second level turbine results in the largest water discharge and the smallest blade angle.

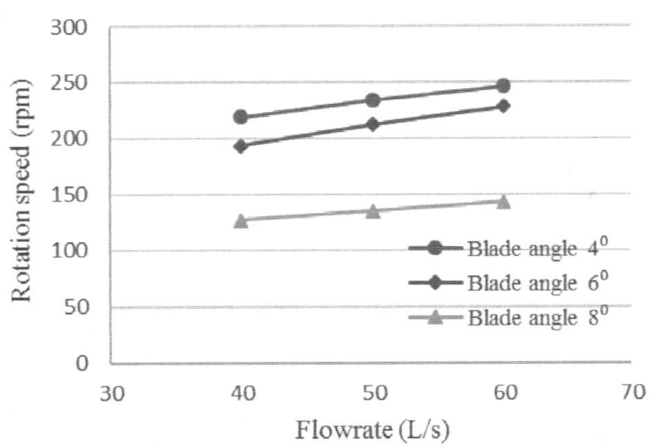

Figure 5. Graph of the effect of blade angle on the second-stage turbine speed rotation.

3.2. Effect of blade angle variations on the torque of turbine

Based on Figure 6, the greatest torque is obtained at 4° blade angle variations with a discharge of 60 L/s and the resulting torque is 0.057 Nm, while the smallest torque is obtained at 8° angle variations with a water discharge of 40 L/s and the resulting torque is 0.019 Nm. The torque produced in the turbine is affected by the discharge and blade angle. As in rotational speed measurements, the greater the water discharge and the smaller the blade angle opening that is applied, the greater the torque produced by the turbine, and vice versa, the smaller the water discharge and the greater the blade angle opening, the smaller the torque velocity of the turbine. This is because crossflow turbines with hydrofoil blades are classified as kinetic turbines, the rotating speed and torque of the turbine is determined by the speed of the water that hits the blade.

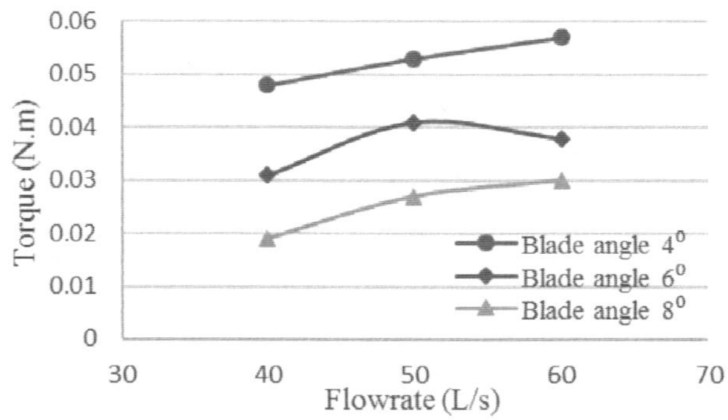

Figure 6. Graph of the effect of blade angle on the first-stage turbine torque.

Based on Figure 7, the largest torque at the second level turbine is obtained at 4° blade angle variations with a discharge of 60 L/s with a value of 0.040 Nm, while the smallest torque is obtained at 8° angle variation with a water flow of 40 L/s with a value of 0.012 Nm. The torque in the second level turbine tends to be smaller than the torque in the first level turbine. This is because the water force from the front has been captured by the first level turbine, resulting in a pressure drop.

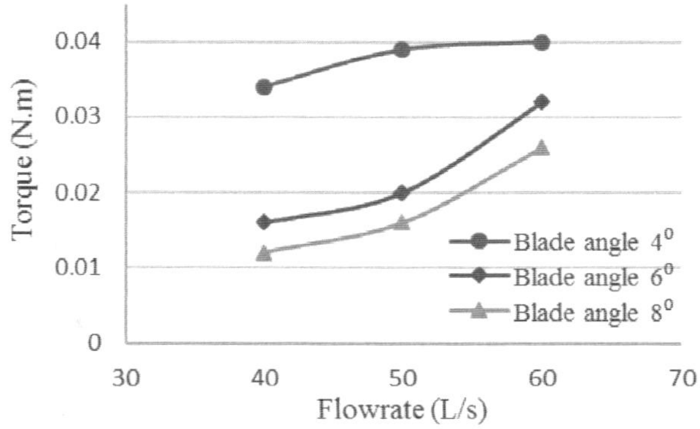

Figure 7. Graph of the effect of blade angle on the second-stage turbine torque.

3.3. Effect of blade angle variations on the total power of turbine

Based on Figure 8, the largest turbine power is 3.03 watts obtained in the use of 4° blade angle variations with a discharge of 60 L/s. While the lowest turbine power is 0.61 watts obtained at the use of 8° blade angle variations with a discharge of 40 L/s. Turbine rotation at blade angle 6° is closer to blade angle 4°, but torsion angle of blade 6° is closer to blade angle 8°. So the 6° turbine power is in the middle between the 4° and 8° turbine power.

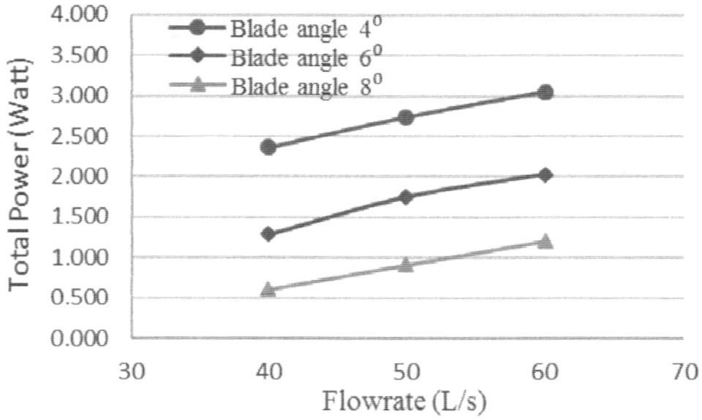

Figure 8. Graph of the effect of blade angle on total power of turbine.

3.4. Effect of blade angle variations on efficiency of turbine

Based on Figure 9, the largest turbine efficiency is 11.15% obtained at the use of a 4° blade angle with a discharge of 40 L/s, while the lowest turbine efficiency is 2.12% obtained at the use of an 8° blade angle with a discharge of 60 L/s. The greater the water discharge the lower the efficiency, this is because the rotation and torque of the kinetic turbine is limited so that the increase in turbine power is not able to compensate for the increase in water power provided [15]. To get greater efficiency, it might be possible to increase the number of turbine levels.

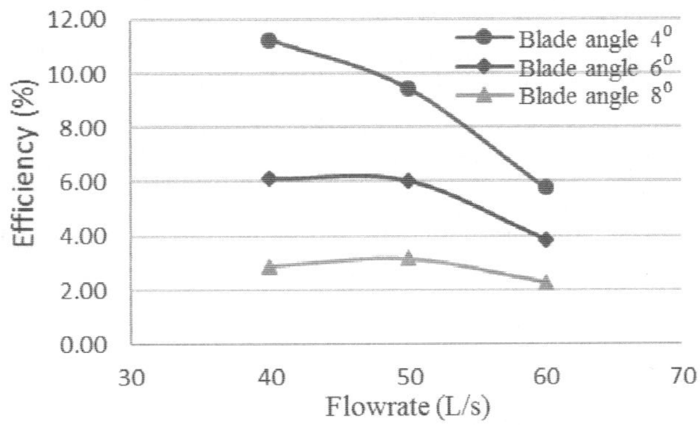

Figure 9. Graph of the effect of blade angle on total turbine efficiency.

Conclusion

The smaller the turbine blade angle, the greater the rotating speed and torque of the turbine produced, so the power and efficiency of the turbine are also greater. The greater the flow of water the greater the rotating speed and torque of the turbine so that the greater the turbine power, but the increase in turbine power is not able to compensate for the increase in water power so that efficiency decreases.

Acknowledgement

Authors would like to thank Pusat Pengembangan Publikasi Ilmiah Universitas Muhammadiyah Sidoarjo (P3I Umsida) Indonesia for supporting the publication of this article.

References

[1] Nandagopal R A and Narasimalu S 2020 Multi-objective optimization of hydrofoil geometry used in horizontal axis tidal turbine blade designed for operation in tropical conditions of South East Asia *Renew. Energy* **146** 166–80
[2] Singh P M and Choi Y D 2013 Shape design and CFD analysis on a 1MW-class horizontal axis tidal current turbine blade *6th International Conference on Pumps and Fans with Compressors and Wind Turbines*
[3] Rantererung C L, Soeparman S, Soenoko R and Wahyudi S 2016 The dual nozzle cross flow turbine performance 1 *ARPN J. Eng. Appl. Sci.* **11** 8538–43
[4] Oyebode O O and Olaoye J O 2019 Comparative Performance Evaluation of Pelton Wheel and Cross Flow Turbines for Power Generation *Eur. Mech. Sci.* **3** 6–12
[5] Chen Z, Nguyen V T T, Inagaki M and Choi Y 2015 Performance of an Open Ducted Type Very Low Head Cross- Flow Turbine *GMSARN Int. J.* **9** 23–8
[6] Anam A, Soenoko R and Widhiyanuriyawan D 2013 Pengaruh Variasi Sudut Input Sudu Mangkok Terhadap Kinerja Turbin Kinetik *J. Rekayasa Mesin* **4** 199–203
[7] Maidangkay A, Soenoko R and Wahyudi S 2014 Pengaruh Sudut Pengarah Aliran dan Jumlah Sudu Radius Berengsel Luar Roda Tunggal terhadap Kinerja Turbin Kinetik *J. Rekayasa Mesin* **5** 149–56
[8] Choi Y Do, Lim J-I, Kim Y-T and Lee Y-H 2008 Performance and Internal Flow Characteristics of a Cross-Flow Hydro Turbine by the Shapes of Nozzle and Runner Blade *J. Fluid Sci. Technol.* **3** 398–409

[9] Mafruddin, Amrul and Amrizal 2006 Studi Eksperimental Sudut Nosel Dan Sudut Sudu Terhadap Kinerja Turbin Cross-flow *J. Tek. Mesin, Univ. Lampung* **5**

[10] Yi S S, Htoo A M and Sein M M 2018 Design of Cross Flow Turbine (Runner and Shaft) *Int. J. Sci. Eng. Technol. Res.* **7** 736–40

[11] Verma D R and Katkade P S D 2018 Horizontal Axis Water Turbine : Generation and Optimization of Green Energy *Int. J. Appl. Eng. Res.* **13** 9–14

[12] Kaunda C S, Kimambo C Z and Nielsen T K 2014 A numerical investigation of flow profile and performance of a low cost Crossflow turbine *Int. J. Energy Environ.* **5** 275–96

[13] Wiranto Arismunandar 1982 *Penggerak Mula Turbin* (Bandung: Penerbit ITB)

[14] Laksmana S C, Fahruddin A and Akbar A 2018 Pengaruh Sudut Pengarah Aliran pada Turbin Air Crossflow tingkat Dua Terhadap Putaran dan Daya *J. Rekayasa Energi Manufaktur* **3** 35–9

[15] Chica E, Aguilar J and Rubio-Clemente A 2019 Analysis of a lift augmented hydrofoil for hydrokinetic turbines *17th International Conference on Renewable Energies and Power Quality (ICREPQ'19)* pp 49–55

The Effect Of The Ratio Of The Hub Diameter (d) To The Diameter Of The Screw (D) To The Performance Of The Archimides Screw

Helmizar[1], Agus Nuramal[1], Novalio Daratha[2], Arif Setiawan[1]

[1]Mechanical Engineering [Universitas Bengkulu]
[2]Electrical Engineering [Universitas Bengkulu]

*Helmizar@unib.ac.id

Abstract. The potency of the ultra-low head hydro resource is quite huge in the world. In Indonesia, only 1.8% of the Pico hydro potency with head of less than 5m is utilized as electric resource [1]. Archimedes screw is one of the promising technology that could be used to extract this low head hydro resource. One prominent factor which is affecting the performance of the Archimedes turbine screw is the ratio of the hub diameter (d) to the diameter of the screw (D). In this paper, the ratio d/D were varied as 0.22, 0.29, 0.33, 0.42. In addition, flow rate (Q), were varied as well as 1.8 l/s, 1.4 l/s, 1.0 l/s. The blade configuration that used in this test is a single blade. The result shows that the highest efficiency is achieved with the diameter ratio of d/D = 0.42 with the eficiency off 40.23%.

1. Introduction

In 2017 the percentage of Indonesians who do not have access to electricity is around 4.65 % [2]. One frequent cause for this problem is the large number of dispersed rural locations with challenging geographical conditions [3]. An additional cause is the uneven distribution of population which has led to an ineffective electrical distribution infrastructure [4]. A micro hydro installation is a solution for this problem. In Indonesia, the amount of installed micro/mini hydro power is around ≈98.47 MW [5], whereas the potential is 450 MW [6]. There are several kinds of micro hydro power converter have been developed nowadays. Archimedes screw is one of the technology developing which is potentially used to exploit small scale hydropower resource. There are several parameters identified play important role to determine the performance of Archimedes screw, such as inclination angle (□□□ diameter of the screw (D), hub diameter (d), etc. This paper, highlights the effect of the ratio of hub diameter (d) to the screw diameter (D). This paper reports screw performance based upon experimental work. Furthermore, the flow rate incoming water was also varied.

The Archimedes screw is also called the Archimedean screw or screw pump, In brief, an Archimedean screw consists of a cylinder containing several continuous helical walls that, when the entire cylinder is rotated on its longitudinal axis, scoop up water at the open lower end and dump it out the upper end [7]. Some scientist stated that this machine was used during the reign of the Assyrian king Sennacherib (704–681B.C.) and that the design was put to use in spectacular fashion to water his palace garden at Nineveh (was an ancient Assyrian city) [8]. However, another scientist said that this machine was invented by Archimedes (287–212 B.C.) [9] [10] [11]. Who invented this machine still under debate see [7], however most people agree to name this machine as an Archimedes Screw. In recent years, the Archimedes Screw has been used working in reverse as a hydropower machine. If water is poured into the top of an Archimedes' screw, the water is lowered within cells which form between the blades and the trough, then it will force the screw to rotate. The rotating shaft which the screw attached then drives an electric generator. Such an installation has the same benefits as using the screw for pumping: the ability to handle very dirty water and widely varying rates of flow at high efficiency. However this machine operates at low rotational speeds, which requires a complex gearbox to connect it to the generator.

Several previous research has been done relating with *Archimedes screw turbine.* [1] Performed experimental work, the inclination angle of the screw (β) was varied, and the effect to the efficiency of the machine investigated. It was found that the Increasing of inclination angle affect to rise the rotational speed of turbine. Yet, it tend to decrease the torque. Furthermore from 3 variations of angle (22°, 30 ° and 40 °), it was found that 22° angle has the highest efficiency.

[12] Reports experimental results, the relationship between torque, rotational speed and power was examined. The downstream water level was varied, and it was found that the higher the downstream water level (the lower the head), the lower power output for the same Torque. In addition, the rotational speed is higher as the downstream water level decreases for the same torque. In addition, Archimedes screw are relatively tolerant to changes in flow and varying water levels, maintaining reasonably high efficiency over relatively wide ranges of these parameters.

The simplification theory of the Archimedes screw was developed by [13]. Based on the idealization, a comparison is made to the hydrostatic pressure wheel. The Efficiency of the Archimedes screw was stated as function of blade geometry (angle, the number of turns and the ratio of inflow depth and radius) and leakage losses. Efficiency increases with increasing number of turns, as well as increasing ratio of inflow depth and radius. Decreasing screw angle also could increase efficiency.

[14] Suggests that increasing the ratio between pitch and diameter (P/D), there is a decrease in volume per bucket, and a proportional decrease in force acting on each bucket. This situation tend to reduce efficiency eventually.

2. Methodology

The experiment was done on irrigation channel in the Batu roto Village, Bengkulu Utara Region. The water from inlet is flowed to the screw via a valve (to control water flow rate) then exits to outlet. The ratio between hub diameter to the screw diameter (d/D) was varied as follow 0.22, 0.29, 0.33, 0.42. The decision to choose these ratio are based on the availability of the material. The dimension of the screw is provided in fig 1. The gap clearance between screw and housing is 5mm whereas, the slope angle of the screw β is 22°.

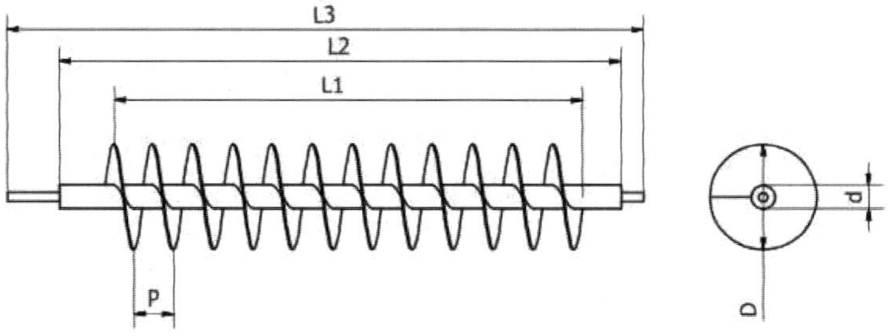

Figure 1. Dimension of the Archimedes Screw

Note
d = 32mm;42mm;48mm;60mm
D = 142mm
P = 54mm
L1= 648mm
L2 = 760mm

L3 = 860mm

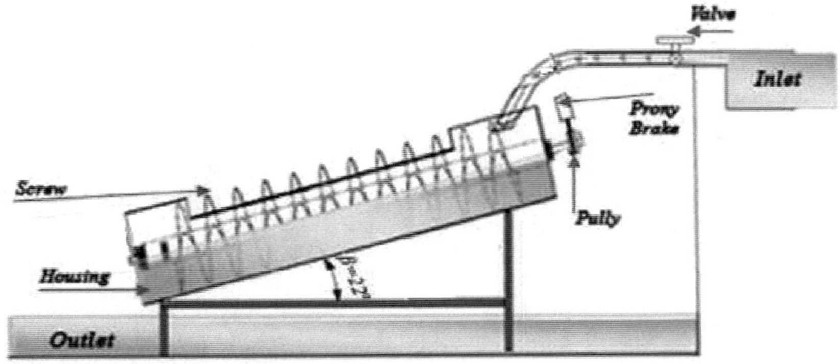

Figure 2. Schematic diagram of the experimental setup.

Figure 3. Experimental apparatus in the field

The mechanical power Ps available at the turbine shaft can be determined by measuring the torque T at a corresponding angular speed ω. The torque is found by measuring the tangential force F on a pony brake with moment arm radius of pulley r.

$$T = F\,r \quad \text{.......................................} 1$$

$$Ps = T\,\omega = \frac{T\,2\,\pi\,n}{60} \quad \text{.................................} 2$$

The efficiency in this case is the ratio between output mechanical power *Ps* and the hydraulic power available, which is calculated as

$$\eta = \frac{Ps}{Pw} \ldots\ldots\ldots\ldots\ldots\ldots\ldots\ldots\ldots\ldots\ldots\ldots\ldots 3$$

$$Pw = \rho\, Q\, g\, h \ldots\ldots\ldots\ldots\ldots\ldots\ldots\ldots\ldots\ldots\ldots 4$$

The torque was measured by using Pony brake. Whereas the rotational speed is determined by using stopwatch for 10 times rotational speed then it is converted into rotational speed. The flow rate of incoming water was varied 1.8 l /s, 1.4 l/s, and 1.0 l/s by using valve control. The effect of variation of the d/D and incoming flow water to the torque, power and efficiency then investigated.

3. Result and Discussion

3.1 The relationship between torque, rotational speed, and flow rate at various of d/D

Figure 4 shows the torque as output of varying rotational speed, flow rate and d/D. it shows that the higher the flow rate, the higher the torque. The highest torque is occurred at the rotational speed of 60rpm for all variations of d/D. All these torques are 0.182Nm, 0.177Nm, 0.176Nm and 0.174 Nm as result of d/D of 0.42, 0.33, 0.29 and 0.22 respectively. Furthermore at any variations of flow rate, the smaller d/D, the torque tends to be smaller as well.

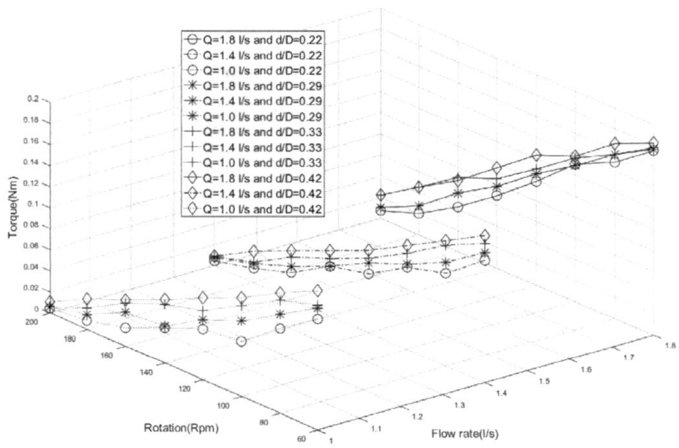

Figure 4. Torque output as function of rotational speed and flow rate at variations of d/D

As d/D is bigger, the arm of the torsional moment occurred on the screw is bigger. Therefore the resultant tangential force on the cell which is filled up with water produces the higher torque as d/D higher. This is alleged to be the cause of the occurrence of greater d / D, the greater the torque produced for various variations of rotation.

In Figure 4 it is also can be seen that the larger the water flow rate, the greater the torque produced in any various of d/D or rotational speed. This phenomenon can be explained as follows, with the increasing of flow rate, the mass of water inside the screw cell is also increasing. Result in the tangential forces (hydrostatic forces within the cells in the tangential direction) are also getting bigger.

Torque Spectrum for d/D=0.42 is shown in figure 5. In general, the torque spectrum pattern for variations d/D=0.42, 0.33, 0.29 and 0.22 is similar to figure 5. The biggest torque always occurs at low rotation (60rpm), largest water discharge (1.8 l/s), and at a small rotational speeds. This pattern is common in various water wheels.

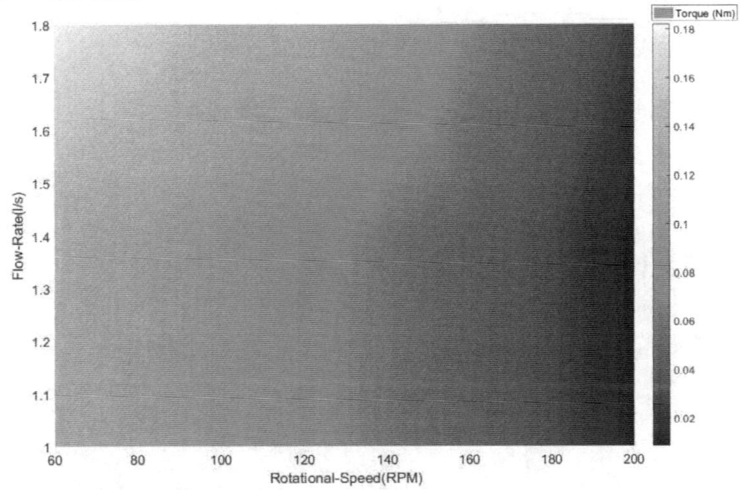

Figure 5 Torque Spectrum Based upon flow rate and rotational speed for d/D=0.42

3.2 The relationship between power output, rotational speed, and flow rate at various of d/D

The graph of power output as varying rotational speed, water flow rate and d/D is presented in figure 6. It shows the tendency of the turbine power to increase as the flow of water increases. This is because a large discharge makes the amount of hydraulic power input greater, therefore the mechanical power output produced is also enlarged. this is similar to what was stated by [15].

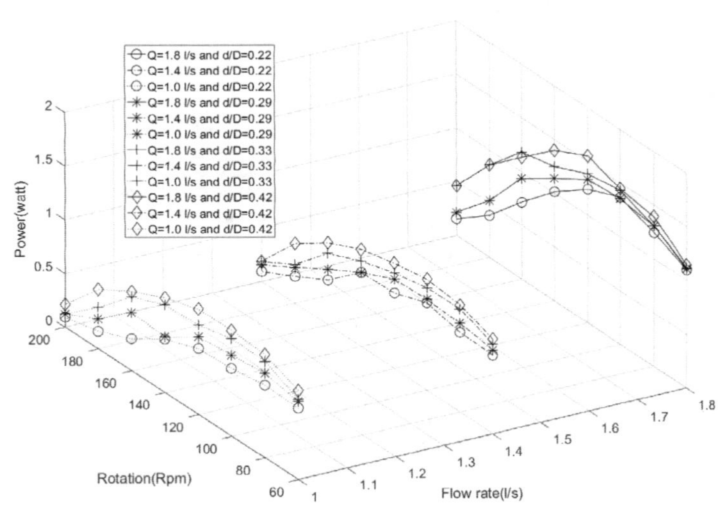

Figure 6. Power output as function of rotational speed and flow rate at variations of d/D

Spectrum power data for d/D=0.42 is shown in figure 7. From the graph, it appears that the largest power output occurs at around 120rpm and the largest discharge (1.8 l/s). The spectrum pattern that occurs at d/D=0.22, 0.29 and 0.33 is similar to what happened at d/D=0.42.

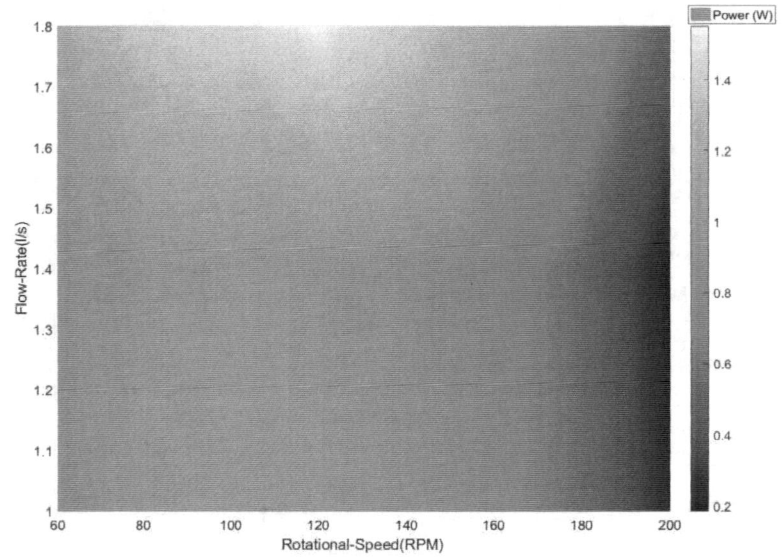

Figure 7. Power output spectrum as function of rotational speed and flow rate at d/D of 0.42

Besides, the bigger the d/D makes the power produced is also enlarged. The greatest power always occurs at d/D=0.42 for various variations of flow rate. At 1.8 l/s discharge the largest output power is 1.60W, occurring at 120 rpm. While for the discharge of 1.4 l/s, the largest power of 1.03W occurs at 100 rpm. Output power of 0.98W is produced at 100 rpm for flow rate of 1.0 l/s.

The power that occurs at d/D=0.42 which is greater than the other diameter ratios which is very closely related to the torque that occurs as explained previously. Where at d/D=0.42 produces the largest torque. this large torque is also closely related to the larger torque arm for larger d/D, which results in a large amount of power in the end.

3.3 The relationship between efficiency, rotational speed, and flow rate at various of d/D

Figure 8 shows a graph of the relationship between efficiency, rotation and flow rate of water entering Archimedes screw. For all discharge variations and screw rotation speeds, the highest efficiency is always achieved at d/D=0.42.

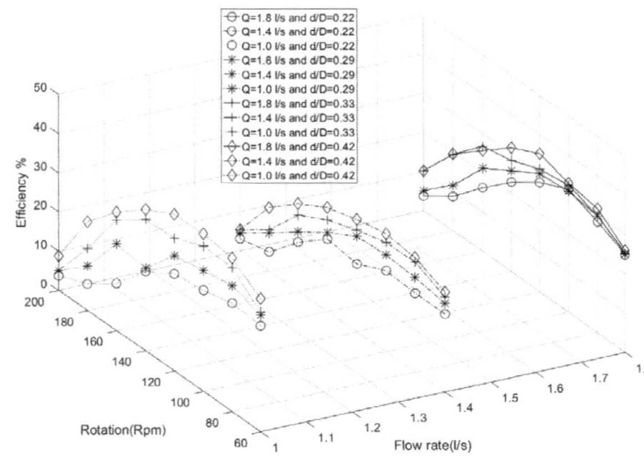

Figure 8. Efficiency as function of rotational speed and flow rate at variations of d/D

At 1.8 l/s flow rate, the highest efficiency occurs at 120 rpm with an efficiency of 35.20%. While the lowest efficiency occurs in variations of screw d/D=0.29 with an efficiency value of 30.73% at 100 rpm. For flow rate of 1.4 l/s the maximum efficiency is 30.08% at 100 rpm while at d/D=0.22, the maximum efficiency value is 20.65% which is also at 100 rpm. Furthermore at the discharge of 1.0 l / s, the efficiency is 40.23% at 100 rpm. While for screw variations of d/D=0.22, the maximum efficiency is 27.82% at 80 rpm.

In general, efficiency increases with an increase in the value of d/D, and this occurs in almost all variations of flow rate. Increasing the d/D ratio means narrowing the cell space filled with water, which means the closer the distance between the hub and the surface of the water in the cell. This tendency is similar to what was conveyed by [13] which states that the closer the water surface to the hub (do), the higher the efficiency, as shown in Figure 9. In the graph it is stated that the greater d0 / R then the efficiency increases (or the more the water level approaches the hub, the more efficiency increases). The results of this study are in accordance with the research conducted (13).

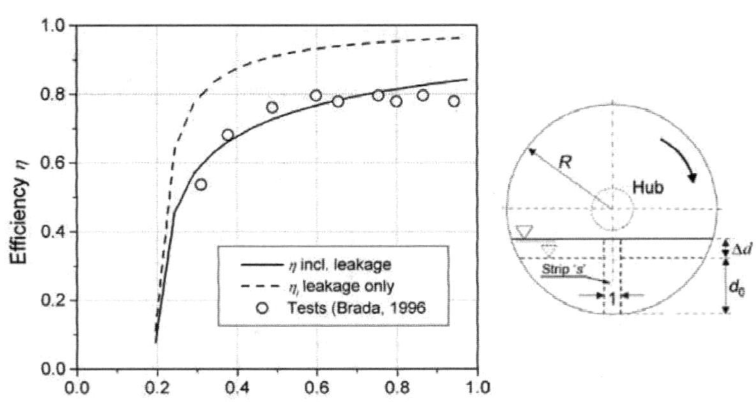

Figure 9. Efficiency graph as a function of do/R [13]

4. Conclusion

This literature review aims to identify and analyze the trends, datasets, methods and frameworks used in the topic of attribute independence assumption assumptions on NB between 2010 and 2018. Based

IOP Conf. Series: Materials Science and Engineering **874** (2020) 012032 doi:10.1088/1757-899X/874/1/012032

on the inclusion and exclusion criteria designed, it shows 71 study studies of attribute independence assumptions on the published NB between January 2010 and December 2018 are investigated in this literature review have been conducted as a review of systematic literature. A systematic literature review is defined as the process of identifying, assessing, and interpreting all available research evidence in order to provide answers to specific research questions.

The results of this study identified three of the most commonly used and influential framework methods in the topic of attribute independence on the NB. They are Menzies et al. Framework, Lessmann et al. Framework, and Song et al. Framework. They are Langley et al [16], Friedman et al [17], and Wu et al [18].

References

[1] N. et al. Erinofiardi, "Experimental study of screw turbine performance based on different angle of inclination," *Energy Procedia*, vol. 110, no. december 2016, pp. 8–13, 2017.

[2] Kementrian Energi dan sumber daya mineral, *Statistik Ketenagalistrikan 2017*, 31st ed. Jakarta: Direktorat Jenderal Ketenagalistrikan, 2018.

[3] Usman Muhammad, "Rural Solar Electrification-Renewable Energy Potential and Distribution for Development in Nigeria," in *World Renewable Energy Forum (WREF) 2012*, 2012.

[4] A. Eberhard, O. Rosnes, M. Shkaratan, and H. Vennemo, "Africa's Power Infrastructure Investment, Integration, Efficiency," Washington DC, 2011.

[5] D. J. Ketenagalistrikan, "Statistik Ketenagalistrikan 2013," 2013.

[6] Kementrian Energi Sumber daya Mineral, "Potensi Energi Baru Terbarukan (EBT) Indonesia," *Kementrian Energi dan Sumber daya Mineral*, 2015. [Online]. Available: http://esdm.go.id/berita/37-umum/1962-potensi-energi-baru-terbarukan-ebt-indonesia.html. [Accessed: 14-May-2015].

[7] S. Dalley and J. P. Oleson, "Sennacherib, Archimedes, and the Water Screw: The Context of Invention in the Ancient World," *Technol. Cult.*, vol. 44, no. 1, pp. 1–26, 2003.

[8] S. Dalley, "Ancient Mesopotamian Gardens and the Identification of the Hanging Gardens of Babylon Resolved," *Gard. Hist.*, vol. 21, no. 1, pp. 1–13, 1993.

[9] A. G. Drachmann, "The Screw of Archimedes," in *Actes du VIIIe Congrès International d'Histoire des Sciences*, 1958.

[10] J. P. Oleson, "Greek and Roman Mechanical Water-Lifting Devices: The History of a Technology," in *Greek and Roman mechanical water-lifting devices: the history of a technology*, Toronto, 1984.

[11] O. Wikander, *Handbook of Ancient Water Technology*. Leiden, 2000.

[12] M. Lyons and W. L. David, "Archimedes Screws for Microhydro Power Generation," in *Proceedings of the ASME 2013 7th International Conference on Energy Sustainability & 11th Fuel Cell Science, Engineering and Technology Conference ESFuelCell2013*, 2013.

[13] G. Müller and J. Senior, "Simplified theory of Archimedean screws Théorie simplifiée," *J. Hydraul. Res.*, vol. 47, no. 5, pp. 666–669, 2009.

[14] M. W. K. Lyons, "Lab Testing and Modeling of Archimedes Screw Turbines," 2014.

[15] and L. Bambang Yulistiyanto, Yul Hizhar, "Effect of Flow Discharge and Shaft Slope of Archimedes Screw Turbine on the Micro Hydro Power Plant," *Din. Tek. SIPIL/*, vol. 12, no. 1, pp. 1–5, 2012.

[16] P. Langley and S. Sage, "Induction of Selective Bayesian Classifiers," *Proc. Tenth Int. Conf. Uncertain. Artif. Intell.*, no. 1990, pp. 399–406, 1994.

[17] N. Friedman, D. Geiger, M. Goldszmidt, G. Provan, P. Langley, and P. Smyth, "Bayesian Network Classifiers *," *Mach. Learn.*, vol. 29, pp. 131–163, 1997.

[18] J. Wu, S. Pan, X. Zhu, Z. Cai, P. Zhang, and C. Zhang, "Self-adaptive attribute weighting for Naive Bayes classification," *Expert Syst. Appl.*, vol. 42, no. 3, pp. 1487–1502, 2015.

IOP Conf. Series: Materials Science and Engineering **874** (2020) 012033 doi:10.1088/1757-899X/874/1/012033

Numerical Studies on Fast Pyrolysis of Palm Kernel Shell in a Fluidized Bed Reactor

Thoharudin[1,2], Yi-Shun Chen[1] and Shu-San Hsiau[1]

[1]Graduate Institute of Energy Engineering, National Central University, Zhongli City, Taoyuan County, 32001, Taiwan ROC
[2]Department of Mechanical Engineering, Universitas Muhammadiyah Yogyakarta, Jl. Brawijaya, Kasihan, Bantul, D.I. Yogyakarta, Indonesia

Email: thoharudin@ft.umy.ac.id

Abstract. Numerical studies using Computational Fluid Dynamics (CFD) of thermal decomposition of Palm Kernel Shell (PKS) have been conducted in a fluidized bed reactor under atmospheric pressure of nitrogen as fluidization gas. To investigate the effect of temperature, the fast pyrolysis temperature was varied in a range from 400 °C to 650 °C. Furthermore, the effect of PKS feeding rate was also studied in a range from 300 g/h to 1500 g/h combined with variations of nitrogen superficial velocities. It was found that the increase of temperature results in the greatening of conversion efficiency and non-condensable gas product yield. The maximum 60.5 % liquid/tar product yield was obtained in 500 °C. The increase of PKS feeding rate led to the reduction of conversion efficiency due to the thermal energy shortage for the decomposition process. However, the increase of PKS feeding rate could reduce the non-condensable gas product with the insignificant change in liquid product yield. The rising superficial velocity of nitrogen favoured the shorter residence time resulting in the increasing tar product yield.

1. Introduction

Palm Kernel Shell (PKS) is abundant biomass residue derived from palm oil production which has a potential resource of renewable energy in Indonesia. In 2015, Indonesia had 8.4 million tons of PKS solid residue which has high carbon content currently used as combustion fuel with very limited efficiency [1]. A promising technology to convert solid biomass into liquid fuel, a useful form of energy, is pyrolysis. Pyrolysis is thermal decomposition of biomass in the absence of oxygen to produce liquid bio-fuel, bio-char and non-condensable gas products. An advanced pyrolysis technology, called as fast pyrolysis, involves fast heating rate and short residence time in medium temperature to maximize the liquid product [2].

Fluidized bed reactor is a popular, widely used and the most successful fast pyrolysis reactor due to many advantages, such as it has continue operational system, simple design and easy scale-up. Moreover, fluidized bed reactor offers a good solid mixing which enables biomass decompose rapidly because of the large surface contact between biomass and bed material as a heat transfer medium. In addition, fluidized bed reactor also provides a short residence time by controlling the fluidization gas flow rate to minimize the secondary cracking of tar which forms non-condensable gas product (NCG) [3].

Pyrolysis of PKS is very attractive to be investigated. High lignin content in PKS makes its bio-oil rich of phenol compound as reported in previous study [4]. Phenol is derived from rich-carbon lignin (Lign-C) and hydroxyl-rich lignin (Lign-OH) decomposed into sinapaldehyde or lumped phenol compound ($C_{11}H_{12}O_4$). Lign-C decomposes into Lign-CC, phenol and other compounds. Moreover,

Lign-CC degrades into phenol and other compounds including char and NCG. In other hand, Lign-OH degrades into lignin and other chemicals, and furthermore lignin decomposes into phenolic compound [5]. The advantage of phenol-rich bio-oil does not only serve as bio-fuel but also shows a great potential source to replace fossil phenols as high-valued chemicals feedstock such as cosmetic, food aromas, resins, pharmaceuticals, pesticides, and etc. currently produced from petrochemical feed stocks extracted from non-renewable resources [6], [7].

Computational fluid dynamics (CFD) is widely used to predict and analyze the phenomena in thermal decomposition process including pyrolysis. Aramideh, et al. [8] conducted numerical study to investigate the productivity of fast pyrolysis using an auger reactor. Jalalifar, et al. [9] investigated numerically the operating factors influencing on the pyrolysis product yields of common biomass pyrolysis in a bubbling fluidized bed reactor. Borello, et al. [10] simulated pyrolysis of sawdust in a packed bed reactor. There are many pyrolysis kinetic schemes provided in the literatures [5], [11], [12]. However, the global kinetic scheme offers a sufficient accurate prediction and simplicity in the use. Therefore, this study aims to investigate numerically the productivity of PKS pyrolysis in a fluidized bed reactor using global kinetic scheme.

2. Methods

A combination of laminar Multi-Fluid Model (MFM) and a chemical kinetic stiff heterogeneous reaction were taken into account using conservation laws of mass, momentum, energy, and species in Eulerian-Granular approach. In general, MFM consists of one gas phase as the primary phase and numbers of solid phases as the secondary phases. These models were solved using a commercial ANSYS Fluent CFD code. In this work, two solid phases were employed, namely biomass and sand. The gas phase consisted of three species, namely tar, non-condensable gas (NCG) and nitrogen. Meanwhile, the biomass phase comprised of nine species, namely cellulose, hemicellulose, lignin, active cellulose, active hemicellulose, active lignin, cellulose char, hemicellulose char, and lignin char. Sand as a bed medium consisted of one species and did not involve in the chemical reaction.

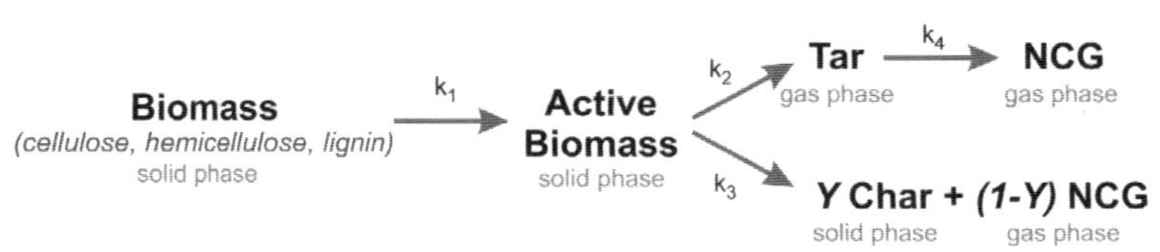

Figure 1. Global kinetic scheme of pyrolysis.

Table 1. Parameters of pyrolysis global kinetic scheme

Component	Reaction	Y of k_3	A (s^{-1})	E ($MJ.kmol^{-1}$)	ΔH ($kJ.kg^{-1}$)
	$k_{1,c}$	-	2.8×10^{19}	242.4	0
Cellulose	$k_{2,c}$	-	3.28×10^{14}	196.5	255
	$k_{3,c}$	0.35	1.3×10^{10}	150.5	-20
	$k_{1,h}$	-	2.1×10^{16}	186.7	0
Hemicellulose	$k_{2,h}$	-	8.75×10^{15}	202.4	255
	$k_{3,h}$	0.6	2.6×10^{11}	145.7	-20
	$k_{1,l}$	-	9.6×10^{8}	107.6	0
Lignin	$k_{2,l}$	-	1.5×10^{9}	143.8	255
	$k_{3,l}$	0.75	7.7×10^{6}	111.4	-20
Tar	k_4	-	4.28×10^{6}	108.0	-42

A global kinetic scheme was considered to solve the pyrolysis simulation. Biomass in a single particle comprised of cellulose, hemicellulose, and lignin, transforms into active biomass by absorbing thermal energy provided by preheated nitrogen and heat source locating along the heater wall. Then, active biomass decomposes into tar, NCG and char. Furthermore, tar degrades into NCG as the consequence of secondary cracking reaction (k4). The reaction scheme and its kinetic parameters are described in Fig. 1 and Table 1 [13], respectively. All reactions are expressed using an Arrhenius form as given in equation 1.

$$k_i = A_i \cdot \exp\left[-E_i/RT\right] \tag{1}$$

Where k_i is the kinetic constant of reaction rate, A_i denotes the Arrhenius pre-exponential constant and E_i represents the activation energy of the chemical reaction. R and T indicate the universal ideal gas constant and temperature, respectively.

The sketch of fluidized bed reactor geometry is illustrated in Fig. 2 which accords to the work conducted by Xue, et al. [14]. 2-D simulations were used to perform the numerical studies due to its low computational cost and a relatively accurate result compared to 3-D simulation [14]. Biomass consisted of cellulose, hemicellulose, and lignin was injected into the reactor at 300 g/h of mass feeding rate and conditioned at 27 °C. A 0.4 mm constant particle diameter of biomass was used in all of the simulations. Pure nitrogen as fluidization gas was preheated until 500 °C and flowed from the bottom of the reactor at a superficial velocity of 0.36 m/s. An electric heater as a thermal energy source was performed as a constant temperature wall maintained at 523 °C. Silica sand employed as a bed medium had a diameter of 0.52 mm, initially heated to 500 °C and packed to a height of 55 mm with a volume fraction of 0.59. The initial sand temperature was set at the same temperature with the preheated nitrogen.

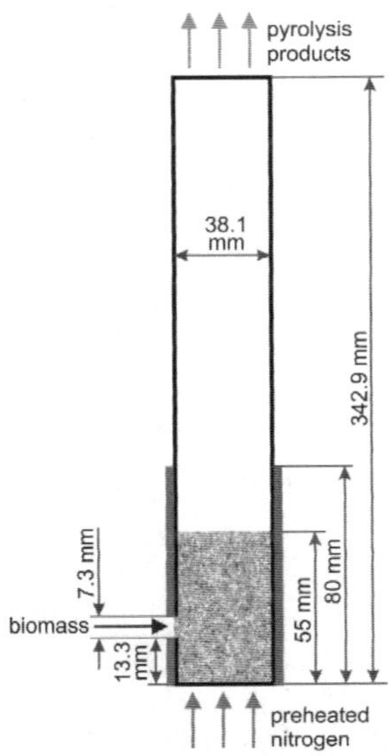

Figure 2. Schematic of fluidized bed reactor.

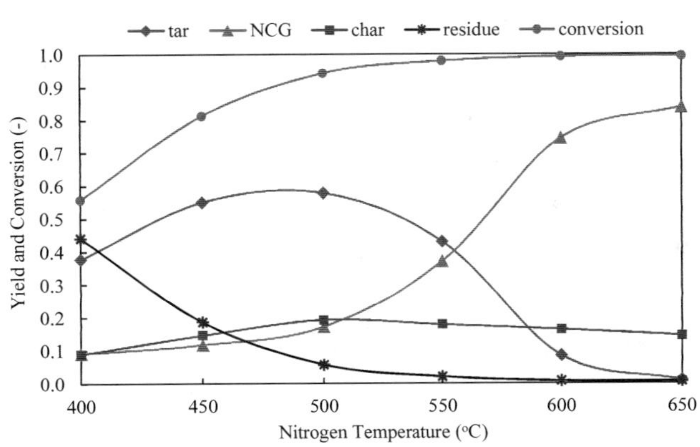

Figure 3. Product yields of PKS pyrolysis for different fluidizing gas temperatures.

The simulation results were compared to the experiment reported by Xue, et al. [14] that used red oak as biomass raw material for pyrolysis in a fluidized bed reactor. Read oak consists of cellulose, hemicellulose, and lignin in the composition ratio of 0.41, 0.32 and 0.27 respectively. This work used PKS as the raw material which has 0.296, 0.232 and 0.472 of cellulose, hemicellulose and lignin, respectively, adapted from a study reported by Abinsa, et al. [15] by neglecting the extractive composition.

The velocity inlet was used as a boundary condition in biomass and preheated nitrogen inlets. One atmospheric pressure was used at the pressure outlet. A no-slip boundary condition was applied to heater wall and adiabatic wall for gas phase and 0.05 of specularity coefficient was implemented for solid phases to both types of wall adopted from Dinh, et al. [16]. Thermo-physical properties of species contributed in the pyrolysis are listed in Table 2 [17].

IOP Conf. Series: Materials Science and Engineering **874** (2020) 012033 doi:10.1088/1757-899X/874/1/012033

The conservation equations were solved using the phase coupled SIMPLE algorithm for the pressure-velocity coupling and the least square cell-based was applied to the pressure based solver. The equations of momentum and energy were calculated using the second order upwind, while the species equation was estimated using the first order upwind method. The volume fraction was discretized using the QUICK algorithm. The first-order implicit scheme for transient formulation was applied with the time step size of 1.25 x 10^{-4} s to 5.0 x 10^{-4} s, depending on the iteration stability. The simulations were performed for a reaction time of 30 s. The yields of the products (tar, NCG, char, residue) and the conversion efficiency were calculated using equation 2 to 6.

Table 2. Thermo-physical properties of pyrolysis species

Phase	Species	Molar Mass (kg.kmol^{-1})	Particle diameter (mm)	Density (kg.m^{-3})	Specific heat (J.kg^{-1}.K^{-1})	Dynamic viscosity (kg.m^{-1}.s^{-1})	Thermal conductivity (W.m^{-1}.K^{-1})
	Nitrogen	28	-	-	1121	3.58×10^{-5}	5.63×10^{-2}
Gas	NCG	30	-	-	1100	3.0×10^{-5}	2.577×10^{-2}
	Tar	100	-	-	2500	3.0×10^{-5}	2.577×10^{-2}
	Cellulose	162					
Biomass	Hemicellulose	132	0.4	400	2300	-	0.3
	Lignin	208					
	Cellulose char	12					
Char	Hemicellulose char	12	0.4	2333	1100	-	0.1
	Lignin char	12					
Sand	Silica	60	0.52	2649	800	-	0.27

$$Yi = \frac{m_i}{m_{biomass}} \tag{2}$$

$$\eta_{conversion} = 1 - Y_{residue} \tag{3}$$

Where Y_i indicates the yield of product species (tar, NCG, char, and residue), m_i is the total mass of the product species including the product which still remains in the reactor and $\eta_{conversion}$ represents the conversion efficiency.

3. Results and Discussions

3.1. Validation Data

The simulation results using red oak as biomass pyrolysis raw material was compared to the experimental data given by Xue, et al. [14], presented in Table 3. The results showed a good relationship between the simulation and experimental red oak pyrolysis data by comparing the product yields distribution and those relative different. Both experiment and simulation product yields of fast pyrolysis were dominated by tar product and followed by NCG and char products with relative differences of 9.7 %, 0.1%, and 17.2%, respectively. Therefore, the simulation model showed a promising capability to investigate different type of biomass (PKS in this work).

Table 3 also shows the comparison between red oak and PKS pyrolysis that indicates pyrolysis of PKS produces a larger amount of char product with the lower products of tar and NCG. It is well known that the composition of biomass has a significant impact on the yields of pyrolysis product. The phenomenon is attributed to the lignin decomposition characteristic. Lumped kinetic scheme reported by Ranzi, et al [5] explained the detailed decomposition of lignin which produces more char compared to cellulose and hemicellulose.

Table 3. Comparison of product yields

Method	Biomass	Tar (%)	NCG (%)	Char (%)	Residue (%)	Ref.
Experiment	Red Oak	71.7 ± 1.4	20.5 ± 1.3	13.0 ± 1.5	-	[14]
Simulation	Red Oak	64.8	18.6	10.8	5.9	This work
Simulation	Palm Kernel Shell	57.4	17.2	19.3	5.7	This work

In this work, PKS was used as the raw material in fast pyrolysis in a fluidized bed reactor. The tar, NCG, and char product yields were 60.5%, 18.0%, and 16.6%, respectively. Compared to the study by Abnisa, et al. [15] which reported that the tar, NCG and char product yields were 47.43%, 17.31%, and 35.26%, respectively. Our work showed higher tar product with the lower char product. This difference was caused by the PKS particle size and the heating rate, whereas our work used smaller particle size and larger heating rate that contributed to the large amount of tar product with the low char product.

3.2. Effect of Fluidizing Gas Temperature

Operating temperature is an essential parameter for the kinetic rate of pyrolysis and it influences the chemical conversion efficiency and its product yields. Biomass consists of cellulose, hemicellulose, and lignin, which have a different characteristic in the decomposition process. Hemicellulose decomposes in a temperature range of 150 to 350 °C, while cellulose degrades at higher temperature ranges, namely 275-350 °C. In the other hand, lignin is a biomass constituent which has the highest temperature decomposition. The decomposition begins at a temperature of 280 °C and it reaches a peak rate at 350-450 °C [18]. PKS is dominated by lignin constituent that strongly influences the conversion efficiency. The higher temperature resulted the elevating conversion efficiency of biomass pyrolysis as presented in Fig. 3. The gradient of conversion efficiency increased sharply in the temperature range of 400 to 500 °C then followed by insignificant increasing conversion at the temperature above 500 °C. The high content of lignin in PKS contributes to the difficulty of PKS to be decomposed. Therefore, PKS needs medium-high temperature to maximize the decomposition efficiency.

Fig. 3 also provides the product yields distribution. The increasing temperature until 500 °C led the greatening of tar, NCG and char products with the decrease of residue or unreacted biomass. Furthermore, the increase of temperature above 500 °C caused to the rising of NCG product significantly with the decrease of tar and char products. High temperature promoted the secondary cracking reaction that converts tar into NCG, therefore the tar product decreases drastically at the temperature above 500 °C. The maximum tar product yield was obtained at 500 °C.

3.3. Effect of PKS feeding Rate

Pyrolysis is a thermal decomposition which means the process needs thermal energy input to carry out the decomposition. Thermal energy not only for heating the biomass, but also it is needed to perform the chemical conversion process. The biomass feeding rate is one of the important parameters that influence the thermal decomposition process. Greater feeding rate of biomass required higher thermal energy input to support the process. Therefore, the increasing feeding rate induced the decreasing temperature inside the reactor due to insufficient thermal energy input as illustrated in Fig. 4. All variations of feeding rate were flowed with 0.36 m/s of nitrogen as fluidizing gas at 500 °C from the bottom side of the reactor. The increasing feeding rates from 300 g/h to 1500 g/h caused the decrement of average reactor temperature from 500 °C to around 400 °C.

The decreasing temperature inside the reactor induced the increasing of both surface heat flux and surface heat transfer coefficient due to the larger temperature difference between the heater wall and biomass as presented in Fig. 5. This phenomenon indicates that the thermal energy input for the decomposition process increased with the raising of PKS feeding rates. Nevertheless, the elevation of thermal energy input into the reactor was still insufficient to maintain the reactor temperature well so that the temperature dropped significantly. As the consequence, the product yields distribution changed associated with the average operating temperature inside the reactor as shown in Fig. 6.

Fig. 6 shows the relationship between PKS feeding rates toward the pyrolysis product yields and its conversion efficiency. The increases of PKS feeding rate contributed to the decreases of conversion efficiency indicated by the rising of unreacted PKS as residue. The temperature drop due to insufficient energy input reduced the decomposition rates which favors the decrement of conversion efficiency and char product. In addition, the greater PKS feeding rates resulted in the decrease of tar and NCG product yields. The decrement was attributed to the lowering the primary decomposition rate of PKS due to the temperature drop inside the reactor.

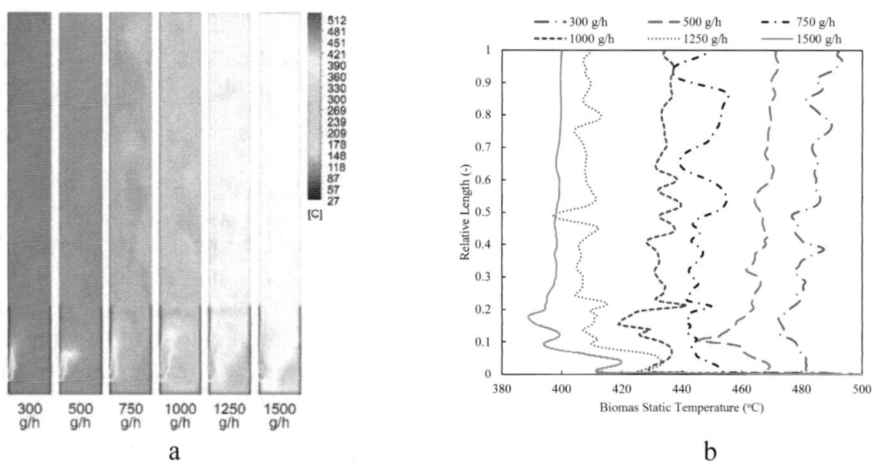

Figure 4. (a) Biomass temperature contour inside the reactor (b) Biomass temperature distribution in the centerline of reactor.

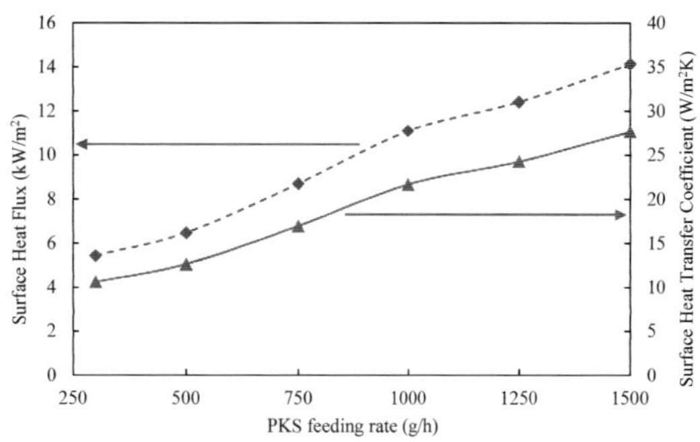

Figure 5. Surface heat flux and surface heat transfer coefficient of heater wall.

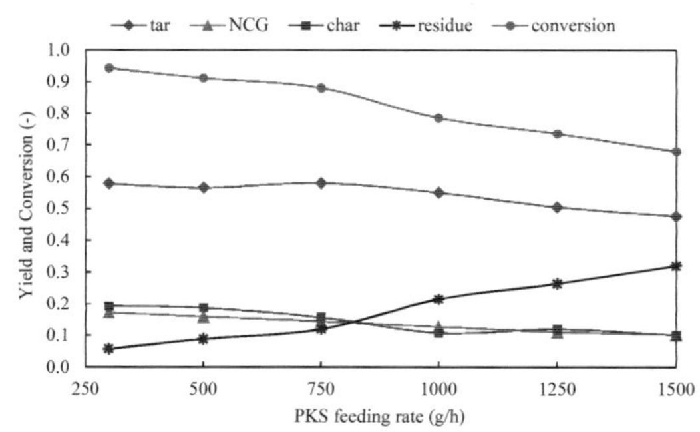

Figure 6. Product yields of PKS pyrolysis for different PKS feeding rates.

3.4. Effect of Fluidizing Gas Superficial Velocity

Nitrogen as fluidizing gas in fluidized bed reactor performs as a heat carrier in the decomposition process, controls the volatile vapour residence time and it has a responsibility to the mixing/segregation between biomass and bed material. Larger superficial velocity of nitrogen indicates the greater energy supply for pyrolysis, shorter residence time, and more rapid mixing between PKS and bed material. Consequently, biomass decomposes rapidly due to the increase of biomass particle heating rate and the tar product elevates due to the minimum rate of secondary cracking reaction. This phenomenon is presented in Fig. 7. The increase of superficial velocity of nitrogen induced to the rising of tar product with the lowering of residue, whereas the char and NCG products showed the insignificant changes in all of feeding rate variations. The superficial velocities of nitrogen were varied i.e. 0.36 m/s, 0.55 m/s, and 0.75 m/s with the estimated residence time of 0.95 s, 0.62 s, and 0.46 s respectively.

Fig. 7 also shows the PKS pyrolysis product yields on the three variations of feeding rates. Those three variations demonstrated the increase of tar product by the elevating nitrogen superficial velocity, whereas the NCG product exhibited insignificant changes. The prominent phenomenon was shown in the residue product where the feeding rates of 1000 g/h and 1500 g/h displaying the drastic decrement with the raising of nitrogen superficial velocity. It also indicates that the increase of fluidizing gas velocity induced the elevating conversion efficiency. This was caused by the energy supply of preheated nitrogen that tends to be used for the primary decomposition rather than secondary cracking reaction resulting a sharp decline in residue product with a slight decrease in NCG product. Because pyrolysis has successive and competitive reactions, the product yields of pyrolysis are influenced by many parameters. Therefore, with the accordant parameters such as medium temperature, appropriate biomass feeding rate and nitrogen superficial velocity, the optimum tar product yield can be achieved.

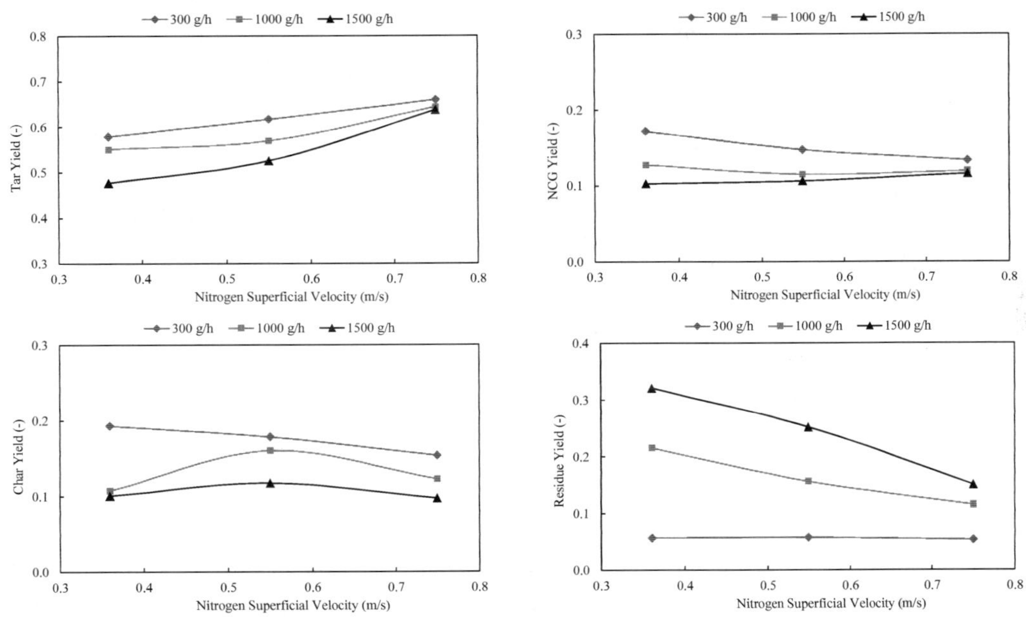

Figure 7. Combination effects of PKS feeding rates and nitrogen superficial velocities on the pyrolysis product yields.

4. Conclusion

In this present study, the global kinetic scheme approach was used to conduct the numerical studies of PKS pyrolysis. Compared to the red oak, pyrolysis of PKS produced larger char product yield with lower tar product yield due to its deferential biomass properties (percentage of cellulose, hemicellulose, and lignin). The increases of nitrogen temperature until 500 oC led to the increases of tar product, while at the above temperature, tar product decreased significantly due to the secondary cracking effect. Therefore, the maximum tar product could be attained in medium temperature (500 °C). The larger PKS feeding rates made the reactor temperature drop that induced the decomposition rate to become low so that the residue/unreacted biomass increased. However, by raising the nitrogen superficial velocity, the residue decreased with the tar product elevated due to bigger thermal energy input from preheated nitrogen even though in a large PKS feeding rate. Therefore, the PKS feeding rate and nitrogen superficial velocity should be maintained in order to achieve the appropriate temperature distribution along the reactor so that the optimum tar product yield could be attained.

References

[1] Hambali E and Rivai M 2017 *The Potential of Palm Oil Waste Biomass in Indonesia in 2020 and 2030* IOP Conf. Ser.: Earth Environ. Sci. vol. 65.

[2] Dhyani V and Bhaskar T 2018 A Comprehensive Review on The Pyrolysis of Lignocellulosic Biomass *Renew. Energy* vol. 129 p. 695.
[3] Bok P J, Choi, S Y, Choi K S, and Jeong W Y 2014 Fast Pyrolysis of Douglas Fir by Using Tilted-Slide Reactor and Characteristics of Biocrude-Oil Fractions *Renew. Energy* vol. 65 p. 7.
[4] Thoharudin, Nadjib M, Santosa T H A, Juliansyah, Zuniardi A, and Shihabudin R 2018 *Properties of co-pyrolysed palm kernel shell and plastic grocery bag with CaO as catalyst* IOP *IOP Conf. Ser.: Earth Environ. Sci.* vol. 209.
[5] Ranzi E, Eliseo; Cuoci A; Faravelli T, Frassoldati A, Migliavacca G, Pierucci S, and Sommariva S 2008 Chemical Kinetics of Biomass Pyrolysis *Energy & Fuels* vol. 22 p. 4292.
[6] Naron D, Collard F-X, Tyhoda L, and Görgens J F 2019 Production of Phenols from Pyrolysis of Sugarcane Bagasse Lignin: Catalyst Screening Using Thermogravimetric Analysis – Thermal Desorption – Gas Chromatography – Mass Spectroscopy *J. Anal. Appl. Pyrolysis* vol. 138 p. 120.
[7] Wang W, Li X, Ye D, Cai L, and Shi S Q 2018 Catalytic Pyrolysis of Larch Sawdust for Phenol-Rich Bio-Oil Using Different Catalysts *Renew. Energy* vol. 121 p. 146.
[8] Aramideh S, Xiong Q, Kong S-C; Brown R C 2015 Numerical Simulation of Biomass Fast Pyrolysis in an Auger Reactor *Fuel* vol. 156 p. 234.
[9] Jalalifar S, Abbassi R, Garaniya V, Hawboldt K, and Ghiji M 2018 Parametric Analysis of Pyrolysis Process on The Product Yields in a Bubbling Fluidized Bed Reactor *Fuel* vol. 234 p. 616.
[10] Borello D, Cedola L, Frangioni G V, Meloni R, Venturini P, De Filippis P, de Caprariis B 2016 Development of a Numerical Model for Biomass Packed Bed Pyrolysis Based on Experimental Validation *Appl. Energy* vol. 164 p. 956.
[11] Di Blasi C 1996 Kinetic and Heat Transfer Control in the Slow and Flash Pyrolysis of Solids *Ind. Eng. Chem. Res.* vol. 33 p. 37.
[12] Miller R S and Bellan J 1997 A Generalized Biomass Pyrolysis Model Based on Superimposed Cellulose, Hemicellulose and Liqnin *Kinetics Combust. Sci. Technol.* vol. 126 p. 97.
[13] Liu B, Papadikis K, Gu S, Fidalgo B, Longhurst P, Li Z, and Kolios A 2017 CFD Modelling of Particle Shrinkage in a Fluidized Bed for Biomass Fast Pyrolysis with Quadrature Method of Moment *Fuel Process. Technol.* vol. 164 p. 51.
[14] Xue Q, Dalluge D, Heindel T J, Fox R O, Brown R C 2012 Experimental Validation and CFD Modeling Study of Biomass Fast Pyrolysis in Fluidized-Bed Reactors Fuel vol. 97 p. 757.
[15] Abnisa F, Arami-Niya A, Wan Daud W M A and Sahu J N 2013 Characterization of Bio-oil and Bio-char from Pyrolysis of Palm Oil Wastes *Bioenergy Res.* vol. 6 p. 830.
[16] Dinh C-B, Liao C-C, and Hsiau S-S 2017 Numerical Study of Hydrodynamics with Surface Heat transfer in a Bubbling Fluidized-Bed Reactor Applied to Fast Pyrolysis of Rice Husk *Adv. Powder Technol.* vol. 28 p. 419.
[17] Lathouwers D and Bellan J 2001 Yield Optimization and Scaling of Fluidized Beds for Tar Production from Biomass *Energy & Fuels* vol. 15 p. 1247.
[18] Basu P 2013 *Biomass Gasification, Pyrolysis, and Torrefaction; Practical Design and Theory* (Amsterdam: Elsevier) pp. 147–176.

IOP Conf. Series: Materials Science and Engineering **874** (2020) 012034 doi:10.1088/1757-899X/874/1/012034

Engineering and Technology Challenges of The Energy Usage: An Estimation for Agro-Industrial In Indonesia

Y H Asnawi[1], A Nuraisyah[1], Zulfiandri[2], V I Sari[1], Rahmawati[1], Gunawan[1], A Rahmana[1]

[1]IPB University, Bogor Indonesia
[2]Esa Unggul University, Indonesia

Abstract. Agro industry is one of strategic industry that supported by a potencial resources from agriculture, fisheries/marine, livestock, farm, and forestry sectors. Energy supply with affordable prices are factors of the agro industry sustainability. The purpose of this study is to ensure the availability of energy in the agriculture industries, the pulp & paper industries, the sugar industries from sugar cane, and the fish canning industries. The stages consist of identifying the needs and availability of raw materials, energy, and formulating the recommendations. A literature study from relevant sources and analysis of efficiency of energy is used to improve the industry competitiveness by using Net Energy Ratio (NER) calculation. Around 38% of the agro industries energy needs are from by-product's biomass. Biomass fuels that usually using by agro industries are palm shells, wood chips, bagasse, black liquor, or depending on the by-products. Aggregately, agro industries energy needs (2017) amounted to 15,711 million TOE. Based on the capacity, the final energy needs of selected agro industries amounted to 25,818 millions TOE that equivalent to 11% from the national final energy demand of 238.8 millions TOE and 30% of the final industrial sector energy needs of 87 millions TOE in 2025.

1. Introduction

The Agro industry is a strategic industry and it is supported by potential resources from farming, fishery/marine, cattle, plantation and forestry. The contribution of the Agro industry toward the PDB of non-oil and gas industry in 2016 is as much as 46.95%, which is the highest compared to other processing industries. The government, the ministry of industry in this case, has responsibility to improve the contribution and growth of the non-oil and gas manufacturing industry for the growth of national economy.

According to the data of Directorate General of Renewable New Energy and Energy Conservation, which is part of the Ministry of Energy and Mineral Resources (ESDM), in the last few years, the growth of the consumption of energy in Indonesia has reached 7% per year. This number is above the growth of the world energy consumption, which is 2.6% per year. The industry sector plays a big part on the high growth, with on average 2.7% increase of energy necessity per year, between 2011-2035. With the demand of energy consumption increasing each year for Industry sector in Indonesia, the number which was 39% in 2010 is projected will jump to 44% by 2030 [1].

In relation to this matter, to push the growth of national economy, the Agro industry is needed to be supported of its growth and roles. For the continuation, it needs to have a form of guarantee for enough supply of raw materials and energy. Some of the strategic Agro industry, such as palm oil, pulp & paper, sugar cane, and canned fish, is expected to be improved of its work and process, in

IOP Conf. Series: Materials Science and Engineering **874** (2020) 012034 doi:10.1088/1757-899X/874/1/012034

relation with the huge potential of Indonesia natural resources and its potential on the national and global market.

However, it is also needed to be understood that along with the growth of the industry sector, it is also necessary for the improvement of resources and energy. In relation with this increase, a full attention on the availability of the resources and energy in affordable prices is required. One of the supports, aside from guaranteeing the availability and prices, is to have the capacity and productivity according to the necessity and quality of an internationally standard product.

The purpose of this activity is to make sure that the availability of energy for agro industry, for the likes of palm oil, pulp & paper, sugar cane, and canned fish. The activities, in details, are as following: 1) To identify the necessity and availability of real energy in Agro industry sector; 2) To identify the challenges related with the fulfillment of the energy necessity in the Agro industry sector.

Hopefully, the result of this identification can be useful to increase the competitiveness of this country's industry, whether it is for facing the intensified global market and of course the local as well.

2. Methodology

In conducting this research, the method of mass and energy balance modelling was used in 3 stages, which are: (1) Stage of identification of necessity and availability of energy, and (2) Stage of formulating the recommendation, in order to face energy fulfillment in relation to the challenges in the Agro industy sector.

Identification stage was done with literature study and data-information gathering from reliable sources. The model of mass and energy balance of the agro industry process was gained through the identification compartment that is related with the production process, the equation of mass balance to determine the input of raw material and supporting matter, as well as the output in form of products and side result between compartment. The balance of mass and energy model is an illustration of the flow of mass and energy in the process of Agro industry. The modeling of the balance of mass and energy uses approach of linier counting based on available data. The approach itself was used to find the critical factor and optimal solution of a relevant mass output with using quantitative model to support decision making. The modeling will result on the ratio (coefficient efficiency) and static variable by using the principality of linear equation.

The analysis stage was done with the approach of the efficiency analysis of the materials and energy to improve its competitiveness. If the produced energy is bigger or equal with the factory necessity, then the factory has the potential for a self-sustaining and efficient energy. The energy efficiency based on the energy is calculated using Net Energy Ratio (NER) [2]. The next stage was the formulation of the recommendation to face the challenges related with the fulfilment of the energy necessity in the Agro industry sector. This was done with expectation that the result of this identification could be useful to increase the competitiveness of the local industry, whether to face the intensifying global trade and of course to fulfill the local demand. The formulation of the recommendation involves aspects of necessity and availability of the Agro industry energy in order to guarantee the supply of the energy resources in competitive prices.

a. Balance of Energy Value

1) The fuel from many sources have different calorie number. The calculation of the amount of the proportion of each component used weight average.

$$Weight\ average = \frac{\sum_i (CVi\ x\ Quantityi)}{\sum_i (Quantityi)} \qquad \boldsymbol{CV{:}Calorific\ Value}\ (e.g. \tfrac{kJ}{m3}) \quad \boldsymbol{Quantity{:}Physical\ quantity}\ (e.g. m3)$$

2) The Percentage of the Energy Necessity per Production [4]

$$\text{Energy Necessity (\%)} = \frac{\boldsymbol{Fuel\ Energy\ (J)}}{\boldsymbol{Total\ Energy\ (J)}} \text{ x } 100$$

3) Fuel Energy [5]

Fuel Energy (J) = ∑BBi (one unit) x FKi (J/one unit)
BB = Volume or mass of the fuel being used (per one unit)
FK = Fuel conversion factor (J/per one unit)

b. Energy Conversion Factor

Conversion factor is used to convert the quantity of the fuel between natural unit (mass or volume) with the energy unit (energy content). The energy conversion factor that was used for the calculation of this research can be seen on Table 1.

The first step to make the mass balance model was to identify the compartment process of the Agro industry production. The equation of mass equilibrium is to determine the input of raw material and additional material, as well as the output between compartments. The equilibrium model was developed into two stages, which is the simple model and complex one. The simple model was built with assumption that the production system of Agro industry is a single compartment made of input and output in the form of product and co-product. The complex model was built according to the stages of the production process of the Agro industry.

Tabel 1. Energy Conversion[6]

Fuel	**Unit**	**Value**
Diesel	MJ/L	35.9
Diesel	MJ/kg	42.9
Coal	MJ/m^3	34,797.30
Coal	MJ/kg	25.75
LPG	MJ/L	24.5
LPG	MJ/kg	46
Natural Gas	MJ/m^3	35.14
Natural Gas	MJ/kg	45.75
Electricity	MJ/kWh	3.6
Biomass	MJ/ton	7,500
Steam	MJ/ton	2,700

The mass equilibrium of the Agro industry production model is gained through compartment identification and the equation of mass equilibrium, which is to determine the input of the raw material and additional material, along with the output between compartment. Simple model assumed that all steps in the process is done in one compartment (Figure 1).

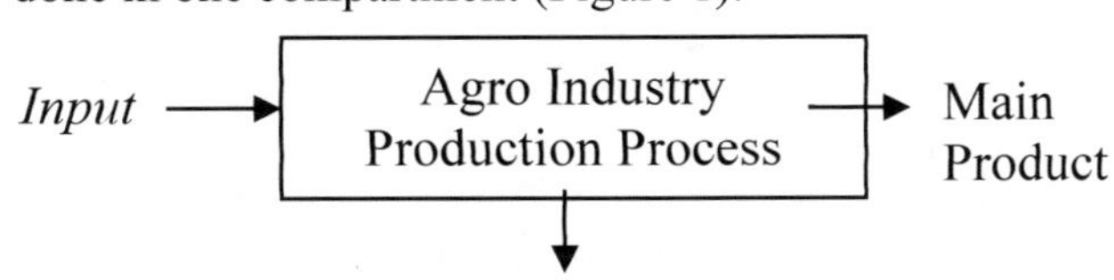

Figure 1. Simple mass equilibrium model[7]

The mass equilibrium would be achieved if the input is the same as the output (product and co-product). The process efficiency was expected to be close to 100%, where all the raw material content is the component of the product that can be converted into renewable-reusable product. The general equation of mass equilibrium with assumption that the side result is waste that can be repurposed:

Input (I) = Product (P) + Side Result (C)[8]

The modelling of mass and energy equilibrium using linear calculation approach based on available data.

$E_n = n \times CV$

Where:

En = Energy
n = Inventory Volume
CV = Calorific Value (energy conversion value)

National Energy Necerssity

1. Energy necessity per ton product (J):

$$ENP\left(\frac{J}{ton}\right) = \frac{Total\ energy\ production\ in\ industry\ (J)}{Total\ Production\ in\ Industry\ (ton)}$$

ENP = necessity per ton product (J/ton)
Total energy production in industry (J)
Total Production in Industry (ton)

2. National Energy Necerssity:
NEN (J) = TNP (ton) x ENP (J/ton)
NEN = National Energy Necerssity (J)
TNP = Total National Production (ton)
ENP = Energy necessity per ton product (J/ton)

3. Result

a. Energy Provision

Industry sector is predicted to be still dominating in the matter of final energy consumption. Indonesia is currently growing to be a developed country, which is indicated with the dominance of the industrial sector in supporting its economic activities. According to the BPPT at Outlook Energi Indonesia 2017, the segment of industrial sector final energy necessity would increase from 29% in 2015 into 43% (basic scenario) or 44% (high scenario) in 2050. The final energy consumption based on type for 2010-2015 was still dominated by Oil type fuel (gasoline, solar, diesel oil, kerosene, burning oil, avtur and avgas) reaching 25%, followed by natural gas (11%), electricity (11%), coal (6.2%) and LPG (4.8%).

1. The necessity for Agro industry energy

In general, the necessity of Agro industry energy is calculated by multiplying the number of productions with the necessity of energy in each Agro industry. The national production is based on the number of national production and installed capacity in 2015. The national production and installed capacity data were acquired from the Ministry of Industry. The calculated necessity for energy is direct energy.

2. The energy necessity based on the source

In this resulted, around 38% of the Agro industry energy necessity comes from biomass, which is produced by co-product of the Agro industry production. The biomass fuel that is used, such as palm oil shell, wood chip, sugar cane pulp, black liquor, all depends on the co-product of the Agro industry. The details of the Agro industry energy based on the energy source can be seen on Table 2. The energy consumption according to the fuel type in Agro industry is shown on Figure 2.

Table 2. The necessity of Agro industry energy based on the energy source

No	Fuel Type	Energy Necessity (one unit)			
		National Production		Installed Capacity	
		(One Unit Fuel)	(TJ)	(One Unit Fuel)	(TJ)
1	Solar (liter)	51,296,726.1	1,841.6	76,078,380.1	2,675.9
2	Natural Gas (m^3)	4,303,229,800.2	151,215.5	8,300,701,592.0	291,686.6
3	Electricity (GWh)	7,242.1	26,071.6	13,862.1	49,903.7
4	Coal (m^3)	9,387,536.0	199,485.1	14,981,452.7	318,355.8
5	Biomass		276,006.0		413,117.9
	Total		654,619.8		1,075,739.90

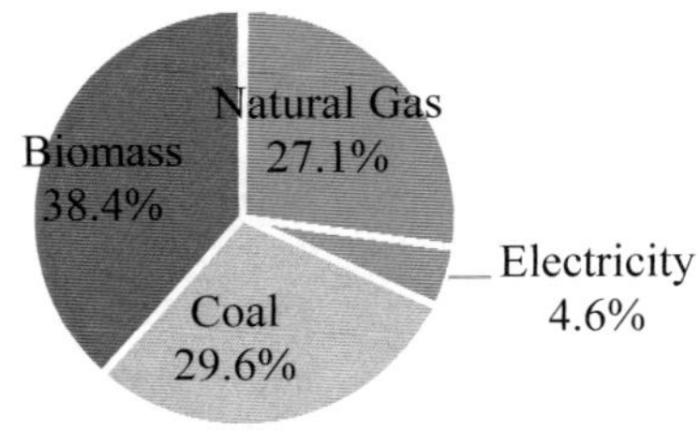

Figure 2. Energy consumption based on the energy type

The national production energy consumption and Agro industry installed capacity were chosen based on the type of fuel as shown on Table 3. The percentage of the chosen energy consumption of the Agro industry installed capacity toward the projection of availability of the national energy in 2025 is shown on Figure 3.

Table 3. Agro Industry Energy Consumption

No	Fuel Type	Energy Necessity (million TOE)		Energy Availability*) (million TOE)
		National Production	Installed Capacity	
1	Solar	0.044	0.064	110 **)
2	Natural Gas	3.629	7.0	62
3	Electricity	7.2 TWh	13.9 TWh	579 TWh
4	Coal	4.788	7.641	105
5	Biomass	6.624	9.915	52
Total (million TOE)		15.711	25.818	

*) Projection for 2020 [9]
**) Availability of fuel energy

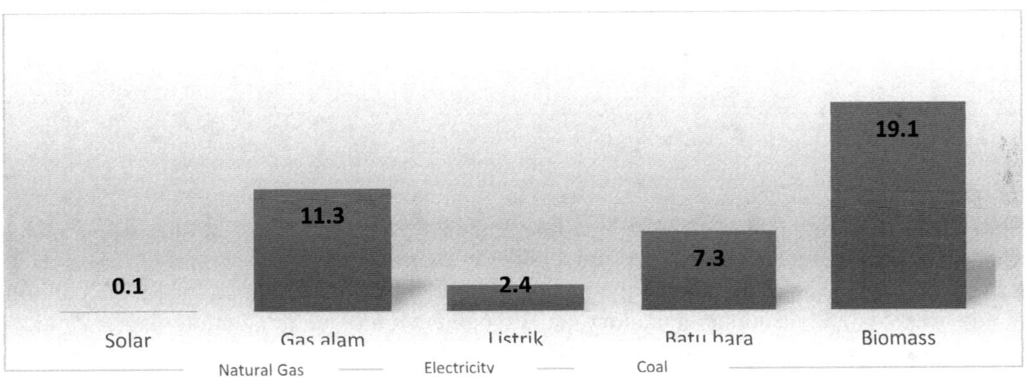

Figure 3. Percentage of the consumption of chosen agro industry installed capacity energy

The necessity of the chosen Agro industry energy for national production (2017) was as much as 15,711 million TOE. Based on the installed capacity. The necessity of chosen Agro industry final energy is as much as 25,818 million TOE, which is equal to 11% of the projected national energy necessity, which itself is 238.8 million TOE and 30% of the industry sector final energy necessity, equaling to 87 million TOE, in the year 2025. The necessity of Agro industry energy for national production without biomass was as much as 9 million TOE or 4.6% of the primary energy without biomass provision in 2015, equally as 195 million TOE. The installed capacity of chosen Agro industry energy without biomass necessity is as much as 16 million TOE, which is 4.8% of the projected energy availability in 2025, or equal as much as 332 million TOE.

3. Total Necessity of Agro Industry Energy

Aggregately, the energy necessity for industry is shown in the table as follow:

Table 4. Energy necessity per ton of the chosen Agro industry product

No	Product	National Energy Necerssity (ton)	Industry installed capacity of energy necessity (TJ)	Energy Necessity: Solar Kilo liter	Solar TJ	Natural Gas m3	Natural Gas TJ	Fuel ton	Fuel TJ	Electricity GWh	Electricity TJ	Coal m3	Coal TJ	Biomass m3	Biomass TJ
1	Biodiesel	30,060,000	74,583.2	0.0	0.0	0.0	0.0	0.0	0.0	1,328.6	4,782.8	0.0	0	4,905,158.7	69,800.4
2	Pulp	11,049,000	407,577.3	34,971.3	1,255.5	530,758,431.8	18,650.9	22,787,678.6	265,476.5	0.0	0.0	5,750,330.1	122,194.5	0	0
3	Paper	23,162,275	376,523.4	21,716.1	779.6	6,510,329,212.5	228,773.0	0.0	0.0	5,994.6	21,580.7	5,900,711.5	125,390.1	0	0
4	Edible Oil and Margarine	26,000,000	58,426.1	2,347.8	84.2	0.0	0.0	0.0	0.0	2,837.6	10,215.2	2,264,780.9	48,126.6	0	0
5	Coffee Powder	238,702.5	500.8	231.9	8.3	10,142,142.1	356.4	0.0	0.0	37.8	136.1	0.0	0.0	0	0
6	Coffee Mix	63,452.6	1,170.7	278.5	10.0	8,504,436.9	298.8	86,504.3	648.8	59.2	213.1	0.0	0.0	0	0
7	Cocoa	891,124	4,231.2	0.0	0.0	79,229,779.1	2,784.1	0.0	0.0	401.9	1,447.0	0.0	0.0	0	0
8	Milk Powder	6,279,433	45,647.0	5,465.7	196.2	1,064,574,689.7	37,409.2	0.0	0.0	2,233.8	8,041.7	0.0	0.0	0	0
9	White Crystal Sugar	3,000,000	77,247.5	1,539.9	55.3	0.000	0.000	10,292,294.7	77,192.2	0.0	0.0	0.0	0.0	0	0
10	Refined Sugar	5,016,000	24,081.5	1,993.6	71.6	0.0	0.0	0.0	0.0	379.3	1,365.3	1,065,630.3	22,644.6	0	0
11	Animal feed	24,000,000	5,806.6	7,533.6	270.5	97,162,899.9	3,414.3	0.0	0.0	589.4	2,121.8	0.0	0.0	0	0
	Total		1,075,795.2	76,078.4	2,731.2	8,300,701,592.0	291,686.6	33,166,477.7	343,317.5	13,862.1	49,903.7	14,981,452.7	318,355.9	4,905,158.7	69,800.4

4. Conclusion

Based on the total energy that is needed, there for the processing of biodiesel took 2.5 GJ energy per ton biodiesel. The energy necessity for processing cooking oil and margarine is 2.25 GJ per ton cooking oil or margarine. The energy necessity on the pulp processing is 36.9 GJ per ton pulp, meanwhile in the paper processing factory, it is 16.3 GJ/ton. The total energy use for the whole process in coffee processing factory is 18.5 GJ/ton coffee mix. As for powdered coffee processing industry, it needs energy as much as 2.1 GJ/ton powdered coffee. The energy necessity per ton product for the cacao processing industry is as much as 4,748.1 MJ/ton cacao powder. The energy necessity in the processing of powdered milk is 7.3 GJ per ton powdered milk. The energy necessity per ton product of rock sugar processing is as much as 25.8 GJ. The energy necessity per ton product that is produced in the processing of the granulated sugar is as much as 4.8 GJ/ton of granulated sugar. Based on the total of weft production that is produced at the cattle weft processing and the total energy that is used at the cattle weft factory, then it can be obtained the energy necessity per ton cattle weft product, which is as big as 241.9 MJ/ton cattle weft. Aggregately, the energy necessity of the selected Agro industry as much as (2017) 15,711 million TOE. Based on the installed capacity, the final energy necessity of the selected Agro industry is as big as 25,818 million TOE, equal to 11% of the projected nationally final necessity energy as much as 238.8 million TOE and 30% of the final energy necessity of the industry sector as much as 87 million TOE in 2025.

Acknowledgement(s)
This research was funded by Directorate General of Agro-Based Industry. Ministry of Industry, The Republic of Indonesia in a program of Agro-Based Industrial Growth and Development Program (2018).

References

[1] Ministry of Energy and Mineral Resources Republic of Indonesia. 2017. Handbook of Energy and Economic Statisic of Indonesia. ISSN 2528 – 3464. Jakarta

[2] Gokcol, C., Dursun, B., Alboyaci, B., & Sunan, E. (2009). Importance of biomass energy as alternative to other sources in Turkey. Energy Policy, 37(2), 424-431.

[3] Pimentel, D. (2003). Ethanol fuels: energy balance, economics, and environmental impacts are negative. Natural resources research, 12(2), 127-134.

[4] Verbong, G., & Loorbach, D. (Eds.). (2012). Governing the energy transition: reality, illusion or necessity?. Routledge.

[5] Spiegel, C. (2007). Designing and building fuel cells (Vol. 87). New York: Mcgraw-hill.

[6] Fariadhie, J. (2009). Perbandingan Briket Tempurung Kelapa Dengan Ampas Tebu, Jerami, dan Batu Bara. Jurusan Teknik Mesin. Universitas Sultan Fatah.

[7] McKendry, P. (2002). Energy production from biomass (part 2): conversion technologies. Bioresource technology, 83(1), 47-54. Doi: 10.1016/S0960-8524(01)00119-5

[8] Edgar, T. F., Himmelblau, D. M., & Lasdon, L. S. (2001). Optimization of chemical processes. McGraw-Hill.

[9] Dewan Energi Nasional. (2016). Outlook Energi Indonesia 2016. ISSN 2527 – 3000. Jakarta.

IOP Conf. Series: Materials Science and Engineering **874** (2020) 012035 doi:10.1088/1757-899X/874/1/012035

Validating The Fully Raster-Based Suitability Model for Sea Cucumber (*Holothuria scabra*)

Bambang Sulistyo[1]*, Dewi Purnama[1], Maya Anggraini[1], and Noviyanti Listyaningrum[2],

[1]University of Bengkulu, Kandang Limun, Bengkulu, Indonesia
[2]Graduate School, University of Gadjah Mada, Bulaksumur, Yogyakarta, Indonesia

* bsulistyo@unib.ac.id

Abstract. This research aimed at validating the fully raster-based suitability modelling for sea cucumber (*Holothuria scabra*) that has been created for the waters of Kiowa Bay, Enggano Island, Bengkulu. To validate the model, eighty-four sea cucumbers were cultivated on two different suitability classes. The field test phase includes setting the cultivation cages; measurement of sea cucumber weight; measurement of water quality parameter; and substrate analysis. Sea cucumbers were put into 2 cultivation cages (Cage 1: *highly suitable*, and Cage 2: *less suitable*), and for each cage, it was divided into 3 sub-cages. Before weighing, the sea cucumber was taken out from the water and then put into a tray layered with tissue, and left for 30 seconds. Measurements were conducted at the beginning of the study (t_0) and then carried out every three weeks until t_3. The data analysis includes the Specific Growth Rate; Daily Growth Rate; Absolute Growth; and Survival Rate. The results showed that the Specific Growth Rate; Daily Growth Rate; Absolute Growth; and Survival Rate at Cage 1 (ie 0.5%; 0.88 gram; 62.28 gram; and 85.71%) better than the same parameters at Cage 2 (ie 0.34%; 0.64 gram; 45.33 grams; and 69.05%). These results indicate that the raster-based data for sea cucumber suitability model has advantages over the sea cucumber suitability model analyzed using vector-based data

1. Introduction

Sea cucumber (*Holothuria Scabra*) is an economic and important marine product which are, usually, obtained from fishing activities. To get the larger number of sea cucumbers, the cultivation is important to do. Enggano Island waters have natural resources that are potential to be utilized, developed and managed as a source of regional economic growth[1]. One of them is the development of sea cucumber cultivation.

Sea cucumber (*holothurians*) is a group of Holothuroidea class marine invertebrates (*Fylum Echinodermata*) that are widespread in marine environments throughout the world, ranging from tidal zones to deep seas, especially in the Indian Ocean and Western Pacific Ocean. Sea cucumbers are slow-moving animals, also an important component in the food chain on coral reefs and their associated ecosystems. Food sources for sea cucumbers mainly consist of organic matter in sand or mud, plankton, coral litter, and detritus found in mud or sand. Besides that some sea cucumbers also eat small organisms such as diatoms, foraminifera, protozoa, nematodes, algae filaments and seaweed[2].

There are found 7 species of sea cucumbers on Enggano Island[3]. One of them is sea cucumber or white sea cucumber . Martoyo stated that sea cucumber is a species that lives in groups that can reach 3 to 10 tails and has a long round shape[4]. Generally, the stomach part of the sea cucumber is whitish-yellow, while the back is gray to blackish with black stripes. When touched, all parts of this sea cucumber feel rough. Many sea cucumbers are found between corals that are still alive or dead and in waters that contain fine sand with fragments of coral or muddy sand.

The selection of location is the first step that determines the success of sea cucumber cultivation. Generally, sea cucumber cultivation is carried out in coastal waters in tidal areas that support rapid water exchange. This is because the location or cultivation of sea cucumbers is a place that directly affects their lives. In other words, to manage marine resources, it requires careful planning[5][6]. When one considers the management of sea cucumber, a suitability map for its farming is needed. This map is generated from a comprehensive and integrated survey. Unfortunately, the collection of such data is time-consuming and also expensive[7]. From those reasons, suitability mapping for the farming of sea cucumber has therefore been generated from models based on the parameters that affect their development.

Generally, modelling related to the suitability determination in one area is expressed as[8] [9] :

$$S = \Sigma\, w_i x_i \quad (1)$$

where S is the suitability, w is the weight of factor i, and x is the potential rating of factor i.

Wulandari have performed an analysis of sea cucumber suitability modelling using vector-based data (say it as the Suitability_Vector, for further discussion) in the waters of Kiowa Bay, Kahyapu village, in the district of Enggano [10]. The analysis used criteria, weighting and scoring which is modified from Marizal[11]. The analysis involves nine parameters that were surveyed at 51 stations: seawater acidity, depth, current velocity, temperature, salinity, brightness, dissolved oxygen concentration, condition of the seafloor, and coastal protection of the area. The results showed that the waters of Kiowa Bay contained 102,477 hectares of *highly suitable*, 62,435 hectares of *suitable*, and 197,991 hectares of *not suitable* for cultivating sea cucumber. This suitability map is the result of analysis using vector-based data and is therefore simplified and consequently uncertain. For example, if one wants to develop the cultivation of sea cucumber using pen-culture techniques with a pen size of 40 meters by 40 meters, there would be uncertainty about where the best location is possible be (in terms of position and coordinates).

DeMers states that the presentation of earth information in the vector-data models assumes the existence of homogeneity in the mapping unit in which classification and simplification are required[12]. These processes are usually carried out subjectively such that there is the possibility of over-simplification and consequent reduction in detail. Thus the results of such analysis are full of uncertainty. Using remote sensing and GIS, those uncertainties can be minimized by applying the fully raster-based suitability modelling which is can be conducted more objectively using established algorithms and mathematical formulas without any simplification required[13]. Using the fully raster-based data, Sulistyo successfully apply them to analyse the erosion and the level of degraded land in Merawu Catchment, Banjarnegara, Central Java, Indonesia[7] [14] [15].

Using the same location, data and parameters as used by Wulandari et al. [10], in 2018 Sulistyo tried to refine the previous model by performing an analysis of sea cucumber suitability modelling using fully raster-based data (say it as the Suitability_Raster, for further discussion)[16].

Instead of using weight and score as it is used when using vector-based data, ie just by selecting its values as presented in tables (which is applicable for the whole and different locations), weighting and scoring when using raster-based data were determined by analysing the PCA (Principal Component Analysis). Through this analysis, the model (or equation) will be different for one location from other locations. For Kiowa Bay, the equation only includes four principal components (PCs) having significant values as[16] :

$$\text{Suitability} = 0.5235 * PC\_1 + 0.2797 * PC\_2 + 0.1329 * PC\_3 + 0.0639 * PC\_4 \quad (2)$$

However, the Suitability_Raster and the Suitability_Vector, are the result of models that must be validated to justify which is the most appropriate, because a model is a subjective approach which does not include all observations, but retains important aspects of reality which remains visible[17].

Refinement of the created model is possible if new information that relates to reality is found[15]. These models have to be validated by installing pen-culture nets at different locations (and for different suitability classes) and farming sea cucumbers to directly observe their growth parameters.

2. Methodology

The research area was located at Kiowa Bay, longitudes between 102.24° and 102.37° east and latitudes between 5.45° and 5.47° south. The study area was approximately 362 hectares (see Figure 1).

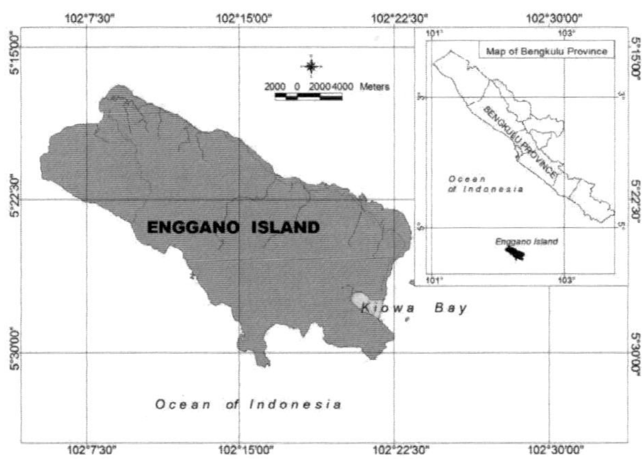

Figure 1. Location of the research in Enggano Island

A flow diagram of the study is presented in Figure 2.

To analyze and handle data, two GIS software packages were used. Meanwhile, some hardware was also required, namely a drafting tablet, equipment for fieldwork including a small boat, binoculars, compass, current meter, refractometer, Secchi disk, Tetra Test Kit O_2, DO meter, pH meter, Thermometer, pipes, waring nets, boards, stakes, ropes, trays, plastic, matches, scissors, ruler, slate board, tissue, battery, digital scales, sacks, nails, flashlights, Korean nylon cords, Global Positioning System, and digital camera.

For ease the discussion, two suitability models are presented in Figure 3 and Figure 4, while its combination is in Figure 5. Figures 5 shows that for the Suitability_Raster, the area which is most suitable for the growth of sea cucumber lies in the northwestern part of the Kiowa Bay, shown in dark blue, which is directly adjacent to the shoreline at a distance from 200 to 600 meters. This location is much more accessible than the location of the most suitable area of the Suitability_Vector. The Suitability_Raster seems more reasonable than those of the Suitability_Vector analysis. Besides, a suitable location in the Suitability_Raster is located far from the open sea such that it can protect sea cucumbers from the instability of marine conditions in Kiowa Bay.

Furthermore, the Suitability_Raster areas with lower a level of suitability (shown gradually by cyan, yellow, reddish yellow and pink) are located in the part of the study area most far away from the shore. Amazingly, for areas that are far from the shoreline, both models of suitability resulted in similar outcomes, namely the suitability value was "*high*" in the Suitability_Raster and "*not suitable*" in the Suitability_Vector.

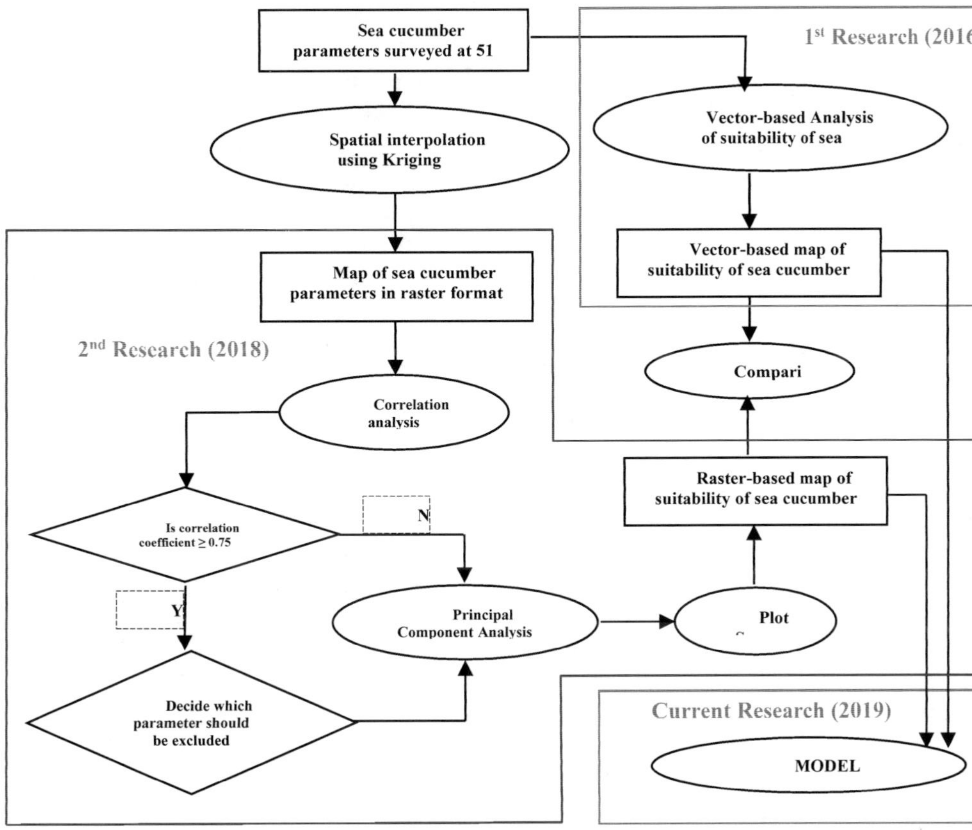

Figure 2. Flow diagram of the study (modified from[16])

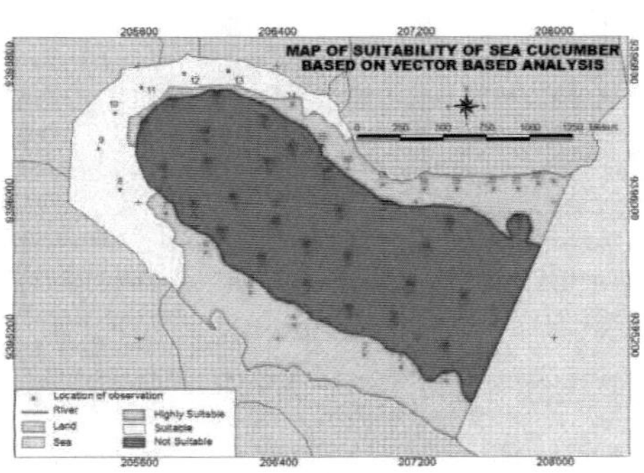

Figure 3. Vector-based map of suitability for sea cucumber

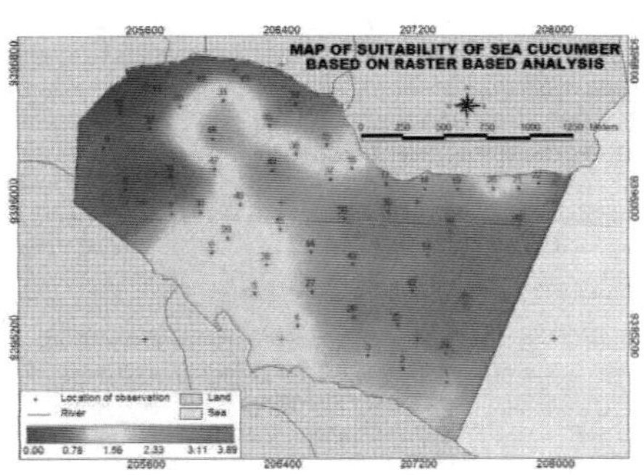

Figure 4. Raster-based map of suitability for sea cucumber

To validate the raster-based suitability model for sea cucumber that has been built, sea cucumbers were cultivated in two cages installed at two different suitability classes (at different locations). Then, their weight is monitored and measured during the period of research. In more detail, the field test phase includes: 1). Preparation for installing a cultivation cage; 2). Measurement of sea cucumber weight; 3) Measurement of water quality parameter; and 4). Substrate analysis.

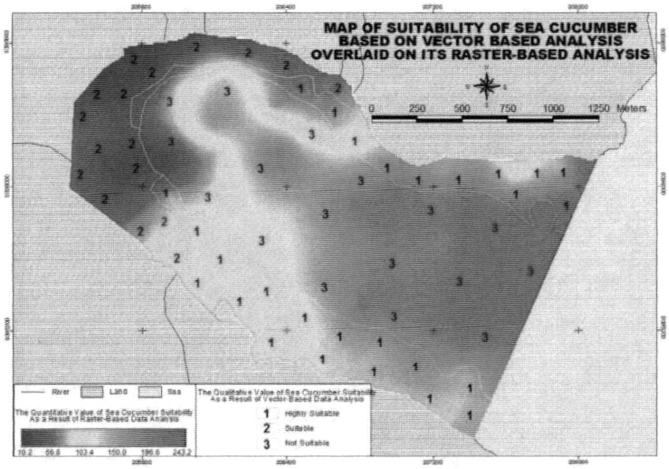

Figure 5. Vector-based map of suitability of sea cucumber (the yellow line) overlaid on the raster-based map

Preparation for installing a cultivation cage

Cage for the cultivation of sea cucumbers is a fence in the form of a fence with a length of 6 meters, a width of 2 meters, and a height of 150 cm which limits the area in the sea to a certain area which is used as a place to cultivate sea cucumbers. Cultivation cage material consists of waring nets, nails, wooden stakes, sacks filled with sand and the dividing wall of the board set 15 cm under seabottom. To measure the weight of sea cucumber, a digital scale with an accuracy of 1000 grams is used. The installation of the cage starts from the immersion of stakes, installation of waring nets, filling the sacks as ballast placed on the edge and inside the cage and fixing the board in waring as adhesive and ballast waring as well (see Figure 6).

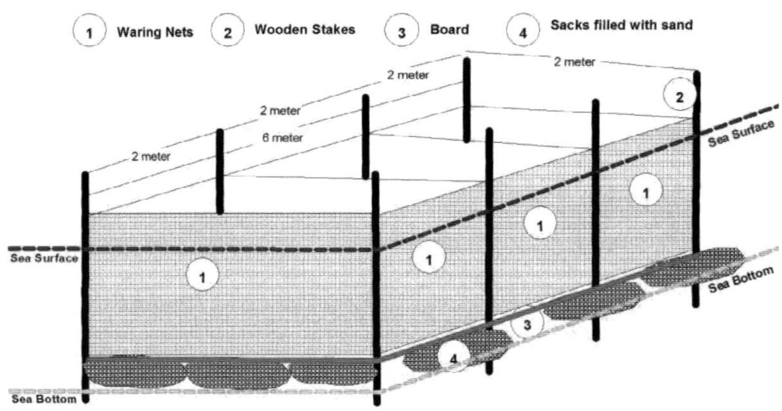

Figure 6. Diagrammatic Construction of Cultivation Cages

Measurement of sea cucumber weight

Eighty-four sea cucumbers were prepared with an initial weight between 60 grams and 266 grams. Sea cucumbers are divided into 2 cultivation cages (each with 42 sea cucumbers) with 3 sub-cages (each with 14 sea cucumbers). In the beginning, all sea cucumbers are weighed (as t_o), then the next measurement is done at every two weeks until t_3. Before weighing the sea cucumber is taken from the water and then put it into a tray filled with tissue and left for 30 seconds, then begins to be weighed.

The treatment in this study is a variation of the location, which consists of 2 treatments (cages) and 3 replications (sub-cages) as follows:

Treatment	The Suitability_Raster	The Suitability_Vector
Cage 1	*Highly Suitable*	*Highly Suitable*
Cage 2	*Less Suitable*	*Highly Suitable*

The analysis used was to compare the different levels of stocking density and survival rates before and after the research activities took place. When two types of treatment are compared, then the average value is used.

Measurement of Water Quality Parameter

Measurement of water quality parameters is carried out during monitoring activities. The parameters measured include physical parameters (depth and temperature) and chemical parameters (salinity, acidity, and dissolved oxygen).

Substrate Analysis

Substrate sampling was carried out every three weeks during the study, then the substrate samples were analyzed in the laboratory to find out the organic matter content in the cultivation cage area. The measurement of the organic content of the substrate was carried out by the ash analysis method, by taking it, weighing 100 grams and putting it in an oven at the temperature of 45°C until it has constant weight (for 3 days). A dry substrate is crushed and put back into the oven and left 30 minutes at really dry temperature. Then weighed again and then is crushed and put into the oven at 45°C for 30 minutes after constant, the substrate was weighed 25 grams and ignited in the furnace with a temperature of 700°C for 3.5 hours. Then the remaining substrate is finally weighed. Analysis of the organic content of the substrate was carried out in the Soil Science Laboratory (at the Department of Soil Science) and the Fisheries Laboratory (at the Department of the Marine Sciences), Bengkulu University.

The data analysis includes:

a. Specific Growth Rate (%)

Specific Growth Rate (SGR) is calculated by the formula[18]:

$$\text{SGR} = \frac{\ln \text{Wt} - \ln \text{Wo}}{\text{t}} \text{ x } 100 \tag{3}$$

Where

Wo : Average weight of individuals at the beginning of the study (gram)
Wt : Average weight of individuals at the end of the study (gram)
t : Research duration (days)

b. Daily Growth Rate (gram) is calculated using the formula from[19]:

$$\text{DGR} = \frac{\text{Wt} - \text{Wo}}{\text{t}} \tag{4}$$

c. Absolute Growth

The absolute growth is calculated by the formula from[20]:

$$\text{AG} = \text{Wt} - \text{Wo} \tag{5}$$

d. Survival Rate (%)

The survival rate (SR) of sea cucumber can be calculated using formula[21], as follows:

$$\text{SR} = \frac{\text{Nt}}{\text{No}} \text{ x } 100 \tag{6}$$

Where

No : Number of individuals at the beginning of the study (tail)
Nt : Number of individuals at the end of the study (tail)

e. Analysis of the physical-chemical and biological data of these waters is aimed at evaluating the level of suitability of the waters in the cultivation cages in the waters of Kiowa Bay, Enggano Island to support the life of sea cucumbers. The results of this environmental observation are then explained descriptively about the effect on the condition of the sea cucumber biology.

f. Substrate Organic Content
Analyse the organic content (OC) of sediments with the following formula[22] :

$$OC = \frac{A-B}{A} \times 100\% \quad (7)$$

Where

A = Constant Weight of Substrate
B = Ash Weight

3. Result and Discussion

The result of growth and survival rate of sea cucumber during the period of research is presented in Table 1.

Table 1. The Result of Growth and Survival Rate of sea cucumber*)

Cage 1

	Wo	No	Wt	Nt	AG	SGR	DGR	SR
Sub-Cage 1	165,79	14	199,57	14	33,79	0,26	0,48	100,00
Sub-Cage 2	143,93	14	190,50	10	46,57	0,39	0,66	71,43
Sub-Cage 3	131,36	14	237,83	12	106,48	0,84	1,50	85,71
Average	147,02		209,30		62,28	0,50	0,88	85,71

Cage 2

	Wo	No	Wt	Nt	AG	SGR	DGR	SR
Sub-Cage 1	160,57	14	248,40	10	87,83	0,61	1,24	71,43
Sub-Cage 2	177,71	14	178,89	9	1,17	0,01	0,02	64,29
Sub-Cage 3	139,71	14	186,70	10	46,99	0,41	0,66	71,43
Average	159,33		204,66		45,33	0,34	0,64	69,05

*)Notation and abbreviations can be seen at the equation or formula given

A. Growth

The high growth in sea cucumber can certainly be influenced by various factors, one of which is environmental factors. Also, another factor is that the high food supply provided during the research period which directly affects the growth of sea cucumber[23].

Based on the results of measurements during the study showed that the treatment of sea cucumber cultivated on Cage 1 resulted in an Absolute Growth of 62.28 gram, SGR of 0.5%, and DGR of 0.88 gram. The high addition of weights at the Cage 1 location is thought to be due to environmental factors and water substrate, where the substrate waters in the Cage 1 location are dominated by seagrass plants and sandy mud. This result is similar to the research of Jasmadi[24], concluding that the growth of sea cucumber in the treatment placed on the sandy substrate resulted in the highest absolute growth. Another study conducted by Louhenapessy[25], from 4 types of substrate used to see the growth rate of sea cucumber, which has the best growth rate is on the mud substrate. Seagrass ecosystems with sandy mud substrate is a habitat for sea cucumber, this ecosystem usually has high organic content and detritus which is the main food for some types of sea cucumber in general[26].

Sandy mud substrate with seagrass plants will ensure the availability of nutrient sources that can meet the needs of sea cucumber. Sabilu states that sea cucumber is thought to be able to use food well, so that the food that is used to survive and maintain the body, is also used to increase the weight that is metabolized and synthesized as meat[27]. The sandy mud area with seagrass plants is the area with the largest distribution of sea cucumber both in the number and the size[28].

Different from the treatment at Cage 1, the treatment at Cage 2 shows a lower rate of Absolute Growth which is 45.33 gram, SGR of 0.34%, and DGR of 0.64 gram. The low addition of weights in this treatment is thought to have a dominant sand substrate with fewer seagrasses, the incompatibility of the environment and natural habitat of sea cucumbers in nature has an impact on the low addition of Absolute Growth, SGR, and DGR of sea cucumbers. This result is in a line with the statement of Yokoyama that sea cucumber will process sediments faster in a finer substrate which is thought to contain a lot of algae, bacteria, fungi, detritus and organic matter[29]. Furthermore, sediment which is located on the surface of the sandy substrate is poor of nutrition, so that sea cucumbers develop a technique of behaviour adaptation to eating by consuming large amounts of sediment[30]. The eating behaviour aims to maximize the absorption of nutrients obtained from organic matter content in sediments as well as efficiency in movement[31].

B. Survival Rate

The survival rate of sea cucumber during the study for Cage 1 was 85.71% and for Cage 2 it was 69.05%. The high survival rate of sea cucumber which is maintained in each treatment is thought to be caused by environmental conditions that support the survival rate of sea cucumbers so that sea cucumbers can maintain their survival rate. Another indication that arises is that the Kiowa Bay area is a productive area capable of producing organic materials. Food for sea cucumber is in the form of detritus and plankton available in the waters of its habitat[32]. Seagrass ecosystems are one of the most productive ecosystems in the shallow sea waters, have an important role in supporting life and the development of bodies, namely as habitat for biota, sediment catchers, nutrient recyclers and so on.

C. Water Quality

The location of the study is in open areas and there are tides so that water fluctuations and water quality may change at any time. The value of water quality in both research locations is presented in Table 2.

Table 2. Results of measurements of water quality during the study

Parameter	Treatment		Optimum Value
	Cage 1	Cage 2	
Temperature (^{0}C)	28.7-28.9	26.9-27.6	24-30 ^{o}C[33]
Salinity (ppt)	30-31	23-25	28-34 ppt[33]
Acidity	7.7	7.7-7.8	7-8.5[21]
Dissolved Oxygen	7.3-7.6	5.0-5.5	> 5ppm[34]
Substrate Organic (%)	5.98	5.18	-

The results of water quality during the study showed that the quality of the waters in the two study areas was still in optimum condition for the growth and survival of sea cucumbers and can still be tolerated by sea cucumbers and is still suitable as a place for cultivating sea cucumbers.

From the results of the above research, especially if one only observed the average value of sea cucumber growth (Specific Growth Rate, Daily Growth Rate, and Absolute Growth) and Survival Rate, it can be seen that Suitability_Raster can better distinguish the results of growth rates rather than the Suitability_Vector. In the Suitability_Raster, different suitability values or class have produced different growth rates and survival rates of sea cucumber, while at the same class of suitability in the Suitability_Vector the result of growth rate is different.

Another advantage of the Suitability_Raster is in pointing the installation place for pen-culture nets in the field[16]. For example, if a pen-culture net with a size of 40 meters by 40 meters (1,600 m^2) was to be installed, the exact location could be easily determined from the Suitability_Raster by selecting 16 adjacent pixels (assuming that the pixel size is 100 m^2), such that the coordinates could be determined precisely. In contrast, this cannot be done using the Suitability_Vector, in which the locating of pen-culture nets with an area of 1,600 m^2 is difficult to locate. In other words, the Suitability_Vector has a high degree of uncertainty compared with the Suitability_Raster in determining exact places. Also, the Suitability_Raster has a higher level of confidence, due to its data inputs are more original than those of the Suitability_Vector, in which data input is simplified causing the variation of earth information is reduced[12].

4. Conclusion

These results indicate that the raster-based data for sea cucumber suitability model has advantages over the sea cucumber suitability model analyzed using vector-based data.

5. Acknowledgments

The authors would like to thank the Chairman of the Institute of Research and Community Service, University of Bengkulu, for facilitating funding and administration for this research. Thanks goes also to Hanifa Assyifaa Siwiristuti, Kamizan Rofik, Surya Marlia Sigito, Risky Suryaman Simbolon, and Muhammad Yusuf Ridho, students of Department of Marine Sciences, Faculty of Agriculture, University of Bengkulu, for assisting us when implementing and collecting data at Kiowa Bay, Kahyapu Village, District of Enggano.

References

[1] Bappeda Provinsi Bengkulu and C.V. Mitra Konsultan, *Laporan Akhir Penyusunan Tata Ruang Dan Potensi Pulau Enggano,* Bappeda Provinsi Bengkulu, Bengkulu, 2004.

[2] L. H. Hyman, *The Invertebrates: Echinodhermata The Ceolomate Bilateria. Vol. IV*. McGraw-Hill Book Company, New York, 1955.

[3] D. Purnama, and D. Hartono, *Studi Jenis Dan Distribusi Teripang Holothuroidea Di Perairan Pulau Enggano*, Lembaga Penelitian Universitas Bengkulu, Bengkulu, 2013.

[4] D. Sartika, "Aspek Biologi Reproduksi Taripang Pasir (Holothuria scabra) di Perairan Pantai Desa Sorue Jaya Kecamatan Soropia Kabupaten Kendari Sulawesi Tenggara," *undergraduate thesis,* Universitas Haluoleo, Kendari, 2002.

[5] A. Anggoro, E. Sumartono, V.P. Siregar, S.B. Agus, D. Purnama, Supriyono, D. A. Puspitosari, T. Listyorini, B. Sulistyo, and Parwito. 2018, "Comparing object-based and pixel-based classification for benthic habitatas mapping in Pari Island," *IOP Conference Series: Journal of Physics: Conf. Series* 1114, 012049

[6] B. Sulistyo, The Accuracy of the Outer Boundary Delineation of Coral Reef Area Derived From the Analyses of Various Vegetation Indices of Satellite Landsat Thematic Mapper, *Biodiversitas,* vol. 18, no 1, pp. 351-358, 2017.

[7] B Sulistyo B, The Effect of choosing three different c factor formulae derived from NDVI on a fully raster-based erosion modeling, 2nd *International Conference of Indonesian Society for Remote Sensing (ICOIRS)*, Published under licence by IOP Publishing Ltd, IOP Conference Series: Earth and Environmental Science, Vol. 47, No 1, 2016.

[8] R. J. Eastman, *Idrisi Andes: Guide to GIS and Image Processing*, Clark Labs, Clark University, Worcester, USA, 2006.

[9] B. Song B and S. Kang, A Method of assigning weights using a ranking and nonhierarchy comparison, *Advances in Decision Sciences,* vol. 1, no. 1, pp. 1-9, 2016.

[10] U. Wulandari, B. Sulistyo and D. Hartono, Aplikasi SIG Untuk Kesesuaian Kawasan Budidaya Teripang Pasir (*Holothuria scabra)* Dengan Metode *Penculture* di Perairan Teluk Kiowa, Desa Kahyapu Kecamatan Enggano, *Jurnal Enggano,* vol. 1, no. 1, pp. 57-73, 2016.
[11] D. Marizal, Y. V. Jaya, and H. Irawan, Aplikasi SIG Untuk Kesesuaian Kawasan Budidaya Teripang *Holothuria scabra* dengan Metode Penculture di Pulau Mantang, Kecamatan Mantang, Kabupaten Bintan. *Marine Science and Fisheries*, vol. 3, no. 6, pp. 1-8, 2012.
[12] M. N. DeMers, *Fundamental of Geographic Information Systems*, John Wiley & Sons, New York, 2008.
[13] D. S. Hadmoko, Toward GIS-based integrated landslide hazard assessment: a critical overview, *The Indonesian Journal of Geography,* vol. 39, no. 1, pp. 87-95, 2007.
[14] B. Sulistyo, T. Gunawan, Hartono, and P. Danoedoro, Toward a fully and absolutely raster-based erosion modeling by using RS and GIS, *The Indonesian Journal of Geography,* vol. 41, no. 2, pp. 149-170, 2009.
[15] B. Sulistyo, T. Gunawan, Hartono, P. Danoedoro, and N. Listyaningrum, Absolute Accuracy of the Erosion Model of DEM-NDVI and it's Modification, *International Journal of Geoinformatics,* vol. 13, no. 2, pp. 23-34, 2017.
[16] B. Sulistyo, D. Purnama, M. Anggraini, D. Hartono, M. D. Wilopo, U. Wulandari, U., and N. Listyaningrum, Refining The Suitability Modeling of Sea Cucumber (*Holothuria scabra*) Using Fully Raster-Based Data, *Forum Geografi,* vol. 31, no. 2, pp. 119-130, 2018.
[17] M. Aral, *Environmental Modeling and Health Risk Analysis (Act/Risks),* Business Media BV, Springer Science, London, 2010.
[18] Murniati, "Pengaruh Tingkat Kombinasi Kotoran Ayam dan Pasir Sebagai Substrat Terhadap Pertumbuhan dan Kelangsungan Hidup Teripang Pasir (*Holothuria scabra, Jager*) yang Diberi Pakan Buatan," *undergraduate thesis*, Universitas Haluoleo, Kendari, 2001.
[19] Supriharyono, *Pengelolaan Ekosistem Terumbu Karang*, Djambatan, Jakarta, 2000.
[20] H. Weatherley, *Growth and Ecology of Fish Population*, Academic Press, New York, London, 1979.
[21] H. Effendi, *Telaahan Kualitas Air Bagi Pengelolaan Sumberdaya dan Lingkungan Perairan,* Kanisius. Yogyakarta, 2003.
[22] Firstyananda, "Komposisi dan Keanekaragaman Makrozoobenthos di Tiga Aliran Sungai," *undergraduate thesis*, Universitas Airlangga, Surabaya, 2011.
[23] M. I. Effendi, *Biologi Perikanan,* Yayasan Pustaka Utama, 1997.
[24] Jasmadi, Growth and Ecological Aspects of Sandfish Holothurian scabra in Pen Culture Lairngangas Coast, South East Moluccas, *Jurnal Ilmu dan Kelautan Tropis,* vol. 10, no. 2, pp. 317-331, 2018
[25] D.G. Louhenapessy, Pengaruh Substrat Berbeda Terhadap Pertumbuhan Teripang Pasir (*Holothuria scabra*), Triton, *Jurnal Manajemen Sumberdaya Perairan,* vol. 9, no. 1, pp. 26-32, 2013.
[26] R. W. Afrely, M. I. Rosyidi, and S. Fajariyah, Keanekaragaman jenis holothuroidea di Zona Intertidal Pantai Pancur Taman Nasional Alas Purwo, *Jurnal Ilmu Dasar,* vol. 16, no. 1, pp. 23-28, 2015.
[27] K. Sabilu, "Pengaruh kombinasi ampas sagu dan kotoran sapi terhadap pertumbuhan teripang pasir (*Holothuria scabra jager*)," *undergraduate thesis,* Universitas Haluoleo, Kendari, 2002.
[28] R. Komala, Keanekaragaman teripang pada ekosistem lamun dan terumbu karang di Pulau Bira Besar, Kepulauan Seribu, Jakarta, *Jurnal Biodiversitas Indonesia,* vol. 1, no. 2, pp. 222-226, 2015.
[29] H. Yokoyama, Growth and Food Source of The Sea Cucumber Apo stichopus Japonicus Cultured Below Fish Cages Potential for Integrated Multi-Trophic Aquaculture. *Aquaculture,* vol. 38, no. 28, pp. 372-375, 2013.
[30] P.G. Navaro, S.G. Sanz, and J.M. Barrio, Feeding and Movement Patterns of the Sea Cucumber Holothuria sanctori. *Marine Biology,* vol. 16, no. 11, pp. 2957-296, 2013.

[31] L.N. Zamora, and A.G. Jeffs, Feeding, Selection, Digestion and Absorption of the Organic Matter from Mussel Waste by Juveniles of the Deposit-Feeding Sea Cucumber, Australostichopus Mollis, *Aquaculture,* vol. 317, no. 1, pp. 223-228, 2011.

[32] R. Wahab, *Teripang di Perairan Pasarwajo, Pulau Buton, Sulawesi Tenggara*, Pusat penelitian Oceanografi-LIPI. Jakarta, 2001.

[33] J.N. Martoyo, Aji, and T.J. M. Winanto, *Seri Agribisnis: Budidaya Teripang,* Jakarta: Penebar Swadaya, Surabaya, 2007.

[34] J.N. Martoyo, Aji, and T.J. M. Winanto, *Budidaya Teripang,* Penebar Swadaya, Jakarta, 2002.

IOP Conf. Series: Materials Science and Engineering **874** (2020) 012036 doi:10.1088/1757-899X/874/1/012036

Identification of Visual and Thermal Comfort in Classroom

Debby Seftyarizki*, Abdul Hamid Hakim, Makmun Reza Razali

Universitas Bengkulu, Bengkulu, Indonesia

* debby.seftyarizki@unib.ac.id

Abstract. As a learning facility which has interaction inside it, a classroom should be comfort in term of visual and thermal. Visual comfort can be obtained by the presence of sunlight through opening. While thermal comfort is the air temperature that influenced by building material used. Room 21 (R.21) of Gedung Kuliah Bersama V (GB V) Universitas Bengkulu (UNIB) is a classroom that mostly used by Architecture Student for receiving lecture from morning until afternoon. To identify the visual and thermal comfort of R.21, computer simulations with Ecotect program was did to get the room lighting and temperature condition in level 60cm above the floor. The result shows that natural light in R.21 GB V UNIB is comfortable enough with 630 lux of average daylight levels. While the temperature in R.21 GB V UNIB is still not comfortable because the average mean radiant temperature is 26,6°C. Visual comfort obtained along with high temperature because R.21 GB V UNIB is located at the corner of the building that have glass window in the east and north of the room. **Keywords**: Visual Comfort, Thermal Comfort, Classroom

1. Introduction

Energy efficiency already became a huge issue in architectural field. According to BPPT, most of energy used in a building is for lighting and air conditioning [1]. This energy is used to obtain the comfortable when doing indoor activity. One of the methods to pursue the efficiency of energy is by using natural resource. Utilize natural lighting and maintaining room temperature through building material also became one of the characteristics of green building according to Green Building Council Indonesia. But, one of the problem by using natural resource is create another question, 'is it comfortable enough?'.

Classroom is a face to face place for transferring knowledge. Since there's an interaction inside it, a classroom should be comfort in term of visual and thermal. Another activity inside a classroom is reading and writing from morning until afternoon. Therefore, the sufficient amount of lighting and appropriate condition of room temperature plays an important factor to obtain the comfortable inside the room. Natural lighting can be obtained by the presence of natural light, while one of the factors that influenced thermal of the room is air temperature. Natural lighting inside the room is mostly coming from the opening or window at the wall. But sometimes, utilization of natural light will also bring the heat along because transparent material can transfer bigger heat into the building. That's why building envelope material also plays a role to transfer the heat from outdoor to inside of the room [2].

The presence of natural lighting through opening can make an interaction between indoor and outdoor that improves the quality of the room. As an example, the renovation of Asian Development

Bank Headquarter building in Manila to bring in the natural light can increase employee productivity [3]. A good quality of the lighting inside the room makes the building user see clearly and comfortable. There're also several research about lighting of classroom, such as measurement of natural light intencity in elementary school with the data of room dimention, floor, ceiling, and wall color, room orentation, opening material and dimention, floor elevation, daylight condition, and date and time simulation using statisic software [4]. Another classroom lighting intencity analysis also used computer simulation to get the daylight factor of architecture leture room with room orientation, room dimension, opening material and dimention, and outdoor components as the variabels [5], and building performance of natural lighting, natural ventilation and room temperature using Ecotect simulation software [6]. Room dimention, building envelope material, and building orientation give an impact to the result of their study. Most of the result study showed that the lighting inside the room is under the standard. It happens because the window is only in one side, east or west opening only on spacious and deep room.

With several facts from studies mentioned above, visual and thermal comfort of classroom in Universitas Bengkulu (UNIB) is need to be identified. There's still no study about visual and thermal comfort of the classroom in Gedung Kuliah Bersama V UNIB. Identification did by seeing the distribution of room temperature and lighting occurred. Because there'll be inequality of room heat and light inside the room, for example, there'll be an area that received too much light or heat while others is received less light or heat. According to SNI 03-2396-2001, a good quality of natural lighting inside a room is light level minimum should be present not only in the area near the window but also on the whole of room, and there's shouldn't be a contras lighting between an area that received too much light or received less light because it can disturbing the vision [7]. According to SNI 03-6572-2001, thermal comfort can be obtain by distribute air temperature evenly. So, this study will try to look the light and temperature in detail on every corner of the room [8]. The final purpose of this study is to known the effectiveness of the classroom as a learning activity so that the learning facility can be more optimize.

2. Methodology

Identification of visual and thermal comfort of classroom in GB V UNIB did by computer software simulation Ecotect. Simulation did to known the room light and temperature condition. Simulation result will show the image that represents the room light and temperature distribution. Qualitative description did to explain the result of simulation. Room 21 (R.21) of Gedung Kuliah Bersama V (GB V) Universitas Bengkulu (UNIB) was chosen as the object study because this classroom is mostly used by Architecture Student for receiving lecture from morning until afternoon. R.21 is located at the corner of the building and faced 5 degrees to north-east. It makes R.21 has 8 windows faced to east and 4 windows faced to north. There's also glass door and small window at the west side of the room face to indoor corridor. R.21 has 8m in wide and 6m in length with wall brick, concrete floor and ceiling, and clear glass and aluminum frame for windows and doors. Simulation software has limitation in providing climate data, whereas no climate data for Bengkulu area. The nearest country that has complete climate data in Ecotect is Malaysia. Therefore, Malaysia climate data are being used is for this simulation along with Bengkulu coordinate. Thermal simulation time did on middle of April and September at 9am, 12pm, and 3pm. It's assumed as the class schedule for first-half academic year and second-half academic year. April is chosen as simulation time because this month is the representation of dry season. Meanwhile September is the representation of rainy season for simulation time. These simulation time considerations are based on SNI 03-6389-2000 about sun radiation data for building energy conservation [9]. Since there's no other building around GB V, obstacle and another external factor are being ignored.

(a) (b) (c)

Figure 1. (a) GB V location, (b) R.21 location, (c) R.21 condition

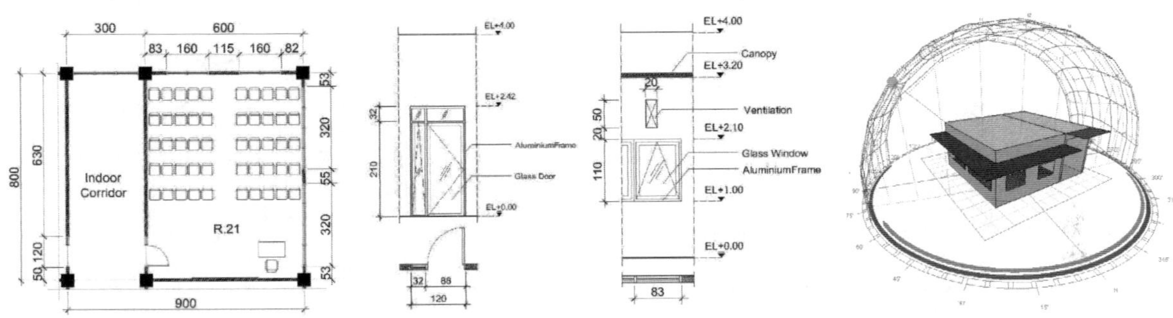

(a) (b) (c)

Figure 2. (a) R.21 layout plan, (b) door & windows detail, (c) Ecotect simulation condition

Several variables also applied for simulation. Indonesia has 10.000 lux for sky illuminance [7], with minimum standard illuminance for classroom is 300-750 lux [10]. Besides light levels, daylight factor also been used to determine the natural light quality inside the room. Day light factor is the ratio of light level inside a building to the light level outside the building, with standard minimum 2% for office room [11]. Thermal Comfort Prediction (PTN) standard for tropical country is more suitable with 0,749 of mean temperature of that place, plus 5,953, because every region in Indonesia has different climate condition so that the thermal comfort for very people on each region is different [12]. Since Bengkulu has 26,75^{O}C mean temperature and 77,8% annual humidity (climate-data.org), 26 ^{O}C will be used as a maximum standard of temperature.

3. Result and Discussion

Discussion of the study did by description of simulation image that indicate by the colors. Image shown is the room plan at level 60cm above the floor that assume as working area. Visual comfort will be identified by explaining the light levels of the room in lux and light factor in percent. Thermal will be identified by explaining the temperature of the room in degree celsius.

A. Visual Comfort

Ecotect simulation for daylight analysis showed that the average value of annual natural light for R.21 GB V UNIB is 630 lux and 6,3% of daylight factor. This annual light is already met the minimum standard of lighting inside a classroom and can be said a well-lit so artificial lighting is not required generally. The highest light in the room is 2000 lux located in A room-corner (indicated by yellow color in figure 3.a) where a glass door faced to indoor corridor exist. It's somehow strange and maybe there's an incorrect data reading in simulation. But, since A corner area is not an important area in the room, so it can be ignored. Another area that received high light is in C corner, which is around 1.000 lux. Most of area in BC side has 800 lux of daylight and 8% of daylight factor. This light level is good enough. But, activity in this area should be careful because can make an uncomfortable condition such as glare when the student doing writing/drawing task on the white paper. Light level at the center of the room is mostly dominate by 550 lux and 4,5% of daylight factor. This area, where mostly students sit and doing activity inside the room, is comfortable enough and can be said effective for learning activity. While AD side area, mostly has 500 lux of daylight and 5% of daylight factor. Even though this area is received the lowest light among the room, it's still meet the minimum standard of light needed in classroom.

Overall, R.21 GB V UNIB can be said that comfortable enough in term of visual. Around 1/4 – 1/3 area of the room is met the minimum standard of classroom lighting. It can be happened because there're a lot of windows in east and north wall, with ratio 1: 4,5 of the room square area. But, one thing to be careful about is the excessive light inside the room because it can lessen the effectiveness of learning activity in classroom. This problem can be solved by proper shading design.

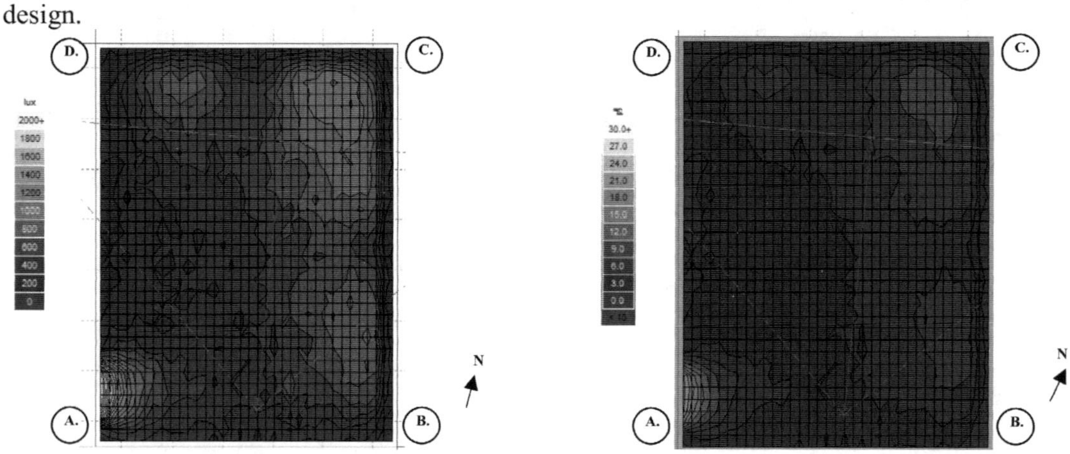

(a) (b)

Figure 3. Daylight Simulation of room plan at el+60cm (a) Daylight level, (b) Daylight factor

B. Thermal Comfort

Ecotect simulation for thermal analysis showed that the average value of mean radiant temperature for R.21 GB V UNIB in April at 9.00 am is 27,65°C, in April at 12.00 pm is 29°C, in April at 3.00 pm is 28,7°C, and in September at 9.00 am is 23,43°C, in September at 12.00 pm is 25,25°C, in September at 3.00 pm is 24,99°C. It can be seen that mean radiant temperature for R.21 GB V UNIB in April is quite high with average temperature 28,45°C for whole day when this class is used for learning activity. While on September, mean radiant temperature for R.21 GB V UNIB is quite low with average temperature 24,56°C. So, average temperature of R.21 GB V UNIB annually is 26,6°C. This result makes R.21 GB V UNIB is not comfortable enough because exceed the Thermal Comfort Prediction (PTN) standard for Bengkulu area. R.21 GB V UNIB received highest heat at midday (12.00 pm) either in April or September. While BC side is the most uncomfortable area because receive the highest heat for whole day. It happened because BC side wall has numbered of glass windows that faced to the east. The highest heat in BC side is 29,3°C on April and 25,7°C on September. So, average temperature in BC side annually is 27,5°C. Mean radiant temperature at the center of the room is mostly dominated by 28°C on April and 24,5°C on September. Average temperature at the center of the room annually is 26,25°C, which is still exceeded the standard. D corner is an area that received lowest heat in the room. Temperature at D corner is 27,3°C on April and 23°C on September. So, average temperature in BC side annually is 25, 15°C, which is comfortable enough. Even though there's several glass windows in CD side wall. But, since this wall faced to the north, it doesn't received so much heat.

Table 1. Mean radiant temperature simulation result

Month	Hour	Temperature
April	9.00 am	27,65°C
April	12.00 pm	29 °C
April	3.00 pm	28,7 °C
September	9.00 am	23,43 °C
September	12.00 pm	25,25°C
September	3.00 pm	24,99 °C

Overall, R.21 GB V UNIB can be said that not comfortable in term of thermal. It can be happened because there're numbered glass windows in east wall that bring a lot of heat for a whole day. To overcome the excessive heat, utilization of natural airflow can give cool effect for building user. Besides that, a proper design of shading or skin façade also can help reducing the heat transfer from outside to inside of the room.

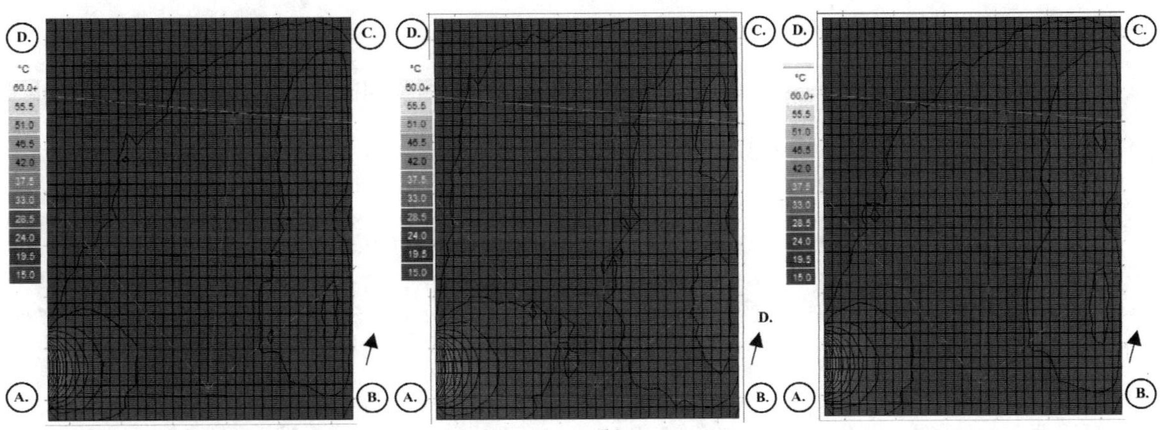

(a) (b)

Figure 4. Thermal Simulation of room plan at el+60cm in April. (a) Mean radiant temperature at 9.00 am, (b) Mean radiant temperature at 12.00 pm, (c) Mean radiant temperature at 3.00 pm

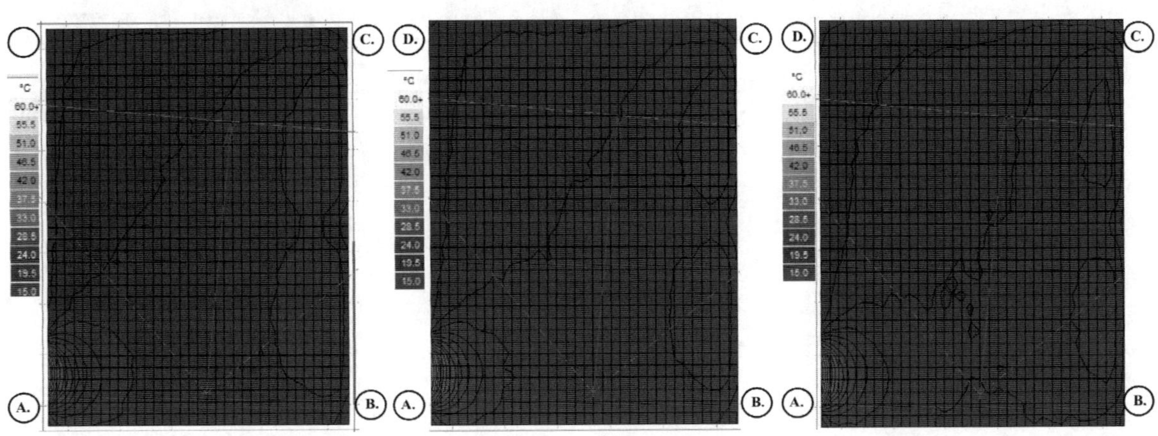

(a) (b)

Figure 5. Thermal Simulation of room plan at el+60cm in September. (a) Mean radiant temperature at 9.00 am, (b) Mean radiant temperature at 12.00 pm, (c) Mean radiant temperature at 3.00 pm

4. Conclusion

From the simulation did above, can be concluded that:

- The presence of numbered glass windows in east wall of R.21 GB V UNIB bring in a lot of light along with heat.
- R.21 GB V UNIB is still not effective enough as a learning facility because the excessive heat inside the room is not comfortable enough in term of thermal. Moreover, although the lighting inside the room is already met the minimum standard, the excessive light inside the room should be careful because can make an uncomfortable condition in term of visual such as glare.

Some recommendation to optimize the comfortable of R.21 GB V UNIB is utilization of natural ventilation to overcome the excessive heat and a proper design of shading or skin façade to reducing the heat transfer and excessive light. This recommendation can be a further research to optimize the effectiveness of R.21 GB V UNIB and another classroom as a learning facility.

References

[1] BPPT, "Perencanaan Efisiensi dan Elastisitas Energi," *Badan Pengkajian dan Penerapan Teknologi,* Jakarta, 2012.

[2] Satwiko, P., "Fisika Bangunan," *Penerbit Andi*, Yogyakarta, 2008.

[3] Kishnani, N., "Greening Asia," *BCI Asia*, Singapore, 2012.

[4] Idrus, I., Hamzah, B., & Mulyadi, R., "Intensitas Pencahayaan Alami Ruang Kelas Sekolah,". *Simposium Nasional RAPI XV – 2016 FT UMS*, pp. 473-479, 2016.

[5] Lisa, N. P., & Nurhaiza., "Analisis Itensitas Pencahayaan Alami pada Ruang Kuliah Prodi Arsitektur Universitas Malikussaleh," *ProsidingTemuIlmiah IPLBI*, pp. 61-66, 2017.

[6] Prasetyo, Y. H., "Kinerja Bangunan Desain Pasif Berdasarkan Simulasi Ecotet Dan Pengukuran Lapangan Studi Kasus : Bangunan Konvensi "Grha Wiksa Praniti" Bandung," *Jurnal Permukiman Vol. 9 No. 1 April*, pp. 41-53, 2014.

[7] Badan Standar Nasional, SNI 03-2396-2001, "Tata Cara Perancangan Sistem Pencahayaan Alami Pada Bangunan Gedung," *Standar Nasional Indonesia*, 2001.

[8] Badan Standar Nasional, SNI 03-6572-2001, "Tata Cara Perancangan Sistem Ventilasi dan Pengkondisian Udara Pada Bangunan Gedung," *Standar Nasional Indonesia*, 2001.

[9] Badan Standar Nasional, SNI 03-6389-2000, "Konservasi Energi Selubung Bangunan Pada Bangunan Gedung," *Standar Nasional Indonesia*, 2000.

[10] Frick, H., Ardiyanto, A., & Darmawan, A., *Ilmu Fisika Bangunan* (PT Kansius Yogyakarta, Yogyakarta, 2008).

[11] Mangkuto, Rizki. A., "Pemodelan dan Simulasi Pencahayaan Alami dalam bangunan di Indonesia," *ITB Press*, Bandung, 2019.

[12] Karyono, H. T., "Arsitektur Tropis," *Penerbit Erlangga*, Jakarta, 2016.

[13] Mangkuto, Rizki. A., "Pemodelan dan Simulasi Pencahayaan Alami dalam bangunan di Indonesia," *ITB Press*, Bandung, 2019.

IOP Conf. Series: Materials Science and Engineering **874** (2020) 012037 doi:10.1088/1757-899X/874/1/012037

Simplified fuzzy logic and arduino-based controller for maximum power tracking in solar panel

Suherman, Mufrihul Fajri, Raja Harahap

Electrical Engineering Department, Universitas Sumatera Utara, Medan-Indonesia

*suherman@usu.ac.id

Abstract. Maximum power tracking for solar panel has been explored since long time ago for increasing solar panel output. Various methods have been implemented from combinational based algorithm to intelligent system, including the use of fuzzy logic. This paper demonstrates the implementation of fuzzy logic on the arduino microcontroller system. A maximum power tracking prototype is developed, and experimental assessments are conducted. Arduino uno is used for the microcontroller, small scale mechanic is developed driven by two electric motors, and 13V solar cell is used as the object. The assessments show that fuzzy logic algorithm is successfully implemented in arduino, the solar panel output is enhanced up to x% and the solar panel efficiency is up to x%.

1. Introduction

Solar panel generates the electrical current by converting photon from sunlight into moveable electrons within silicon based material. Sunlight is the most promising energy source as it is sustainable and available whole the year. Solar panel requires materials that catch sunlight, enable to transfer this energy to alleviate electron moving through its band gap to a conductive band. Positive-negative (PN) junction is generally employed to enable electron excitation. Study mentioned that the possible converted energy from sunlight achieves 7.45 x 10^{17} kWh [1].The solar panel system has been used worldwide to replace the conventional energy source. Solar panel is expected to be the main energy source in the future. However, challenges should be dealt with as solar panel efficiency is still low [2].

Existing work have been performed to enhance solar panel efficiency, including material enhancement [3], cooling system implementation [4], the maximum power tracking method [5] and DC to AC conversion [6]. Solar cells that are made of methylammonium lead triiodide (MAPbI3) exerted efficiencies greater than 20% [3]. But the stability is not well constructed for long term. Some issues including the chemical stability, thermal stability, and current–voltage hysteresis are in research interest. Meanwhile, active and passive cooling systems are choices to compensate the surface temperature increment in solar panel [4]. Active cooling system applied air flow and/or water sprinkle on surface or at the back of solar panel. Active needs electrical energy to drive pump or fan. Passive cooling system decreased surface temperature by using heat-sink or water to absorb the heat. Extracting solar panel maximum power is the most interesting research in solar PV system [5]. Maximum power point tracking (MPPT) is the key for efficient solar cell system. MPPT can be performed either by using timeworn conventional methods or soft computing algorithms. The method performance varies in terms of tracking speed, complexity, and partial shading performance. In addition, MPPT is also used for other renewable energy sources, such as wind turbine [7]. Since solar cell output is direct current (DC), inverter is needed to convert DC to AC. The DC-AC waveform conversion is often costly, poor quality and harmonic contained [6]. Therefore, inverter research to support solar cell efficiency is also important.

Since many soft computing methods have been explored by many researchers, this paper focuses on reporting how fuzzy logic implemented in arduino for MPPT application. This paper gives an alternative guide besides reports that are provided by some papers [8-10]. Fuzzy logic components include fuzzification interface, knowledge base, inference mechanism and defuzzification interface. Those components objective is to solve control problem systematically. However, in some cases, it can be simplified to some multiple IF THEN ELSE operations.

2. Methodologies

In order to implement fuzzy logic for MPPT, hardware is designed. Hardware consists of four light dependant resistors (LDRs), voltage sensors, LCD display and an arduino nano as the controller. Circuit is shown in Figure 1. Four light sensors; LDRs act as voltage divider, are located in the East and West right and left panel movement, in the North and South for up and down panel movement. LDR in series to 7500 ohm resistor produces voltage decrement as sunlight stronger. This sensing signal is sent to arduino nano through analog pin. Sensor values are digitized by the internal analog to digital converter (ADC) as inputs for fuzzification using fuzzy logic rules to determine which servomotor responses. LDR is separated from one another so that it receives light only from direct exposure to sun. Program occurrence is reported to LCD display. The solar panel prototype has a 13 Volt peak voltage.

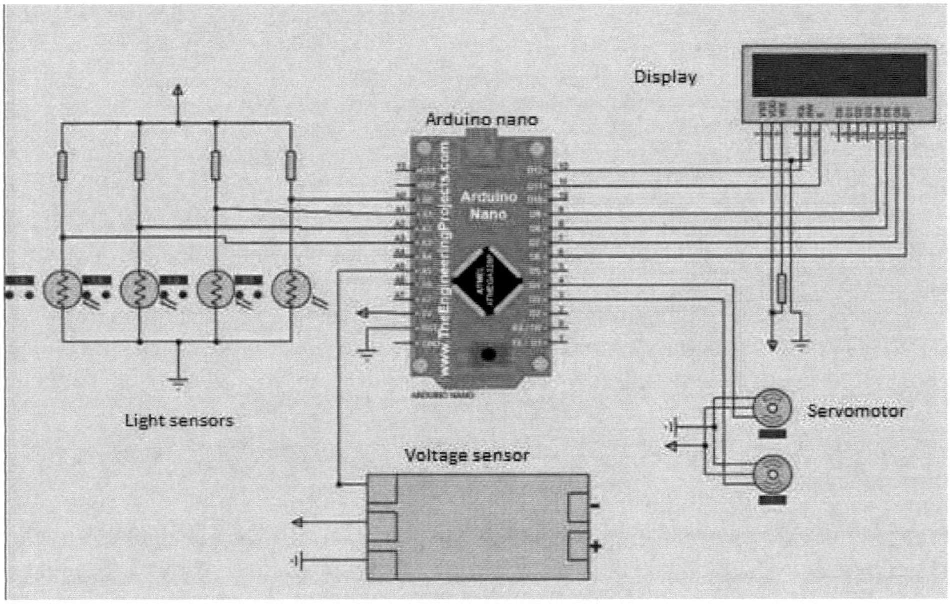

Figure 1. MPPT circuit

Table 1 shows the servomotor function that moves panel up and down with vertical mechanic and moves right and left with horizontal mechanic. Servomotor size can be enlarged depending on the size of solar panel. Additional motor driver is required when larger motor size installed. In defuzzification stage, vertical and horizontal motor moves step by step by one degree. Figure 2 shows software design for horizontal motor. The vertical motor uses the same design using North and South sensors. Arduino nano code is shown in the following page.

Table 1. Servomotor and sensor positions

Servomotor	Sensors	Function
Vertical	North and South	Moving panel up and down
Horizontal	East West	Moving panel right and left

IOP Conf. Series: Materials Science and Engineering **874** (2020) 012037 doi:10.1088/1757-899X/874/1/012037

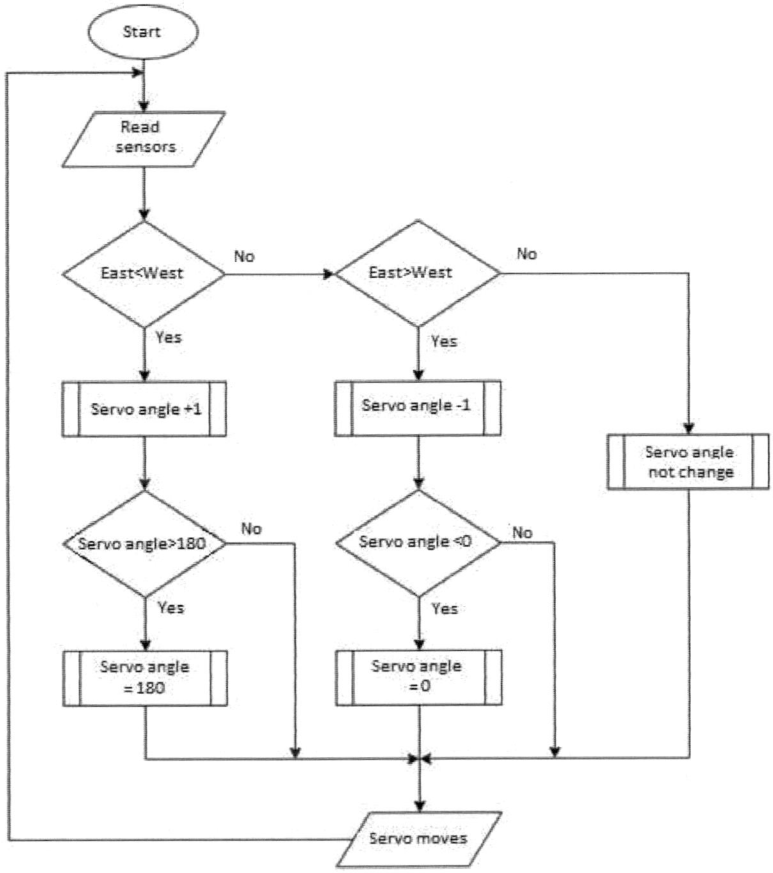

Figure 2. Software design

```
#include <Servo.h>
#include <LiquidCrystal.h>
#define voltage_input A5
LiquidCrystal lcd(12, 11, 9, 8, 7, 6);
Servo horizontal; Servo vertical; int servoh = 90; Servo vertical; int servov = 90; int time_ = 10; int tol = 1;

void setup() {
  lcd.begin(16, 2); lcd.setCursor(0,0); lcd.print("=====START====="); delay(1000); lcd.clear(); Serial.begin(9600);
  horizontal.attach(4); vertical.attach(3); pinMode(A0,INPUT); pinMode(A1,INPUT); pinMode(A2,INPUT);
  pinMode(A3,INPUT); pinMode(A5,INPUT);
}

void loop() {
  int down = (analogRead(A3)/tol); int up = (analogRead(A2)/tol); int right = (analogRead(A1)/tol); int left = (analogRead(A0)/tol);
  Serial.print("\nright="); Serial.print(right); Serial.print("        left="); Serial.print(left); Serial.print("              up=");
  Serial.print(down); Serial.print("                  down="); Serial.print(up);
  if (up < down){ servov = ++servov; if (servov > 180){ servov = 180;}
  }else if (up > down){servov= --servov; if (servov < 0){ servov = 0;}
  }else if (up = down){}
  vertical.write(servov);

  if (left < right){servoh = --servoh; if (servoh < 0){servoh = 0;}
  }else if (left > right){servoh = ++servoh; if (servoh > 180){servoh = 180;}
  }else if (left = right){}
  horizontal.write(servoh);
  delay(time_);
  read();
}

void read(){
 int read = analogRead(voltage_input);  float voltage = read * (5.0 / 1024.0) * 4.6;
 Serial.print("\nvoltage    "); Serial.println(voltage,3); Serial.print("A5 = ");Serial.print(read); Serial.println(" ");
 lcd.clear(); lcd.setCursor(0,0); lcd.print(" VOLTAGE :    "); lcd.setCursor(6,1); lcd.print(voltage);
 delay(200);
}
```

IOP Conf. Series: Materials Science and Engineering **874** (2020) 012037 doi:10.1088/1757-899X/874/1/012037

3. Assessment results

Figure 3 shows the MPPT impact to output voltage, current and power of the solar panel, compared to static solar panel. Increments on those parameters are obvious. Voltages, current and power are consistently increased in every hour, especially in the morning and afternoon. This is caused by static panel put horizontally so that output increments between 11.30 to 13.30 at noon are less that morning and afternoon.

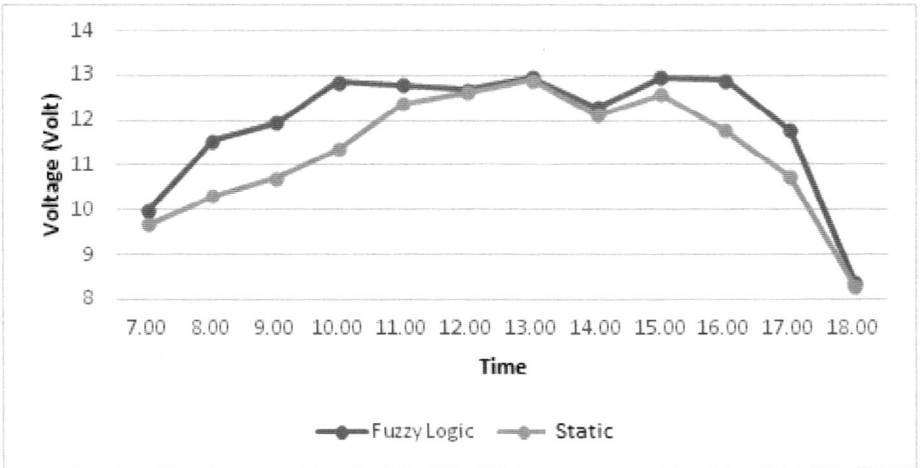

(a) Voltage

(b) Current

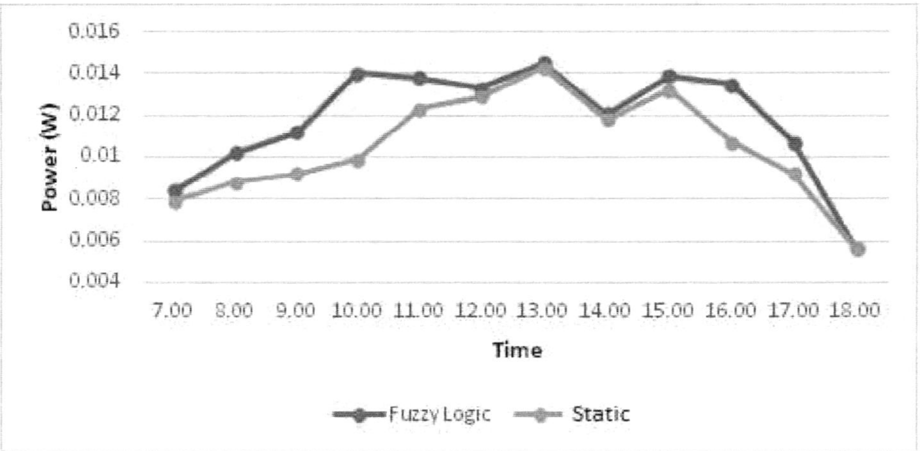

(c) Power

Figure 3. Fuzzy logic-based MPPT versus static solar panel

Voltage increases up to 1.5 V, current increases up to 0.21 mA and power increases up to 4.1 mW. In percentage, compared to static panel, voltage, current and power increments achieve 5.65%, 6.10% and 12.24% subsequently. In this simple fuzzy implementation, motor moves one degree in each program looping, so that the time required to set MPPT is quite tremendous, more than 10 s (Figure 4). This duration can be suppressed, by adding more degree for each movement. However, MPPT may not be optimal.

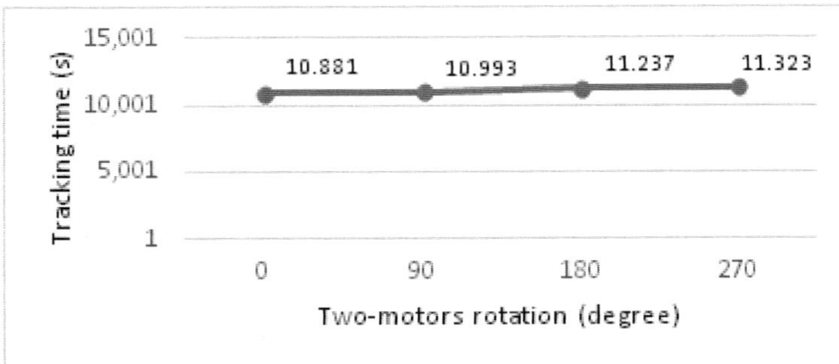

Figure 4. Tracking time

4. Conclusions

This paper has discussed the implementation of fuzzy logic for MPPT of solar panel. Hardware requires at least 4 light sensors located separately in the East, West, North and South. Fuzzyfication and defuzzification processes are simply denoted by using IF THEN ELSE rules to determine servo motor movements. Assessment shows that voltage, current and power increments achieve 5.65%, 6.10% and 12.24% subsequently compared to static panel. Future work should consider more detail fuzzification and defuzzification processes to achieve optimal results for bigger solar panel.

References

[1] Chopra, K. L., & Das, S. R. (1983). Why thin film solar cells?. In Thin film solar cells (pp. 1-18). Springer, Boston, MA.

[2] Suherman, S., Sunarno, A., Hasan, S., & Harahap, R. (2020). Water and Heat-sink Cooling System for Increasing the Solar Cell Performances. EAI Endorsed Transactions on Energy Web, 7(26).

[3] Ming, W., Yang, D., Li, T., Zhang, L., & Du, M. H. (2018). Formation and diffusion of metal impurities in perovskite solar cell material CH3NH3PbI3: implications on solar cell degradation and choice of electrode. Advanced Science, 5(2), 1700662.

[4] Oropeza-Perez, I., Østergaa rd, P. A. (2018) Active and passive cooling methods for dwellings: A review. Renewable and Sustainable Energy Reviews, 82, 531-544.

[5] Ram, J. P., Babu, T. S., & Rajasekar, N. (2017). A comprehensive review on solar PV maximum power point tracking techniques. Renewable and Sustainable Energy Reviews, 67, 826-847.

[6] Hutabarat, M. A., Hasan, S., Rambe, A. H. (2020). Design and simulation hybrid filter for 17 level multilevel inverter. Bulletin of Electrical Engineering and Informatics, 9(3), 886-897.

[7] Ardiles, R., Hasan, S., Pane, Z. (2019, November). Maximizing Power Tracking of an Offshore Wind Turbine Using Perturb and Observe Method. In Journal of Physics: Conference Series (Vol. 1373, No. 1, p. 012008). IOP Publishing.

[8] Fannakh, M., Elhafyani, M. L., & Zouggar, S. (2018). Hardware implementation of the fuzzy logic MPPT in an Arduino card using a Simulink support package for PV application. IET Renewable Power Generation, 13(3), 510-518.

[9] Bhattacharya, T., & Sarkar, P. (2018, March). ARDUINO UNO based packed U cell inverter for photovoltaic application. In 2018 International Symposium on Devices, Circuits and Systems (ISDCS) (pp. 1-6). IEEE.

[10] Ammar, A. G., Mohammed, A. S., Sarab, A. C., Craciunescu, A., & FADEL, M. A. (2018, October). Experimental installation of photovoltaic MPPT controller using Arduino board. In 2018 International Conference on Applied and Theoretical Electricity (ICATE) (pp. 1-6). IEEE.

IOP Conf. Series: Materials Science and Engineering **874** (2020) 012038 doi:10.1088/1757-899X/874/1/012038

Frequency selection for ground monitoring using propagation characteristics

Suherman, Ali Hanafiah, Naemah Mubarakah, Erwin Wijayanto

Electrical Engineering Department, Universitas Sumatera Utara, Medan-Indonesia

*suherman@usu.ac.id

Abstract. Ground monitoring can be used for observing ground structure changes or other specific objectives. Rather than using expensive ground penetration radar, the ground propagation characteristics may be able to describe the structure changes. Ground propagation can be evaluated by using a simple radio transmitter and receiver. This paper employs a vector network analyzer (VNA) to explore ground propagation characteristics. As the available VNA works from 500 MHz to 1 Ghz, the paper proposes how to select suitable frequency for ground monitoring. Tangential comparison between mathematical model and direct measurement is proposed. The result shows that the selected frequency exerts promising propagation characteristic even dough signal diffraction causes irregular signal strength at the receiver.

1. Introduction

Underground monitoring is useful to observe ground structure even dough object detection or finding is the most frequent applications [1]. The ground structure was initially conducted by the electromagnetic induction that is able to detect metallic materials presence in the ground [2]. Electromagnetic induction makes use radiated electromagnetic field and senses the remaining field left over on the detected object. Ground penetration radar or GPR [3] is the most popular method for observing underground object by transmitting radio signal into the object and receiving the reflected signal to determine object pattern. Ground penetration radar is like other surface radar but uses higher power as signal penetrates to a more dense material than of the surface radar. Time domain reflectometer (TDR) [4] and capacitance measurement [5] can also be use to explore ground content.

As GPR is intended to explore object shape rather than ground structure changes, GPR is too expensive for ground monitoring [6]. This paper proposes ground monitoring using a separated radio transmitter and receiver. Although the same radio schema has been used for through the earth (TTE) radio, which is for communication purpose [7], this paper focuses the usage for structure monitoring. Some extended radio facilities such as noise cancelation [8] are not discussed. Instead, the study uses the existing vector network analyzer. This paper presents the initial step which is frequency selection techniques based on tangential comparison between mathematical model and measurement results.

2. Research Method

In order to select the best frequency for monitoring ground structure, mathematical model is employed and direct measurement was performed. The approximated mathematical model for underground propagation uses Friis equation as plotted in Equation 1 [9]:

$$P_r(dBm) = P_t(dBm) + G_r(dB) + G_t(dB) - L_0(dB) \quad (1)$$

The Lo denotes propagation loss which varies to distance (d in km) and frequency (f in MHz).

$$L_0(dB) = 32.4 + 20\log(d) + 20\log(f) \quad (2)$$

Additional factor (L_p) is added to correct the Friis equation which is mainly used for surface propagation. The received power is approximated by Equation 3.

$$P_r = P_t + G_r + G_t - L_p \quad (3)$$

L_p comprises of space loss $L_0$and ground loss L_s. L_s determines losses based on attenuation constat α (1/m)and phase shift, β (radian/m). The corrected loss is expressed in Equation 4 [9], while value α and β are given by Equation 5 and 6 [9]. Magnetic permeability μ, and permittivity ℇ are selected as in [10].

$$Ls = 8.69\alpha d + 154 - 20\log(f)(Hz) + 20\log(\beta) \quad (4)$$

$$\alpha = \omega\sqrt{\frac{\mu\varepsilon'}{2}[\sqrt{1+\left(\frac{\varepsilon'}{\varepsilon'\prime}\right)^2} - 1} \quad (5)$$

$$\beta = \omega\sqrt{\frac{\mu\varepsilon'}{2}[\sqrt{1+\left(\frac{\varepsilon'}{\varepsilon'\prime}\right)^2} + 1} \quad (6)$$

Measurement was conducted by using scenario plotted in Figure 1. Depth and distance are adjusted to the limit of the employed VNA (Figure 2). Frequency selection is also based on VNA capability. Object size is a box made of metal sheet with size 20 cm x 20 cm x 20 cm, located at the middle measurement path.

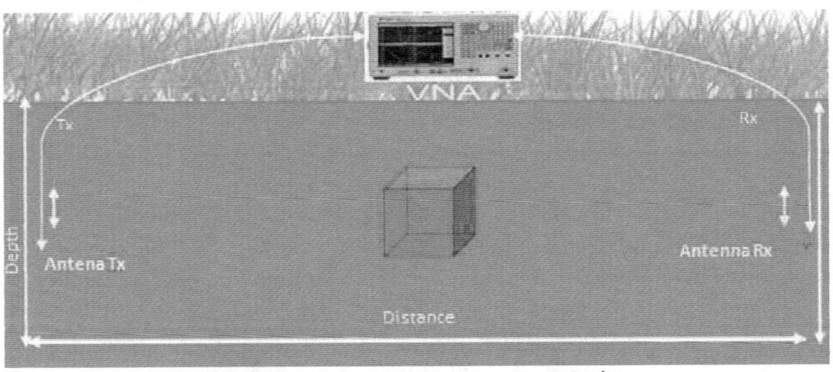

Figure 1. The measurement scenario

The proposed frequency selection is by finding the linier trend of mathematical model and the measurement result. Tangential values were extracted from the linear trend. Since negative tangent is expected, only negative tangential values from both mathematical model and results determine the selected frequency.

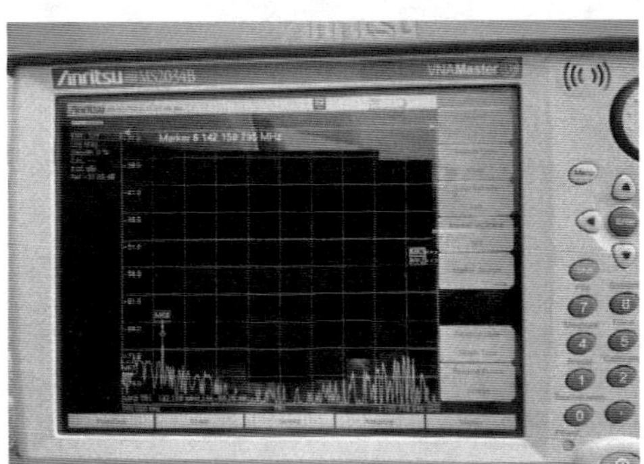

Figure 2. VNA for measurement

3. Modelling and measurement results

Figure 3a shows the loss and the received signal patterns generated by the mathematical model. Loss in lower frequency band increases to distance; while in higher frequencies tend to be flat. By using normalized graph, the received signal is inversely proportional to losses as plotted in Figure 3b.

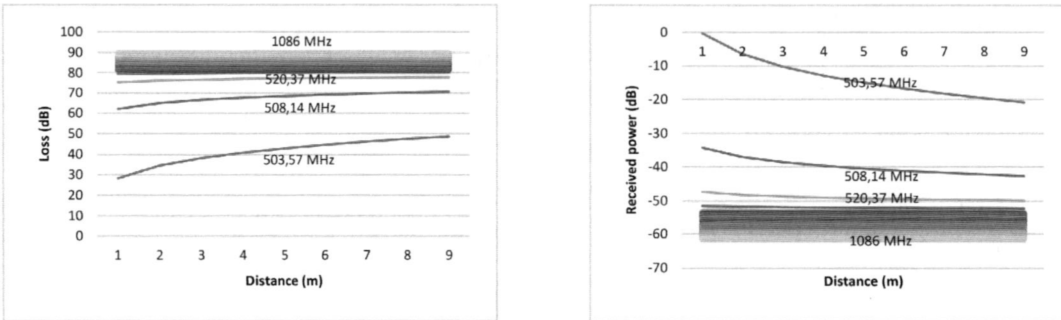

(a) (b)

Figure 3. Propagation loss and received signal generated by mathematical model

Figure 4a shows the sample of measurement results. Amplitudes of received signals vary to working frequency. Frequency band lower than 600 MHz gets the best signal strength followed by band higher than 1 GHz. This variation is influenced by underground scattering. The samples of received signal to distance is shown in Figure 4b. Unlike surface propagation, signal decrease inconsistently to distance.

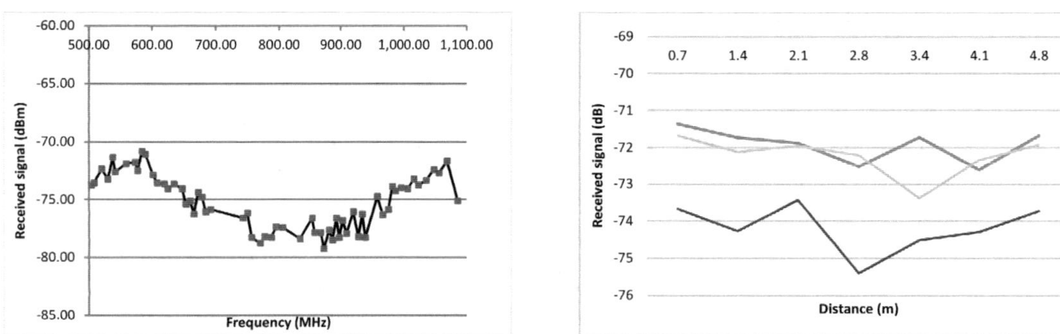

(a) (b)

Figure 4. Received signal level to frequency and distance

The comparison of tangential values between mathematical model and measurement is shown in Figure 5. Tangential value of the model changes from minus to zero, while measured tangential values changes irregularly.

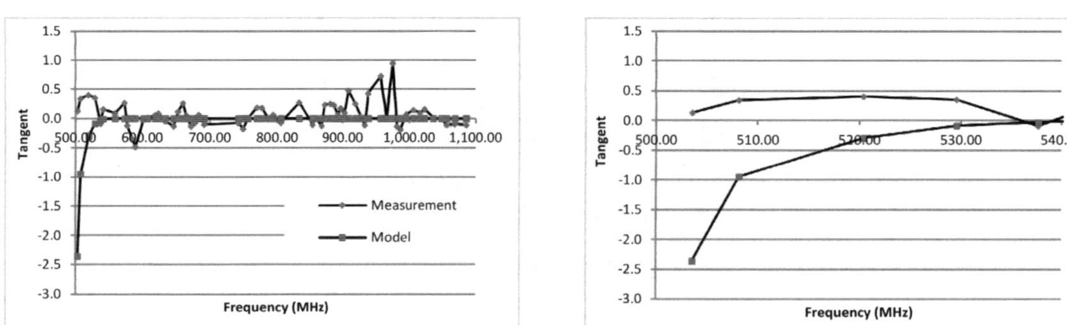

(a) (b)

Figure 5. Tangential based selection

Since both tangential values for the selected frequency should be in negative region, frequency of 537.69 MHz is selected as best signal. Figure 6 shows the received signal sample of frequency 537.69 MHz that changes to distance. This frequency will be used for the ground monitoring.

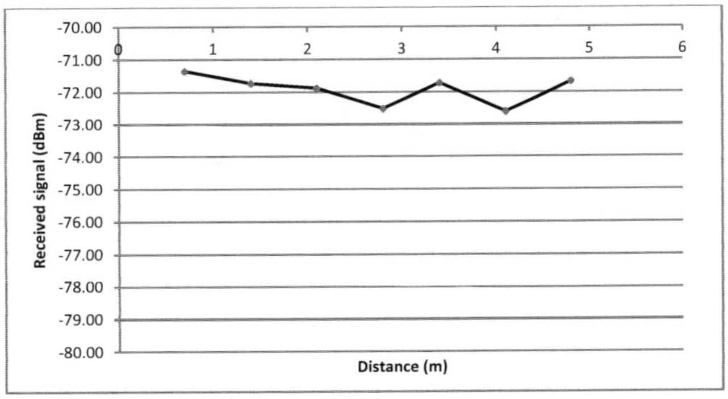

Figure 6. Received signal pattern of the selected frequency

4. Conclusions

This paper has presented frequency selection method for underground monitoring system based on propagation characteristics. The proposed method made use of the tangential value comparison between mathematical model and direct measurement. As VNA used in this experiment has frequency range of 500MHz to 1 GHz, the selected frequency is 537.69 MHz. The selected frequency has decreasing trend of signal strength to distance. Future work will discuss the use of this frequency for underground monitoring.

Acknowledgement

This research has been supported by DRPM DIKTI under the scheme of Penelitian Dasar Unggulan Perguruan Tinggi research grant 2019.

References

[1] Conyers L B. 2016. Interpreting ground-penetrating radar for archaeology", Routledge.
[2] Yongbo W and ChenT. 2018. Deep buried metal pipeline detection using frequency domain electromagnetic method. In AGU Fall Meeting.
[3] Katerina Z and Tomas C. 2019. Application of ground penetrating radar methods in soil studies: A review. Geoderma (343) 116-129.
[4] Burkes K W, Elham B M, and Ramtin H. 2015. Water tree detection in underground cables using time domain reflectometry. IEEE Power and Energy Technology Systems Journal (2) 2 53-62.
[5] Liu Y H, Yunzhi H, Rui T, and Beibei W. 2015. Application of interdigital capacitive sensors for detecting power cable insulation damage. IEEE International Conference on Mechatronics and Automation (ICMA), pp. 1795-1799.
[6] Torrione P A, Morton K D, Sakaguchi R and Collins L M, 2014. Histograms of oriented gradients for landmine detection in ground-penetrating radar data. IEEE Transactions on Geoscience and Remote Sensing, 52(3) pp.1539-1550.
[7] Carreño J, Silva L, Neves S, Aguayo L, Braga A J, Barreto A N and Garcia L U. 2016. Through-The-Earth (TTE) Communications for Underground Mines. Journal of Communication and Information Systems, 31(1).
[8] Ralchenko M, Svilans M, Claire S, and Roper M. 2010. Finite-difference time-domain modelling of through-the-Earth radio signal propagation. Computers & Geosciences 85: 184-195.
[9] G. S. et al. 2010. To the rescue!. World Coal 12.
[10] Ian F A. 2009. Signal propagation techniques for wireless underground communication networks. United States: Elsevier.

AUTHOR INDEX

Institute of Physics
Dirac House, Temple Back
Bristol BS1 6BE UK

CURRAN ASSOCIATES INC.
proceedings.com

ISSN: 1757-8981
ISBN 978-1-7138-3175-4